DEFENDING

SCIENCE

—within reason

DEFENDING
SCIENCE
—within reason

BETWEEN SCIENTISM
AND CYNICISM

SUSAN HAACK

Prometheus Books

59 John Glenn Drive
Amherst, New York 14228-2197

Published 2003 by Prometheus Books

Inquiries should be addressed to
Prometheus Books
59 John Glenn Drive
Amherst, New York 14228–2197
VOICE: 716–691–0133, ext. 207
FAX: 716–564–2711
WWW.PROMETHEUSBOOKS.COM

07 06 05 04 03   5 4 3 2 1

Library of Congress Cataloging-in-Publication Data

Haack, Susan.
　　Defending science—within reason : between scientism and cynicism / by Susan Haack.
　　　　p. cm.
　　ISBN 1–59102–117–0 (cloth : alk. paper)
　　1. Creative ability in science. 2. Science—Methodology. 3. Science—Philosophy.
I. Title.
Q172.5.C74H33 2003
501—dc21

2003012541

Printed in Canada on acid-free paper

When one turns to the magnificent edifice of the physical sciences, and sees how it was reared; what thousands of disinterested moral lives of men lie buried in its mere foundations; what patience and postponement, what choking down of preference, what submission to the icy laws of outer fact are wrought into its very stones and mortar; how absolutely impersonal it stands in its vast augustness—then how besotted and contemptible seems every sentimentalist who comes blowing his smoke-wreaths, and pretending to decide things from out of his private dream!

William James, "The Will to Believe"

There are no scientific methods which alone lead to knowledge! We have to tackle things experimentally, now angry with them and now kind, and be successively just, passionate and cold with them. One person addresses things as a policeman, a second as a father confessor, a third as an inquisitive wanderer. Something can be wrung from them now with sympathy, now with force; reverence for their secrets will take one person forward, indiscretion and roguishness in revealing their secrets will do the same for another. We investigators are, like all conquerors, seafarers, adventurers, of an audacious morality and must reconcile ourselves to being considered on the whole evil.

Friedrich Nietzsche, *Daybreak*

Falsehood is so easy, truth so difficult. . . . [E]ven when you have no motive to be false, it is very hard to say the exact truth. . . .

George Eliot, *Adam Bede*

CONTENTS

PREFACE

My title speaks of "Defending Science"; but, though every now and then you will hear the rumble of a distant skirmish, or smell a whiff of gunpowder, this book is not intended as another salvo in the so-called "Science Wars." Rather, its purpose is to articulate a new, and hopefully a true, understanding of what science is and does. Discussions of the Old Deferentialism, with its focus on the "logic of science," on structure, rationality, and objectivity, and of the New Cynicism, with its focus on power, politics, and rhetoric—and of the deep cultural currents of admiration for and uneasiness about science of which they are manifestations—serve only as background to this constructive project.

My title speaks of defending science "Within Reason," and the play on the two meanings is intentional. I shall defend the pretensions of science to tell us how the world is, but in only quite a modest, qualified way ("within reason" in its colloquial sense), and from the perspective of a more general understanding of human cognitive capacities and limitations, and our place as inquirers in the world ("within reason" in a more philosophical sense). Science has managed to discover a great deal about the world and how it works, but it is a thoroughly human enterprise, messy, fallible, and fumbling; and rather than using a uniquely rational method unavailable to other inquirers, it is continuous with the most ordinary of empirical inquiry, "nothing more than a refinement of our everyday thinking," as Einstein once put it. There is no distinctive, timeless "scientific method," only the modes of inference and procedures common to all serious inquiry, and the multifarious "helps" the sciences have gradually devised to refine our natural human cognitive capacities: to amplify the senses, stretch the imagination, extend reasoning power, and sustain respect for evidence.

For a while I toyed with the idea of beginning: "There's no such thing as scientific method, and this is a book about it."[1] But that would have been too clever by half; or rather, it would have been half-right at best. For, once key ideas about scientific evidence and scientific inquiry had begun to come into focus, and I had learned enough of the history of molecular biology to illustrate those ideas from real-life scientific episodes, I glimpsed new ways of approaching difficult but fascinating questions far beyond my original agenda: about the differences between science and literature, the tensions between science and religion, the interactions of science with the law; and about the place of science in our lives, its value, its dangers, its limits, and even the possibility of its eventual annihilation, culmination, or completion. No doubt that's why this now seems to me the most Pragmatist of my books: influenced here by Peirce, there by James, its approach to the social sciences informed by Mead's work, its concern with science and values by Dewey's; and above all, liberated by the example of this rich tradition from the uneasy reluctance of analytic philosophy to stray beyond strictly linguistic, logical, or conceptual questions.

I came to see more clearly that science is valuable not only for the "magnificent edifice" of knowledge built over centuries by many generations of scientists, not only for the technological developments that have made our lives longer and more comfortable, but as a manifestation of the human talent for inquiry at its limited, imperfect, but sometimes remarkable best. I came to grasp more firmly that, though writers inquire, and scientists write, the word "literature" does not refer, like the word "science," to a federation of kinds of inquiry, but to a federation of kinds of writing; and so to understand how pointless and unnecessary it is to worry about whether science or literature is more valuable.

How did I get involved in a project as vast, as demanding, as overwhelming as this turned out to be? For the usual reasons; or at least, *my* usual reasons. I thought—given the ideas I had sketched in *Evidence and Inquiry* about the place of the sciences within empirical inquiry generally, and the couple of essays in *Manifesto of a Passionate Moderate* in which I had taken on some extravagances of self-styled "cultural critics" of science—enough of the surrounding crossword entries were completed, enough letters supplied in as-yet uncompleted entries; so that it should be, not easy exactly, but well within the realm of the feasible to say something useful about scientific knowledge and scientific inquiry, and about the place of science in our culture. As usual, when I began the work I had no idea what I was in for.

Spelling everything out proved almost as hard as thinking it all through. I have done my best to be as direct as possible, to eschew unnecessary technicalities, and to avoid the dreadful muddy blandness that pervades so much contemporary academic prose. But doubtless—as some readers will think that my ideas are too radical, and others that they are not radical enough; as some will complain that I spend too little time on the niceties of recent philosophy of science, or of the new "Science Studies," and others that I spend too much; and as some will reproach me for devoting too little attention to arcane details of quantum mechanics, and others for devoting too little attention to ethical issues about stem-cell research—some will find my style too dryly analytic, and others will find it too exuberantly literary, or too wryly playful. What can I say, except that George Eliot was right: "even when you have no motive to be false, it is very hard to say the exact truth"; and even harder when you *have* a motive, such as the polite reluctance to give offense that I needed to overcome to say forthrightly that the scientific and the religious world-pictures really are incompatible, really can't be reconciled.

As I write this Preface—almost the last step in a long journey of many false starts and wrong turnings, in which occasional moments of illumination and exhilaration had to compensate for long stretches of near-despair and a constant sense of my inadequacies—I think of Eliot again, this time reflecting, many years after its publication, on her *Romola*: "There is no book about which I more thoroughly feel that I swear by every sentence as having been written with my best blood."[2] Unfashionable as such Victorian seriousness is in today's academy, it captures my feelings about this book quite precisely.

NOTES

Note: In endnotes, books and articles are referred to by short titles; full details are to be found in the bibliography.

1. Adapted from Steven Shapin's nice line in *The Scientific Revolution*, p. 1: "There was no such thing as the Scientific Revolution, and this is a book about it."

2. My source is Andrew Sanders' introduction to the Penguin edition of *Romola*, p. 10.

ACKNOWLEDGMENTS

C hapter 1 of this book, "Neither Sacred nor a Confidence Trick," is a much revised and adapted version of "Defendiendo la ciencia—dentro de la razon," translated by Wenceslao González, published in Málaga, Spain, in *Filosofía actual de la ciencia*, ed. Paulo Martínez Friere, *Contrastes*, Supplemento 3, 37–56, in 1998. A modified English version, "Defending Science—Within Reason," was published in Santa Caterina, Brazil, in *Principia* 3, no. 2, 187–211, in 1999; and a Chinese translation by Chen Bo in Beijing, in *Studies in the Dialectic of Nature* 17, no. 5, 11–19, in 2001. An abridged version of chapter 3, "Clues to the Puzzle of Scientific Evidence," appeared in *Principia* 5, nos. 1–2, 253–81, in 2002. Chapter 8, "Stronger Than Fiction," is a much expanded and modified version of "Science, Literature, and the 'Literature of Science'," published in *Partisan Review* 67, no. 4, 640–47, and in *Occasional Papers* 47, American Council of Learned Societies, 45–56, both in the fall of 2000; it also includes material from "Misinterpretation and the 'Rhetoric of Science': Or, What Was the Color of the Horse?", published in *Texts and Their Interpretation*, *Catholic Philosophical Quarterly*, 1998, 69–91. Chapter 9, "Entangled in the Bramble Bush," draws substantially on "An Epistemologist in the Bramble Bush: At the Supreme Court with Mr. Joiner," published in the *Journal of Health Politics, Policy, and Law* 26, no. 2, 217–48, in April 2001, and reprinted electronically in *Philosophy, Science, and Law* in the fall of the same year.

Various versions of chapters of the book formed the basis of my lectures as Phi Beta Kappa Romanell Professor, delivered at the University of Miami in 1997; of my Gail Stine Memorial Lecture at Wayne State University, and my Bermann lectures at Umeå University, Sweden, in 1999; of my Cowling lectures at Carleton College, my Gilbert Ryle lectures at Trent University, Canada, my Spenser-Leavitt lectures at Union College, and my Gustav Bergman lecture at the University of Iowa, in 2000; of my Henri J. Renard lecture at Creighton University, and a public lecture at Michigan State University supported by the Templeton Foundation, in 2001; and of my lectures as Lansdowne Professor at the University of Victoria, British Columbia, and the lecture I gave on the occasion of my receiving the Faculty Senate Distinguished Scholar Award at the University of Miami, in 2002.

And parts of the book have been presented in philosophy departments, at conferences, and as public lectures at universities around the world, from Madrid, Santiago de Compostela, and La Coruña in Spain, Lund, Stockholm, and Uppsala in Sweden, Oslo and Bergen in Norway, to Florianópolis in Brazil, as well as at universities and colleges across the U.S. and Canada. An abridged version of chapter 3 was presented, for example, at the International Colloquium on Logic, Epistemology, and Philosophy of Science organized by the editors of *Principia* in Florianópolis in 2001, and at the "Mind and Language" seminar at NYU, the department of philosophy at VPI, and the European Congress for Analytic Philosophy in Lund in 2003. The short paper that eventually grew into chapter 8 was presented at the annual meeting of the American Council of Learned Societies in 1999; and an early version of the full chapter was presented at the Seminar on Issues in Science and the Humanities at the Whitney Humanities Center at Yale later that year, at the Forum for Contemporary Thought at the University of Virginia in 2000, and as a public lecture organized by the Department of English at the University of Georgia in 2001. The shorter paper that formed the basis of chapter 9, originally written for a conference organized by the School of Law and the Department of Philosophy at the University of North Carolina, Chapel Hill, was presented in the Law Schools at Penn, Virginia, Maryland, Iowa, Boston, Creighton, and the College of William and Mary; the full chapter was presented at Case Western Reserve Law School, and, with some thoughts about differences between the culture of the law in the United States and Canada, at Dalhousie Law School in Nova Scotia.

I learned a great deal, on all these occasions, from listeners' comments and questions; as I did from a small army of helpful correspondents, among them not only philosophers but also scientists, historians of science, engineers, economists, and legal and literary scholars; and from the students who asked good questions, brought articles and clippings they thought might interest me, or just looked appropriately baffled when I said something incomprehensible. I am grateful to all the many people who helped, in different ways, to make this a better book than it would have been had I struggled with it entirely alone—not least to the librarians of the University of Miami School of Law, and especially Virginia Templeton, for bibliographical assistance. I am especially grateful to Mark Migotti, who read the whole thing in typescript, raised good questions and made useful suggestions, and even stuck with me through the final process of hunting down repetitive passages and fixing twisted and broken sentences; and, as always, to Howard Burdick—because the usual but-for-whoms don't begin to cover it.

I

NEITHER SACRED NOR A CONFIDENCE TRICK

The Critical Common-Sensist Manifesto

> That men should rush with violence from one extreme, without going more or less into the contrary extreme, is not to be expected from the weakness of human nature.
>
> —Thomas Reid, *Essays on the Intellectual Powers*[1]

Attitudes to science range all the way from uncritical admiration at one extreme, through distrust, resentment, and envy, to denigration and outright hostility at the other. We are confused about what science can and what it can't do, and about how it does what it does; about how science differs from literature or art; about whether science is really a threat to religion; about the role of science in society and the role of society in science. And we are ambivalent about the value of science. We admire its theoretical achievements, and welcome technological developments that improve our lives; but we are disappointed when hoped-for results are not speedily forthcoming, dismayed when scientific discoveries threaten cherished beliefs about ourselves and our place in the universe, distrustful of what we perceive as scientists' arrogance or elitism, disturbed by the enormous cost of scientific research, and disillusioned when we read of scientific fraud, misconduct, or incompetence.

Complicated as they are, the confusions can be classified into two broad kinds, the scientistic and the anti-scientific. Scientism is an exaggerated kind of deference towards science, an excessive readiness to accept as authoritative any claim made by the sciences, and to dismiss every kind of criticism of science or its practitioners as anti-scientific prejudice. Anti-science is an exaggerated kind of suspicion of science, an excessive readiness to see the interests of the powerful at work in every scientific claim, and to accept every kind of criticism of science or its practitioners as undermining its pretensions to tell us how the world is. The problem, of course, is to say when the deference, or the suspicion, is "excessive."

Disentangling the confusions is made harder by an awkward duality of usage. Sometimes the word "science" is used simply as a way of referring to certain disciplines: physics, chemistry, biology, and so forth, usually also anthropology and psychology, sometimes also sociology, economics, and so on. But often—perhaps more often than not—"science" and its cognates are used honorifically: advertisers urge us get our clothes cleaner with new, scientific, Wizzo;[2] teachers of critical thinking urge us to reason scientifically, to use the scientific method; juries are more willing to believe a witness when told that what he offers is scientific evidence; astrology, water-divining, homeopathy or chiropractic or acupuncture are dismissed as pseudo-sciences; skeptical of this or that claim, people complain that it lacks a scientific explanation, or demand scientific proof. And so on. "Scientific" has become an all-purpose term of epistemic praise, meaning "strong, reliable, *good*." No wonder, then, that psychologists and sociologists and economists are sometimes so zealous in insisting on their right to the title. No wonder, either, that practitioners in other areas—"Management Science," "Library Science," "Military Science," even "Mortuary Science"—are so keen to claim it.

In view of the impressive successes of the natural sciences, this honorific usage is understandable enough. But it is thoroughly unfortunate. It obscures the otherwise obvious fact that not all and not only practitioners of disciplines classified as sciences are honest, thorough, successful inquirers; when plenty of scientists are lazy, incompetent, unimaginative, unlucky, or dishonest, while plenty of historians, journalists, detectives, etc., are good inquirers. It tempts us into a fruitless preoccupation with the problem of demarcating *real* science from pretenders. It encourages too thoughtlessly uncritical an attitude to the disciplines classified as sciences, which in turn provokes envy of those disciplines, and encourages a kind of scientism—inappropriate mimicry, by practitioners of other disciplines, of the manner, the technical terminology, the mathematics, etc., of the natural sciences. And it provokes resentment of the disciplines so classified, which encourages anti-scientific attitudes. Sometimes you can even see the envy and the resentment working together: for example, with those self-styled ethnomethodologists who undertake "laboratory studies" of science, observing, as they would say, part of the industrial complex in the business of the production of inscriptions;[3] or—however grudgingly, you have to admit the rhetorical brilliance of this self-description—with "creation science."[4] *And*, most to the present purpose, this honorific usage stands in the way of a straightforward acknowledgment that science—science, that is, in the descriptive sense—is neither sacred nor a confidence trick.

Science is not sacred: like all human enterprises, it is thoroughly fallible, imperfect, uneven in its achievements, often fumbling, sometimes corrupt, and of course incomplete. Neither, however, is it a confidence trick: the natural sciences, at any rate, have surely been among the most successful of human enterprises. The core of what needs to be sorted out concerns the nature and conditions of scientific knowledge, evidence, and inquiry; it is epistemological. (No, I haven't forgotten Jonathan Rauch's wry observation: "If you want to empty the room at a cocktail party, say 'epistemology'";[5] but the word is pretty well indispensable for my purposes because, unlike "theory of knowledge," it has adjectival and adverbial forms.) What we need is an understanding of inquiry in the sciences which is, in the ordinary, non-technical sense of the word, realistic, neither overestimating nor underestimating what the sciences can do.

What we have, however, is a confusing Babel of competing, unsatisfactory accounts of the epistemology of science. How did we come to such a pass?

FROM THE OLD DEFERENTIALISM TO THE NEW CYNICISM

Once upon a time—the phrase is a warning that what follows will be cartoon history—the epistemological *bona fides* of good empirical science needed to be defended against the rival claims of sacred scripture or a priori metaphysics. In due course it came to be thought that science enjoys a peculiar epistemological authority because of its uniquely objective and rational method of inquiry. In the wake of the extraordinary successes of the new, modern logic, successive efforts to articulate the "logic of science" gave rise to umpteen competing versions of what I call the "Old Deferentialism":[6] science progresses inductively, by accumulating true or probably true theories confirmed by empirical evidence, by observed facts; or deductively, by testing theories against basic statements and, as falsified conjectures are replaced by corroborated ones, improving the verisimilitude of its theories; or instrumentally, by developing theories which, though not themselves capable of truth, are efficient instruments of prediction; or, etc., etc. There were numerous obstacles: Humean skepticism about induction; the paradoxes of confirmation; the "new riddle of induction" posed by Goodman's "grue"; Russell Hanson's and others' thesis of the theory-dependence of observation; Quine's thesis of the underdetermination of theories even by all possible observational evidence.[7] But these, though acknowledged as tough, were assumed to be superable, or avoidable.

It is tempting to describe these problems in Kuhnian terms, as anomalies facing the Old Deferentialist paradigm just as a rival was beginning to stir. Kuhn himself, it soon became apparent, hadn't intended radically to undermine the pretensions of science to be a rational enterprise. But most readers of *The Structure of Scientific Revolutions*, missing many subtleties and many ambiguities, heard only: science progresses, or "progresses," not by the accumulation of well-confirmed truths, or even by the elimination of conjectures shown to be false, but by revolutionary upheavals in a cataclysmic process the history of which is afterwards written by the winning side; there are no neutral standards of evidence, only the incommensurable standards of different paradigms; the success of a scientific revolution, like the success of a political revolution, depends on propaganda and control of resources; a scientist's shift of allegiance to a new paradigm is less like a rational change of mind than a religious conversion—a conversion after which things look so different to him that we might almost say he lives "in a different world."

Even so, a quarter of a century ago now, when Paul Feyerabend proclaimed that there is no scientific method, that appeals to "rationality" and "evidence" are no more than rhetorical bullying, that science is not superior to, only better entrenched than, astrology or voodoo, he was widely regarded—he described himself—as the "court jester" of the philosophy of science. Post-Kuhnian Deferentialists, adding "incommensurability" and "meaning-variance" to their list of obstacles to be overcome, modified and adapted older approaches; but remained convinced not only of the rationality of the scientific enterprise, but also of the power of formal, logical methods to account for it.

But then radical sociologists, feminists, and multiculturalists, radical literary theorists, rhetoricians, and semiologists, and philosophers outside strictly philosophy-of-science circles, began to turn their attention to science. Proponents of this new almost-orthodoxy, though they disagreed among themselves on the finer points, were unanimous in insisting that the supposed ideal of honest inquiry, respect for evidence, concern for truth, is a kind of illusion, a smokescreen disguising the operations of power, politics, and rhetoric. Insofar as they were concerned at all with the problems that preoccupied mainstream philosophy of science—theory-dependence, underdetermination, incommensurability and the rest—they proclaimed them insuperable, further confirmation that the epistemological pretensions of the sciences are indefensible. Appeal to "facts" or "evidence" or "rationality," they maintained, is nothing but ideological humbug disguising the exclusion of this or that oppressed group. Science is largely or wholly a matter of interests, social negotiation, or of myth-making, the production of inscriptions or narratives; not only does it have no peculiar epistemic authority and no uniquely rational method, but it is really, like all purported "inquiry," just politics. We arrived, in short, at the New Cynicism.[8]

Feyerabend, who seems in retrospect the paradigm Old Cynic, promised to free us of "the tyranny of . . . such abstract concepts as 'truth,' 'reality,' or 'objectivity'." Now New Cynics like Harry Collins assure us that "the natural world has a small or non-existent role in the construction of scientific knowledge";[9] and Kenneth Gergen that the validity of theoretical propositions in the sciences "is in no way affected by factual evidence."[10] Ruth Hubbard urges that "[f]eminist science must insist on the political nature and content of scientific work";[11] and Sandra Harding asks why it isn't "as illuminating and honest to refer to Newton's laws as 'Newton's rape manual' as it is to call them 'Newton's mechanics'."[12] Ethnomethodologist of science Bruno Latour announces that "[a]ll this business about rationality and irrationality is the result of an attack by someone on associations that stand in the way";[13] and rhetorician of science Steve Fuller proposes "a 'shallow' conception . . . that locates the authoritative character of science, not in an esoteric set of skills or a special understanding of reality, but in the appeals to its form of knowledge that *others* feel they must make to legitimate their own activities."[14] Richard Rorty informs readers that "[t]he only sense in which science is exemplary is that it is a model of human solidarity,"[15] and Stanley Fish that "the distinction between baseball and science is not finally so firm."[16] Reacting against the scientism towards which the Old Deferentialism sometimes veered uncomfortably close, rushing with violence into the opposite extreme, the New Cynics take an unmistakably anti-scientific tone.

Perhaps it is no wonder that many scientists came to regard philosophy of science as at best irrelevant—"about as useful to scientists as ornithology is to birds."[17] With the exception of a few enthusiastic Popperians, working scientists seem to have been mostly unaware of or indifferent to Old Deferentialist aspirations to offer them advice on how to proceed. But in 1987—provoked by Alan Chalmers' observation, at the beginning of his popular introduction to philosophy of science, that "[w]e start off confused and end up confused on a higher level"[18]—physicists Theocharis and Psimopoulos published an impassioned critique of "betrayers of the truth": Popper, Kuhn, Lakatos, and Feyerabend, "the worst enemy of science." And as the influence of New Cynicism grew, more scientists were moved to defend the honor of their enterprise: Paul Gross and

Norman Levitt in *Higher Superstition*; Max Perutz denouncing rhetoric of science as "a piece of humbug masquerading as an academic discipline"; Sheldon Glashow deconstructing his letter of invitation to a conference on The End of Science; Alan Sokal parodying post-modernist "cultural critique" of science; and Steven Weinberg replying to the "cultural adversaries" of science.[19] Sometimes they made a good showing; but, not surprisingly, they didn't aspire to supply the realistic account of the epistemology of science required for an adequate defense against the extravagances of the New Cynicism.

Mainstream philosophy of science, meanwhile, had become increasingly specialized and splintered. Though many philosophers of science ignored the New Cynicism, a few tackled it head-on: John Fox criticizing self-styled "ethnomethodologists of science"; Larry Laudan in running battles with proponents of the "Strong Programme" in sociology of science; Mario Bunge protesting the influence of Romanticism and subjectivism; Noretta Koertge wrestling with the social constructivists.[20] Sometimes they made a good showing too. And with an increased readiness to accommodate social aspects of science,[21] a turn towards one or another form of naturalism, and, since Bas Van Fraassen's influential defense of constructive empiricism, something of a retreat from strong forms of scientific realism, there were efforts to acknowledge the elements of truth at which the New Cynicism gestures, without succumbing to its extravagances. But some mainstream philosophers of science—especially, perhaps, those most anxious not to offend the feminists—seem to have gone too far in the direction of the New Cynicism: as when Ronald Giere suggested that there might be something in the Nazi concept of "Jewish physics," as in feminist complaints of the masculinity of science;[22] and many others, still clinging to the Old Deferentialist reliance on formal, logical methods as sufficient to articulate the epistemology of science, seem not to have gone far enough. In my opinion, anyway, the realistic understanding we need still eludes us.[23]

Still, as I articulate my Critical Common-Sensist account—which I believe can correct the over-optimism of the Old Deferentialism without succumbing to the factitious despair of the New Cynicism—I won't forget the cautionary story of the student who is said to have observed in his Introduction to Philosophy examination that while some philosophers believe that God exists, and some philosophers believe that God does not exist, "the truth, as so often, lies somewhere in between." Though there are elements of truth in both the Old Deferentialism and the New Cynicism, a crude split-the-difference approach won't do; for the truth, as Oscar Wilde so nicely put it, "is seldom pure and never simple."[24]

CRITICAL COMMON-SENSISM

Critical Common-Sensism acknowledges, like the Old Deferentialism, that there are objective standards of better and worse evidence and of better and worse conducted inquiry; but proposes a more flexible and less formal understanding of what those standards are. Critical Common-Sensism acknowledges, like the New Cynicism, that observation and theory are inter-dependent, that scientific vocabulary shifts and changes meaning, and that science is a deeply social enterprise; but sees these, not as obstacles to an understanding of how the sciences have achieved their remarkable successes, but as part of such an understanding.

The core standards of good evidence and well-conducted inquiry are not internal to the sciences, but common to empirical inquiry of every kind. In judging where science has succeeded and where it has failed, in what areas and at what times it has done better and in what worse, we are appealing to the standards by which we judge the solidity of empirical beliefs, or the rigor and thoroughness of empirical inquiry, generally. Often, to be sure, only a specialist can judge the weight of evidence or the thoroughness of precautions against experimental error, etc.; for such judgments require a broad and detailed knowledge of background theory, and a familiarity with technical vocabulary, not easily available to the lay person. Nevertheless, respect for evidence, care in weighing it, and persistence in seeking it out, so far from being exclusively scientific desiderata, are the standards by which we judge *all* inquirers, detectives, historians, investigative journalists, etc., as well as scientists. In short, the sciences are not epistemologically privileged.

They are, however, epistemologically distinguished; the natural sciences at least, fallible and imperfect as they are, have succeeded remarkably well by the core epistemological standards of all serious inquiry. But distinction, unlike privilege, has to be earned; and the natural sciences have earned, not our uncritical deference, but our tempered respect.

The evidence for scientific claims and theories—consisting, not simply of the "data" or "observation statements" of the Old Deferentialism, but of experiential evidence and reasons working together—is like empirical evidence generally, but even more complex and ramifying. Think of the controversy over that four-billion-year-old meteorite discovered in Antarctica, thought to have come from Mars about 11,000 years ago, and found to contain what might possibly be fossilized bacteria droppings. Some space scientists think this is evidence of early Martian life; others think the bacterial traces might have been picked up while the meteorite was in Antarctica; others again believe that what look like fossilized bacteria droppings might be merely artifacts of the instrumentation.

How do they know that the meteorite's giving off these gases when heated indicates that it comes from Mars? that it is about 4 billion years old? that this is what fossilized bacteria droppings look like? Like crossword entries, reasons ramify in all directions.

How reasonable a crossword entry is depends on how well it is supported by its clue and any already-completed intersecting entries; on how reasonable those other entries are, independent of the entry in question; and on how much of the crossword has been completed. Similarly, what makes evidence stronger or weaker, a claim more or less warranted, depends on how supportive the evidence is; on how secure it is, independent of the claim in question; and on how much of the relevant evidence it includes.

While judgments of evidential quality are perspectival, dependent on background beliefs about, for instance, what evidence is relevant to what, evidential quality itself is objective; the determinants of evidential quality are not subjective or context-dependent, as New Cynics suppose. Neither, however, are they purely logical, as Old Deferentialists thought; they are *worldly*, depending both on scientists' interactions with particular things and events in the world, and on the relation of scientific language to kinds and categories of things. Observation interacts with background beliefs, as clues interact with already-completed crossword entries: what observations are taken to be relevant, and what is noticed when observations are made, depends on background assumptions; and the workings of instruments of observation, such as the mass spectrometer and ultra-high resolution transmission microscope on which those space scientists relied, depend on other scientific theories.

Scientific inquiry is continuous with the most ordinary of everyday empirical inquiry. There is no mode of inference, no "scientific method," exclusive to the sciences and guaranteed to produce true, probably true, more nearly true, or more empirically adequate, results. As Percy Bridgman put it, "the scientific method, as far as it is a method, is nothing more than doing one's damnedest with one's mind, no holds barred."[25] And, as far as it is a method, it is what historians or detectives or investigative journalists or the rest of us do when we really want to find something out: make an informed conjecture about the possible explanation of a puzzling phenomenon, check how it stands up to the best evidence we can get, and then use our judgment whether to accept it, more or less tentatively, or modify, refine, or replace it.

Inquiry is difficult and demanding, and we very often go wrong. Sometimes the problem is a failure of will; we don't really want to know badly enough to go to all the trouble of finding out, or we really *don't* want to know, and go to a lot of trouble *not* to find out. And even with the best will in the world, we often fail.

Our senses, our imaginations, and our intellects are limited; we can't always see, or guess, or reason, well enough. The remarkable successes of the natural sciences are due, not to a uniquely rational scientific method, but to the vast range of "helps" to inquiry devised by generations of scientists to overcome natural human limitations. Instruments of observation extend sensory reach; models and metaphors stretch imaginative powers; linguistic and conceptual innovations enable a vocabulary that better represents real kinds of thing and event; techniques of mathematical and statistical modelling enable complex reasoning; the cooperative and competitive engagement of many people in a great mesh of sub-communities within and across generations permits division of labor and pooling of evidence, and—though very fallibly and imperfectly, to be sure—has helped keep most scientists, most of the time, reasonably honest.

As you and I can complete the crossword faster and more accurately if your knowledge of Shakespeare and quick ear for puns complements my knowledge of popular music and quick eye for anagrams, so scientific inquiry is advanced by complementary talents. And interactions among scientists, both within and across generations, are essential not only to the division of scientific labor, but also to the sharing of scientific evidence, to a delicate mesh of reasonable confidence in others' competence and honesty. Moreover, scientific claims are better and worse warranted, and there is a large grey area where opinions may reasonably differ about whether a claim is yet sufficiently warranted to put in the textbooks, or should be subjected to further tests, assessed more carefully relative to an alternative, or what. There can no more be rules for when a theory should be accepted and when rejected than there could be rules for when to ink in a crossword entry and when to rub it out; "the" best procedure is for different scientists, some bolder, some more cautious, to proceed differently.

Scientific theories are (normally) either true or else false. A scientific claim or theory is true just in case things are as it says; e.g., if it says that DNA is a double-helical, backbone-out macromolecule with like-with-unlike base pairs, it is true just in case DNA *is* a double-helical, backbone-out macromolecule with like-with-unlike base pairs. Truth, like evidential quality, is objective; i.e., a claim is true or false, as evidence is better or worse, independent of whether anybody, or everybody, believes it to be so. But there is no guarantee that every scientist is entirely objective, i.e., is a completely unbiased and disinterested truth-seeker, immune to prejudice and partisanship; far from it. Nevertheless, the natural sciences have managed, by and large and in the long run, to overcome individual biases by means of an institutionalized commitment to mutual disclosure and scrutiny, and by competition between partisans of rival approaches.

A popular stereotype sees "the scientist" as objective in the sense, not merely of being free of bias or prejudice, but as unemotional, unimaginative, stolid, a paradigmatically convergent thinker. Perhaps some scientists are like this; but not, thank goodness, of all them. "Thank goodness," because imagination is essential to successful scientific inquiry, and a passionate obsession with this or that problem, even a passionate commitment to the truth of this or that elegant but as yet unsupported conjecture, or a passionate desire to best a rival, can contribute to the progress of science. Another stereotype, this time perhaps more philosophical than popular, sees "the scientist" as an essentially critical thinker, refusing to take anything on authority. A systematic commitment to testing, checking, and mutual disclosure and scrutiny *is* one of the things that has contributed to the success of natural-scientific inquiry; but this is, and must be, combined with the institutionalized authority of well-warranted results. Not that crossword entries once inked-in never have to be revised; but only by taking some for granted is it possible to isolate one variable at a time, or to tackle a new problem with the help of others' solutions to older problems.

Not all scientific theories are well supported by good evidence. Most get discarded as the evidence turns out against them; nearly all, at some stage of their career, are only tenuously supported speculations; and doubtless some get accepted, even entrenched, on flimsy evidence. Nevertheless, the natural sciences have come up with deep, broad, and explanatory theories which are well anchored in experience and interlock surprisingly with each other; and, as plausibly filling in long, much-intersected crossword entries greatly improves the prospects of completing more of the puzzle, these successes have enabled further successes. Progress in the sciences is ragged and uneven, and each step, like each crossword entry, is fallible and revisable. But each genuine advance potentially enables others, as a robust crossword entry does—"nothing succeeds like success" is the phrase that comes to mind.

Earlier, I stressed the complicated interactions among individuals that enable division of scientific labor and pooling of scientific evidence. But of course science is a social enterprise in another sense as well: it is a social institution embedded in the larger society, both affecting and affected by other social institutions; subject to political and cultural forces, it both influences and is influenced by the beliefs and values of that larger society. Earlier, I stressed the ways in which the social character of natural-scientific inquiry has contributed to its success. But of course both the internal organization and the external environment of science may hinder, as well as enable, good, fruitful, inquiry.

The disasters of Soviet and Nazi science remind us how grossly inquiry can be distorted and hindered when scientists seek to make a case for politically desired conclusions rather than to find out how things really are. Less melodramatic, but still disturbing, among the other potential hindrances that come immediately to mind are the necessity to spend large amounts of time and energy on obtaining funds, and to impress whoever supplies them, in due course, with your success; dependence for resources on bodies with an interest in the results coming out this way rather than that, or in rivals being denied access to them; pressure to solve problems perceived as socially urgent rather than those most susceptible of solution in the present state of the field; a volume of publications so large as to impede rather than enable communication. It would be less than candid not to admit that this list does not encourage complacency about the present condition of science. Once, important scientific advances could be made with the help of a candle and a piece of string; but it seems scientists have made most of those advances. As science proceeds, more and more expensive equipment is needed to obtain more and more recherché observations; and the more science depends for resources on governments and large industrial concerns, the worse the danger of some of the hindrances listed. While scientific techniques and instruments grow ever more sophisticated, the mechanisms needed to sustain intellectual integrity are strained.

Where the social sciences are concerned, it isn't so easy to think of examples of discoveries analogous to plausibly filled-in, long, much-intersected crossword entries, nor to be so sure that real progress has been made—it's no wonder that some people deny that the social "sciences" are really sciences at all. Social scientists' enthusiasm for the mathematical techniques and the cooperative and competitive inquiry that have helped the natural sciences to build on earlier successes has sometimes encouraged a kind of affected, mathematized obscurity, and prematurely gregarious (or dogmatically factional) thinking. More immediately to the present point, where the social sciences are concerned some of the prejudices apt to get in the way of honest inquiry are political as well as professional. Where the physical sciences are concerned, given the manifest irrelevance of sex, race, or class to the content of physical theory, the New Cynics' complaints of pervasive sexism, racism, classism, etc., seem far-fetched. But where the human and social sciences are concerned, given the manifest relevance of sex, race, or class to the content of some theories, political and professional preconceptions can come together, and the complaints seem only exaggerated.

Perhaps it is because the social sciences are especially susceptible to the influence of political prejudice and bias that sociologists of science seem especially susceptible to a dreadful argument ubiquitous among New Cynics—the "Passes-for Fallacy":[26] what has passed for, i.e., what has been accepted by scientists as, known fact or objective evidence or honest inquiry, etc., has sometimes turned out to be no such thing; *therefore* the notions of known fact, objective evidence, honest inquiry etc., are ideological humbug. The premise is true; but the conclusion obviously doesn't follow. Indeed, this argument is not only fallacious, but also self-undermining: for if the conclusion were true, the premise could not be a known fact for which objective evidence had been discovered by honest inquiry. Dreadful as this argument is, however, it has played a significant role in encouraging the recent alliance of radical sociology of science with the New Cynicism.

As a result, many sociologists of science have been cool or even hostile not only towards mainstream philosophy of science, but also to science itself; and this has reinforced the disinclination already felt by some mainstream philosophers of science, and by scientists themselves, to take sociology of science seriously as a potential ally in the task of understanding the scientific enterprise. This unfortunate quarrel between epistemology and sociology has obscured the potential for a sensible sociology of science to contribute to our understanding of what arrangements encourage, and what discourage, good, thorough, honest inquiry, efficient communication of results, effective testing and criticism.

Since both scientism and anti-scientific attitudes have their roots in misunderstandings of the character and limits of scientific inquiry and scientific knowledge, up till now I have focused on epistemological issues. But I don't at all mean to deny the legitimacy or denigrate the importance of those difficult ethical, social, and political questions about the role of science in society: who should decide, and how, what research a government should fund? who should control, and how, the power for good and evil unleashed by scientific discoveries?

As this suggests, the vexed question of science and values *is* vexed, in part, because of its many ambiguities. Scientific inquiry is a kind of inquiry; so epistemological values, chief among them creativity and respect for evidence, are necessarily relevant—which is not to say that scientific inquiry always or inevitably exemplifies such values. But moral and political values are relevant too; it is legitimate to ask, for instance, whether some ways of obtaining evidence are morally unacceptable, or whether, and if so how, and by whom, access to and applications of potentially explosive scientific results should be controlled.

Some New Cynics suggest that the fact that scientific discoveries can be put to bad uses is a reason for doubting the *bona fides* of those discoveries; and some hint that those of us who believe that science has made many true discoveries, or even that there is such a thing as objective truth, reveal themselves to be morally deficient in some way. But it isn't enough to point out the obvious confusion, nor to protest the blatant moral one-up-personship. It is essential, also, to articulate sober answers to those difficult questions about the role of science in society: to point out, inter alia, that only by honest, thorough inquiry can we find out what means of achieving desired social changes would be effective. And, as always, it is essential to avoid the exaggerations of the scientistic party as well as the extravagances of the anti-science crowd: to point out, inter alia, that decisions about what ways of handling the power that scientific knowledge of the world gives us are wise or just, are not themselves technical questions that may responsibly be left to scientists alone to answer.

THE ROAD AHEAD

This has been, of course, only the sketchiest of preliminary sketches, drawn in the broadest strokes. Questions crowd in thick and fast: about the place of the sciences within empirical inquiry generally, about the flaws in a narrowly logical conception of rationality, about the nature, structure, and pooling of scientific evidence, about the long, failed search for the "scientific method," about the ideas of truth, objectivity, and progress; about the differences, and the similarities, between the natural and the social sciences; about the differences, and the similarities, between the role of metaphors and linguistic innovations in science and in literature; about the potential relevance of sociology and rhetoric of science to its epistemology; about the tensions between science and religion; about the role of science in society—its interactions with the law, for example, or its relations with industry; and about whether politics can, or should, be kept out of science.

The road ahead is long, and steep in places. Like Ishmael, "I try all things; I achieve what I can."[27]

NOTES

1. Reid, *Essays on the Intellectual Powers* (1785), 6:4:xvi. Reid was the founder of the Scottish school of Common Sense philosophy. "Critical Common-Sensism" was C. S. Peirce's phrase for his adaptation of Reid's ideas, and I have borrowed the phrase for my adaptation of Peirce's.

2. After I had written this paragraph I found the following in Barzun, *Science: The Glorious Entertainment*: "sometimes the word ['science'] degenerates into a vague honorific, synonymous with the advertiser's 'reliable' or 'guaranteed'" (p. 14); and the comments on the honorific use of "science" in Chalmers, introduction to *What Is This Thing Called Science?* and in McCloskey, *Knowledge and Persuasion in Economics*, pp. 56ff.

3. See chapter 7, pp. 190–94, for details.

4. See chapter 11, pp. 272–74.

5. Rauch, *Kindly Inquisitors*, p. 35.

6. An idea I first introduced in "Knowledge and Propaganda: Reflections of an Old Feminist," in 1993. The same year, referring to the same genre of ideas about science, Kitcher wrote in *The Advancement of Science* of "Legend," the old, over-optimistic picture of scientific knowledge and inquiry.

7. See chapter 2 for details.

8. Though this particular version was new—hence, "the New Cynicism"—the attitude of distrust and hostility to inquiry in general, and the sciences in particular, is familiar from older manifestations.

9. Collins, "Stages in the Empirical Programme of Relativism," p. 3.

10. Gergen, "Feminist Critique of Science and the Challenge of Social Epistemology," p. 37.

11. Hubbard, "Some Thoughts About the Masculinity of the Natural Sciences," p. 13.

12. Harding, *The Science Question in Feminism*, p. 113.

13. Latour, *Science in Action*, p. 205.

14. Fuller, *Philosophy, Rhetoric, and the End of Knowledge*, p. xx.

15. Rorty, "Science as Solidarity," p. 46.

16. Fish, "Professor Sokal's Bad Joke," p. 82.

17. Unfortunately Weinberg, who quotes this observation in *Facing Up*, p. 8, says he can't remember where he heard it.

18. Chalmers, introduction to *What Is This Thing Called Science?* p. xvi in the first (1976) edition, p. xix in the second (1982) edition (I cannot find this sentence in the introduction to the third [1999] edition). In context, it seems the observation was intended to be mildly self-deprecatory.

19. Theocharis and Psimopoulos, "Where Science Has Gone Wrong"; Gross and Levitt, *Higher Superstition*; Perutz, "A Pioneer Defended"; Glashow, "The Death of Science!?"; Sokal, "Transgressing the Boundaries"; Weinberg, *Facing Up*.

20. Fox, "The Ethnomethodology of Science"; Laudan, "The Pseudo-Science of Science"; Koertge, "Wrestling With the Social Constructor."

21. See, e.g., the articles collected in McMullin, ed., *The Social Dimensions of Science*.

22. Giere, "The Feminism Question in the Philosophy of Science," p. 12.

23. Though we would disagree on many of the details, I agree with Chalmers that the failure of the mainstream philosophy of science to offer an adequate defense has "play[ed] into the hands of the anti-science movement" (*Science and Its Fabrication*, p. 8).

24. Actually, it was Algernon, in Wilde's *The Importance of Being Earnest* (1895).

25. Bridgman, *Reflections of a Physicist*, Philosophical Library, New York, 1955, p. 535.

26. A phrase I first introduced in 1993 in "Knowledge and Propaganda: Reflections of an Old Feminist"; see also "Staying for an Answer."

27. Melville, *Moby-Dick*, p. 335.

2

NAIL SOUP

A Brief, Opinionated History
of the Old Deferentialism

> I do not think that one can hope to understand [science]
> unless one appreciates that . . . , however formal its symbolism
> may sometimes become, it is not an exercise in logic. When
> some philosophers talk about the logic of scientific investiga-
> tion . . . I can only suppose that they speak metaphorically.
>
> —Henry Harris, "Rationality in Service"[1]

Believing, rightly, that science is, in some sense, a rational undertaking, Old Deferentialists assumed, wrongly, that the new, formal logic would suffice to articulate its epistemological core; believing, rightly, that the Old Deferentialism had failed, New Cynics concluded, wrongly, that the epistemological pretensions of the sciences are indefensible. Of course, this excessively brisk diagnosis simplifies what is really a whole complex tangle of issues; but it serves its purpose if it directs our attention to the assimilation of the rational to the narrowly logical taken for granted by Deferentialists, and unchallenged by Cynics.[2]

This assimilation was encouraged by the rise of modern, mathematical logic, and by the consequent shift of usage to which my reference to a "narrowly logical" conception of rationality alludes. In the twentieth century an older, ampler conception of logic as the theory of whatever is good in the way of reasoning, was displaced by a narrower conception of logic as the formal theory of validity.

Rather as, thanks to the success of a particular brand, "hoover" became a generic word for vacuum-cleaning, and "xerox" for photocopying, the word "logic" gradually shed its older, broader scope, and took on its modern, narrower reference, setting the stage for a confusion of the reasonable or rational—the logical, in the broad sense—with the formally, narrowly logical.[3]

In fact, Logical Positivism was called "logical," in contradistinction to earlier forms of plain Positivism, because of its reliance on the remarkable advances made by Boole, Peirce, Frege, Russell, etc., in formal deductive logic. These formal innovations, initially motivated in part by the desire to understand the foundations of mathematics, seemed the very model of rigor; and understandably encouraged the hope—especially among philosophers preoccupied with physics, the most mathematized of the sciences—that the new logic, or something very much like it, would suffice to answer central questions about the foundations of science. According to the earliest formulations of Logical Positivism, there are only two kinds of meaningful statement: the analytic, including the statements of logic and mathematics, and the empirically verifiable, including the statements of empirical science. Anything else is, cognitively speaking, nonsense, an expression of emotion at best. Much of traditional philosophy—metaphysics, ethics, aesthetics—was discarded, along with theology, as meaningless verbiage, or bad poetry. If philosophy was not to be abandoned altogether, it had to be reinvented; and so it was, as the "logic of science." "To pursue philosophy," wrote Carnap, "can only be to clarify the concepts and sentences of science by logical analysis. The instrument for this is the new logic."[4]

From the beginning, however, there was a problem about the status of theoretical statements. The original idea was that theoretical terms are simply abbreviations of congeries of observational terms, and theoretical statements reducible to observational statements by means of correspondence rules linking the two vocabularies. But this idea soon ran into difficulties, and it had to be acknowledged that scientific theories are not conclusively verifiable. Some liberalization of the original verificationist account was needed.

Three main responses emerged. Instrumentalism reconstrued unrestricted generalizations and theoretical "statements" as falling outside the scope of the verifiability criterion of meaning; they are not really statements at all, but rules for deriving observation statements. Inductivism, in the various forms proposed by Reichenbach, Carnap, and Hempel, weakened the original requirement of verifiability by allowing statements to qualify as empirically meaningful provided they could be probabilified or confirmed by observational evidence.[5] The goal was an inductive logic analogous to the more familiar deductive systems, but formalizing a relation of confirmation or probabilification rather than logical implication. Popper's deductivism, by contrast, replaced verifiability as a criterion of meaning by falsifiability as a criterion of the scientific. According to Popper, scientific theories can no more be probabilified or confirmed than they can be verified; but they can be falsified—and the logic of falsification is the deductive rule of *modus tollens*, from "if *p* then *q*" and "not-*q*" to infer "not-*p*."

I set instrumentalism aside for now,[6] to focus on the inductivist and deductivist responses. Chronologically as well as diagnostically, it makes sense to begin with deductivism, the purest form of the narrowly logical approach: though it didn't appear in English translation until 1959, the original, German edition of Popper's *Logik der Forschung*, published in 1934, was well known to Carnap, Hempel, etc.; and, though deductivism faces special difficulties of its own, inductivism is no less vulnerable to the underlying problems created by the assimilation of the rational and the narrowly logical.

A narrowly logical model of the epistemology of science suggests a picture of scientific evidence and inference as involving chains of statements in a fixed and predetermined vocabulary, anchored in observation statements the truth of which can be taken for granted, and linked by relations of deductive implication, or analogous but weaker relations of inductive support. It is ill suited to acknowledge that the warrant of so-called "observation statements" depends in part on background beliefs as well as on observation; it has no place for conceptual innovation, for shifts of meaning or vocabulary; it can't allow that evidential support might depend, not just on form, but on content, on the relation of statements to the world; it is severely strained by issues about comprehensiveness of evidence; and it makes the social character of science look incidental at best, embarrassing at worst.

No wonder, then, that as I began to glimpse the flaws in the Old Deferentialism, I kept remembering the fine old joke to which my title alludes: A peddlar shows up in a village bringing with him a large nail, which he claims makes excellent soup if you just stir hot water with it. The villagers duly supply hot water; stirring away, the peddlar explains that while plain nail soup is good, nail soup with onions is even better. The villagers duly supply onions; stirring away, the peddlar explains that while nail soup with onions is very good, nail soup with onions and carrots is better yet. The villagers duly supply carrots, . . . and so on. And no wonder, either, that as I noticed how the skeptical consequences of Popper's deductivism are quietly elided into a more plausible but no longer purely deductivist fallibilism, and how semantic and pragmatic elements are quietly slipped into Hempel's purportedly purely syntactic account of confirmation, a shrewd observation of J. L. Austin's kept echoing in my head: that in every important philosophical thinker, "there's the part where he says it, and the part where he takes it back"—the part where Old Deferentialists offer narrowly logical explanations of the epistemology of science, and the part where they slip extra-logical ingredients into the soup.

THE OLD DEFERENTIALISM, DEDUCTIVIST STYLE

Popper's approach is called "Deductivism" because it holds that the only logic involved in science is deductive; "Falsificationism" because it holds that the hallmark of a scientific theory is falsifiability, and the essence of scientific method bold conjecture and severe test; and "Critical Rationalism" because it holds that the rationality of science lies in the susceptibility of scientific theories to criticism. It also invites the label "Logical Negativism": "logical" because, despite his disagreements with the old Positivists and their heirs, Popper is no less in the grip of the assimilation of the objective and the logical than they; "negativism" because, replacing verification by falsification, he turns the Logical Positivist picture upside down. But hard-line deductivism turns out to be, not the appealingly fallibilist philosophy of science Popper leads you to expect, but a covert skepticism; and this thin deductivist soup can be transformed into the nourishing minestrone described on the menu only by slipping in non-deductive, and ultimately extra-logical, ingredients.

Let me begin at the beginning, with the part where Popper says it. The first moment of Falsificationism, like the first moment of Verificationism, is a criterion of demarcation. But Popper is concerned to demarcate, not the meaningful from the meaningless, but real science, such as Einstein's theory of relativity, from non-scientific disciplines like pure mathematics or history, from metaphysics and pre-scientific myths, and from such pseudo-sciences as Freud's and Adler's psychoanalytic theories and Marx's "scientific socialism."[7]

Curiously enough, Popper acknowledged from the beginning that his criterion of demarcation is a "convention"; and in 1959, in the new Introduction to the English edition of *The Logic of Scientific Discovery*, he even affirms that scientific knowledge is continuous with common-sense knowledge.[8] Nevertheless, the demarcation criterion—to qualify as scientific, a theory must be, not verifiable, but falsifiable, i.e., potentially capable of being shown to be false if it is false—is central to his philosophy of science. Popper sometimes says things like, *"it must be possible for an empirical scientific system to be refuted by experience"*;[9] but the official formulation of falsifiability is logical, couched in terms of relations among statements: an empirical scientific theory is a universal generalization incompatible with some basic statement, i.e., a singular statement reporting the occurrence of an observable event at a specified place and time.[10]

According to Popper, we have known since Hume that induction is unjustifiable; there can be no inductive logic. There is only deduction of basic statements from a conjecture, and, when a potentially falsifying basic statement is accepted as true, deduction of the negation of the conjecture. The rationality of science lies, not in a method for accumulating verified or well-confirmed or probable theories, but in the elimination of error as bold, *im*probable conjectures are subjected to severe tests. Hence Popper's methodological supplement to his formal criterion: scientific theories should not be protected from falsification by ad hoc or "conventionalist" modifications.[11]

The English translation of the title of Popper's *Logik der Forschung, The Logic of Scientific Discovery*, is notoriously misleading, since according to Popper there *is no* logic of scientific discovery; for questions about how theories are arrived at are psychological or sociological, i.e., causal, not logical.[12] Nor, however, would it be right to say that Popper's focus is on the logic of scientific justification; for he insists that scientific theories are never justified. At best, a theory is "corroborated," i.e., tested but not yet falsified. Degree of corroboration depends on the number and severity of the tests passed; it is relative to a time and, Popper tells us, depends in part on the sincerity of efforts to falsify the theory. And that a theory is corroborated, to however high a degree, doesn't show that it is rational to believe it, or that it is true, or even that it is probable; indeed, the testability of a hypothesis is *inversely* related to its degree of logical probability. Nor is corroboration a measure of verisimilitude or nearness to the truth, but only an indicator of how the verisimilitude of a theory *appears*, relative to other theories, at a time.

All this is troubling enough; but there is an even more serious problem: it isn't clear that Popper's account really allows that any scientific conjecture is ever shown to be false. A conjecture is "falsified," he says, when a basic statement incompatible with it is accepted. But he maintains that the acceptance of basic statements is a matter of convention, a decision on the part of the scientific community.[13] For logical relations can hold only among statements, not between statements and events; and so basic statements can never be justified by a scientist's observations—no more than by his thumping the table. Recalling with a wry smile that Popper had complained that the old Positivist Otto Neurath "unwittingly throws empiricism overboard," I notice the judiciously placed scare quotes with which he disguises the fact that he has done precisely that: "basic statements must be testable, inter-subjectively, by 'observation'."[14] Popper suggests that his conventionalism is harmless, because it is at the level of basic statements; such statements are testable, he points out, and the right decision is to stop with those which are "especially easy to test."[15] But how, one wonders? and why should we stop there? (Even so loyal a Popperian as John Watkins feels obliged to ask: so why don't scientists make one last effort and *actually test* those basic statements? Why make a decision to accept the statement, "there's a hippopotamus in the garage," when we could just go and look?)[16]

As David Stove observes, though he continues to use the words, Popper has apparently stripped "knowledge," "discovery," etc., of their connotation of success;[17] for the "discoveries" to which his "objective, scientific knowledge" refers are only conjectures which scientists haven't yet managed to falsify. And, as Alan Olding points out in a review of Stove's book, Popper has apparently stripped failure-words like "falsify," of their essential connotation, too; for a theory which is "falsified," in his sense, may not be false.[18] This is closet skepticism.

But then there are the parts where Popper takes it back; such as "Philosophy of Science: A Personal Report," published in 1957, where Popper several times describes hypotheses which have been tested but not yet falsified as (not "corroborated," but) "confirmed." However, in a footnote added in 1959 to the English edition of the *Logic of Scientific Discovery* Popper reports that, early on, Carnap had mistranslated his word "*Bewährung*" by "confirmed," and that for a while, thinking the issue merely verbal, he had not only let this unhappy translation go, but even occasionally used "confirm" himself; but, he continues, this had been a bad mistake on his part, since it conveyed the false impression that a theory's having been corroborated means that it is probably true, or that it is rational to believe it.[19]

Elsewhere, Popper concedes that theory's having been corroborated means that it is rational to prefer it, in a pragmatic sense, as the basis for action; but he goes on to add that this doesn't mean that the preference is "rational" in the sense of being based on good reasons for thinking the theory will be successful in the future—there *can be no* "good reasons" in this sense.[20] So it seems that all this "concession" amounts to is that, in deciding how to act, we had best go with theories we don't so far know to be false.

The fear that acknowledging an epistemological role for scientists' observations must lead to an objectionable subjectivism is found not only in Popper's earlier work, but also in his later advocacy of "epistemology without a knowing subject." But there are also passages, early and late, where Popper seems to take back his denial of the relevance of observation to the justification of basic statements. Already in *The Logic of Scientific Discovery* you find a nice analogy depicting science as resting, not on the firm ground of infallible observation statements, but on piles driven into a swamp,[21] and an argument to the effect that even the simplest basic statement, like "Here is a glass of water," is really theory-impregnated; which suggests, not conventionalism, but a fallibilist conception of basic statements as partially justified by scientists' observations. And much later, in response to criticism from Anthony Quinton and A. J. Ayer, apparently eschewing his earlier denial of the relevance of experience to justification, it is this much more plausible fallibilist position that Popper defends.[22] He doesn't explain, however, how it could be accommodated within an approach which dismisses the idea of supportive-but-not-conclusive evidence, and which, being purely logical, can accommodate only relations among statements.

Popper's admirers—not only philosophers, but also distinguished scientists, and even some justices of the U.S. Supreme Court—sometimes run together the parts where he says it and the parts where he takes it back, coming away with the impression that he has supplied a crisp criterion of demarcation and a tough-minded account of scientific method which, acknowledging the constraint of experience, allows that theories which have survived testing are thereby confirmed. I once heard Sir Hermann Bondi, declaring himself a strong Popperian, describe cosmology as having become a science in 1826, when Wilhelm Olbers made the first falsifiable cosmological conjecture; a conjecture which, Bondi continued, was subsequently roundly falsified, and replaced by a new conjecture which is by now *well confirmed by observational evidence*.[23] This is much like the Supreme Court's misreading of Popper in *Daubert*, where the falsifiability criterion of demarcation is proposed as an indication of genuinely scientific, *and hence reliable*, testimony.[24]

But the parts where Popper says it really can't be reconciled with the parts where he takes it back—any more than his marvelous analogy likening scientific knowledge to a medieval cathedral built over the centuries by generations of workers can be squared with his official account, which makes science more like an endless building project in which one structure is torn down and replaced by another which in due course is itself torn down and replaced, . . . and so on.

THE OLD DEFERENTIALISM, INDUCTIVIST STYLE

Insofar as the problems with Popper's approach arise from his deductivism, an inductivist approach might resolve them; insofar as they result from his commitment to the narrowly logical model, however, it is unlikely that an approach in terms of a formal logic of confirmation would do significantly better. As we shall see, it doesn't.

Carl Hempel is no less committed than Popper to the idea that logic is the key to understanding the rationality of the scientific enterprise; indeed, an entry in the *Philosopher's Lexicon* defines "hempel-mindedness"—unkindly, to be sure, but also shrewdly—as a tendency to assume that all philosophical problems can be represented in the language of first-order logic. And unlike Reichenbach or Carnap, Hempel doesn't attempt to explain confirmation by appeal to the calculus of probabilities. In fact, because of his conception of the "hypothetico-deductive" method of science, and his account of scientific theories as deductive structures, he has sometimes been classified as a deductivist.[25] But Popper specifically identifies him as one of those inductivists with whom he disagrees;[26] and this seems appropriate in view of the fact that Hempel not only criticizes Popper's falsificationist criterion of demarcation, but also devotes several papers to articulating the "logic of confirmation."

When Hempel announces that such a logic is "one of the most urgent desiderata of the present methodology of empirical science," and even that it will explain "what determines the soundness of a hypothesis . . . the way it stands up when tested, i.e., when confronted by relevant observations,"[27] it sounds as if he expects logic to resolve the core issues in the epistemology of science. However, he acknowledges that the logic of confirmation won't tell us anything about how new theories or concepts are arrived at—that is a psychological matter; nor will it tell us when to accept a hypothesis or when to reject it—that may not be formalizable. So far, perhaps, so good. But before long Hempel has taken back even the relatively modest claim that the logic of confirmation will explain when a hypothesis is sound; shortly thereafter he retracts even the minimal claim that a purely syntactic theory will explain under what conditions evidence supports a hypothesis; and very soon he slips the first non-logical ingredients into his purportedly syntactic soup.

Hempel distinguishes a relative and an absolute sense of "confirm": in the relative sense, confirmation is the relation between a hypothesis and the evidence statements that support it; in the absolute sense, confirmation is the property a hypothesis has when it is supported by evidence itself solid. Hempel's "logic of confirmation" is a logic only of *relative* confirmation. For an explanation of absolute confirmation we would require, in addition, an account of what makes observation statements reliable.[28]

Hempel assumes that scientific hypotheses and observation statements can be expressed in a specifiable "language of science," including "a clearly delimited observational vocabulary of terms designating more or less [*sic*] observable attributes of things and events." An observation statement is one that either asserts or denies that a given object has an observable property. What properties are observable, however, is relative to the instruments of observation used; and, Hempel continues, "the convention" [*sic*] should be that observation statements are "not irrevocable."[29] Unlike Popper, Hempel doesn't deny that observation is relevant to the soundness of scientific theories; instead, relegating the problem of reliable observation to pragmatics, he severely restricts the scope of his purely logical account.

Moreover, in the course of Hempel's discussion of the "raven paradox," it soon transpires that more, and more than logical, ingredients are needed even for his account of relative confirmation. Hempel first argues that Nicod's criterion—that any instance of a universal generalization confirms it—though not a necessary condition of confirmation, is sufficient. But, together with the Equivalence Condition—that whatever confirms a hypothesis confirms any logically equivalent hypothesis—Nicod's sufficient condition of confirmation gives rise to an apparent paradox. By Nicod's criterion "x is non-black and x is a non-raven" confirms "all non-black things are non-ravens"; so, by the Equivalence Condition, it also confirms the logically equivalent "all ravens are black." But this implies that an observation of a white shoe or a red herring confirms "all ravens are black"; which is counter-intuitive, to put it mildly. Hempel insists, however, that it is not this result, but the intuition it offends, that is mistaken. We find the result paradoxical only because, instead of focusing strictly on the relation of hypothesis H to evidence E, we sneak in background information about the number of non-black things versus the number of ravens around; and thus "fail to observe the methodological fiction . . . that we have no evidence for H other than that included in E."[30]

And Hempel's account of relative confirmation also turns out to require semantic concepts. Hempel first defines the development of hypothesis H for a finite class of individuals, C: a statement which says what H would say if there existed only those objects which are elements of C. Then he defines direct confirmation: an observation report O directly confirms H if O entails the development of H for the class of objects mentioned in O. And finally he defines confirmation, simpliciter: O confirms H if H is entailed by a class of sentences each of which is directly confirmed by O.[31] Though it was presented in a paper entitled "A Purely Syntactical Definition of Confirmation," Hempel later describes this (using "satisfies" in the technical sense of Tarski's semantic conception of truth)[32] as the "satisfaction criterion," because its basic idea is that a hypothesis is confirmed by an observation report if it is satisfied in the finite class of those individuals mentioned in the report.

In any case, by the time of his "Postscript on Confirmation" (1964), Hempel was ready to concede to his critics that his definition was too narrow,[33] and that it might have been unwise to begin, as he had, with a categorical conception of (relative) confirmation, rather than first defining degrees of confirmation. But the most significant concession comes with his reaction to "quite a different aspect of the problem":[34] Nelson Goodman's "grue" paradox.[35] x is grue just in case either it is examined before t, and is green, or it is not examined before t, and is blue. Adapting Goodman's example to his own "ornithological paradigm," Hempel realizes that his criterion entails that a raven observed before t and found to be black confirms "all ravens are blite," which implies that all ravens not examined before t are white—which is strongly counter-intuitive. Unlike the raven paradox, Hempel admits, the grue paradox is real. Following Goodman's suggestion that a predicate is projectible just in case it is entrenched, i.e., has been used in previously projected generalizations, he concedes that a purely syntactic criterion of confirmation presupposes hypotheses formulated in projectible terms, and that such terms "cannot be singled out by syntactical means alone."[36] But this is just a backhanded way of saying that confirmation is not a purely logical notion; which is, in effect, to take it *all* back.

More than a quarter of a century later, Hempel was ready to concede that "certain ideas advanced by Kuhn seem . . . important and illuminating." Now, denying that the goal of science is to discover truths about the world, he claims that "logical considerations" show that the fact that scientific theories conform to empirical evidence "has no bearing at all on the question of their truth."[37] Instead of following through on the insight to which the "grue" paradox brought him so tantalizingly close—that there is an extra-logical element in supportiveness of evidence, that not only the form but also the content of scientific language matters—by this time Hempel apparently abandons his earlier recognition that observation statements are fallible, and forgets his earlier distinction between relative and absolute confirmation, the most nourishing ingredients he had earlier slipped into the soup.

Presumably informed by our readiness to speak of this or that scientific claim as "probable" or "likely," another style of inductivism looked to the mathematical calculus of probabilities as the basis for the "logic of science." The earliest version was due to Hans Reichenbach, who proposed a probabilistic revision of the verificationist criterion according to which a statement is meaningful only if it is possible to determine its degree of probability. "Probable," Reichenbach held, has only one interpretation, the frequentist: the probability of an event is the limit of the relative frequency of that type of event in a given infinite sequence. According to the straight rule of induction, when the relative frequency of a property in the sample available to us is n, in the absence of special knowledge of the limit of that sequence we should posit that the relative frequency will approach a limit of approximately n as the sequence continues. And according to Reichenbach's pragmatic justification of induction—though we don't, and can't, know whether sequences in nature actually have limits—if they do, then if we continue to follow the straight rule we shall discover this order.[38]

Like Reichenbach, Rudolf Carnap thinks the verification principle should be liberalized. Like Reichenbach, Carnap takes a probabilistic approach to inductive logic. Unlike Reichenbach, however,[39] Carnap distinguishes two concepts of probability (an essential step forward, though only the first step towards distinguishing the many uses of "probable").[40] The frequency concept, Carnap argues, probability$_1$, applies to classes of events, and is inherently unsuitable as an explication of the relation of evidence to scientific hypotheses; for this purpose, we need a logical concept, probability$_2$, representing a relation between statements analogous to but weaker than logical implication.[41]

Like Hempel, Carnap acknowledges that the logic of confirmation isn't a complete meal, just one course; inductive logic can no more constitute a complete epistemology of science than deductive logic can constitute a complete epistemology of mathematics. He is concerned only with "the logical aspect of confirmation," as a measure of evidential support;[42] applying this measure will also require a "methodology"—including a Total Evidence Requirement, to the effect that judgments of the degree of confirmation of a hypothesis should be based on all the available evidence.[43] Carnap's conception of logic, however, is broader than Hempel's; relying, not exclusively on form, but on analytic meaning relations among predicates, it is overtly semantic.[44]

Evidence E deductively implies hypothesis H just in case all possible states of affairs in which E obtains are states of affairs in which H obtains. Analogously, the degree to which E inductively supports H depends on the extent to which the possible states of affairs in which both E and H obtain overlap with the states of affairs in which E obtains.[45] Working in a fixed language with one or more families of mutually incompatible predicates such as "green," "blue," "black," "white," etc., Carnap defines a measure function (representing the size of the relevant sets of possible states of affairs), and then characterizes the degree to which E confirms H, the probability$_2$ of H given E, as the ratio of the measure of the conjunction of H and E to the measure of E. On certain assumptions about logical features of the measure and confirmation functions, Carnap shows, probability$_2$ satisfies standard theorems of the probability calculus.

Actually, like Hempel, Carnap soon admits that his inductive logic isn't a complete account even of evidential support. Since it applies only to languages of an artificially simple structure, it can't aspire to provide a measure of, say, the degree of confirmation of Einstein's theory by the observational evidence available when it was first formulated, or by the observational evidence available in 1919; such evidence is "immensely extensive," and Einstein's theory is not stated in accordance with "the rigorous standards of modern logic"—the defect, apparently, is in Einstein's formulation, not in modern logic! Still, Carnap assures us, inductive logic can go beyond the degree of confirmation of the hypothesis that all swans are white by the evidence that x is a swan and is white to "more complex evidence"—such as, that y is a swan and is either small or white (gosh).[46]

Moreover, despite its austerity, some ingredients of Carnap's soup are hard to swallow. Each and every positive instance equally increases the degree of confirmation of a restricted general hypothesis—when yet another black raven from a temperate zone, surely, ought not to raise the degree of confirmation of "all ravens are black" as much as a black specimen from the Arctic or the Antarctic. (Carnap admits that variation of instances ought somehow to be accommodated.) As the size of the domain increases, the increment of confirmation afforded by each positive instance is less; so, as determined by c^*, the confirmation function Carnap favors, the degree of confirmation of an unrestrictedly universal hypothesis by any amount of evidence is zero, since with unrestricted generalizations there will always remain indefinitely many instances still unobserved. (Carnap suggests that predictions rationally based on past evidence don't need to rely on unrestricted generalizations.) Efforts to fix the flaws in Carnap's approach,[47] still based on the assumption that degree of confirmation is analytically grounded in meaning-relations within a fixed and predetermined language, had at best a limited success.

Though Carnap never took it *all* back, as Hempel seemed to, he writes with great shrewdness and an unexpected dry wit of the risks of producing "a theory which is wonderful to look at in its exactness, symmetry and formal elegance, yet woefully inadequate for the task . . . for which it is intended"; this, he continues, "is a warning directed at [myself] by [my] critical super-ego."[48]

THE KUHNIAN REVOLUTION

Thomas Kuhn's is a paradigm of the "naturalistic" intellectual temperament that Carnap contrasts with his own—and sees as running the opposite risk: a theory "rich in details but weak in power of explanation."[49] So it is understandable why readers of *The Structure of Scientific Revolutions*, accustomed to the severely restricted, narrowly logical diet recommended by the Old Deferentialism, were apt either to binge indiscriminately, or else to spurn Kuhn's offerings as hopelessly indigestible. Kuhn offers welcome relief from the simplification that takes "all swans are white" as the standard of a scientific statement, assumes a fixed scientific language, and gives the distinction of discovery versus justification a central place. But he seems more than a little discouraging on questions of evidence and method; so, though rich in historical ingredients, his soup left discerning diners hungry for epistemological meat.

Let me begin with the part where Kuhn says it—with the radical picture that has led many of his admirers, and nearly all his critics, to see him as the father of the New Cynicism. There is no such confrontation of theory and observation as naive philosophers of science take for granted. In periods of normal science, the reigning paradigm (a conglomeration of theory, exemplary experiments, instrumentation, etc.) is deployed in the resolution of the countless difficulties which every paradigm always faces, and any failure is attributed to the practitioner. But if persistently inexplicable anomalies cause a crisis, and if a rival paradigm is available, there may be a revolution, large or small, in which one paradigm is displaced by another, a new and incommensurable way of seeing and conceptualizing the world. There are no paradigm-independent standards of evidence, no neutral observations, no crucial experiments, no constancy of meaning between paradigms; rather, in something like a *Gestalt* switch, converts to the new paradigm come to see the world differently, literally as well as metaphorically. Proponents of rival paradigms don't so much disagree as talk past each other in mutual incomprehension. A scientist's shift of allegiance to a new paradigm is more like a religious conversion than a rationally defensible change of mind—a conversion so drastic that he not only sees the world differently, but may even be said to be "responding to a different world."[50]

As for the progress of science, it isn't mythical, exactly, but it isn't what it seems. In periods of normal science, sure, there is progress—but it is in the eye of the beholder; how else could those normal scientists view their work? And revolutionary episodes will be seen as progressive too, for it will be the history perceived by the winning side that will find its way into the textbooks. So either way the history of science will be *seen as* progressive; but "does it really help," Kuhn asks, "to imagine that there is some one full, objective, true account of nature and that the proper measure of scientific achievement is the extent to which it brings us closer to that ultimate goal?"[51]—evidently a question expecting the answer, "No." In the final chapter of *The Structure of Scientific Revolutions*, noting that he has thus far felt the need for the word "truth" only once,[52] Kuhn suggests that the history of science is better thought of as evolution-from than as progress-towards.

But then there are the parts where he takes it back. The electrical paradigm due to Franklin enormously increased the "effectiveness and efficiency" of research, "providing evidence for a societal version [*sic*] of Francis Bacon's acute methodological dictum; 'Truth emerges more readily from error than from confusion'." At least some of the puzzle-solving achievement of normal science "always proves to be permanent." There are certain rules and commitments "without which no man is a scientist": he must be "concerned to understand the world and to extend the precision and scope" with which science orders it, he must "scrutinize . . . some aspect of nature in great empirical detail," and "if that scrutiny reveals pockets of apparent disorder, these must challenge him to a new refinement of his observational techniques or to a further articulation of his theories."[53]

Moreover, the radical-sounding phrases that frightened Kuhn's critics are almost always hedged, hinting at a less radical view. When paradigms change, "[i]t is rather as if the professional community had suddenly been transported to another planet. . . . Of course, *nothing of quite that sort does occur.* . . . Nevertheless . . . *we may want to say* that after a revolution scientists are responding to a different world."[54] "[T]he principle of economy will urge us to say that after discovering oxygen Lavoisier worked in a different world. I shall inquire in a moment about the possibility of avoiding *this strange locution.*"[55] And: however things may appear to him, "the scientist after a revolution is *still looking at the same world.*"[56]

And in later writings Kuhn insisted that he had been misunderstood. In 1970, speculating very suggestively about the role of conceptual innovation in science, and observing that "though logic is a powerful . . . tool of scientific inquiry, one can have sound knowledge in forms to which logic can scarcely be applied," Kuhn suggests that Popper isn't so far wrong about revolutionary science, but has ignored the more cumulative processes of normal science; thus, apparently, envisaging both falsification and accumulation after all.[57] In 1983, he avers that his concern had always been as much epistemological as sociological: "what is it about what scientists do, I have been asking, that makes their output knowledge?" The recent preoccupation of sociologists of science with social and economic interests, he continues, ignoring such cognitive interests as love of truth, "often seems disaster"; though he rather spoils the effect by adding, after "love of truth," "fear of the unknown, if one prefers."[58] In 1993 he suggests that the incommensurability of paradigms is not after all a matter of mutual untranslatability—it's just that new vocabulary is acquired directly, not by translation; and, in a startlingly modest reconstrual of his earlier, radical, and incoherent claims, writes of a pluralism of "*professional*" worlds (my italics).[59] "I never meant," he wrote to me in 1995, "to impugn the rationality of science."

TWO NEW DEFERENTIALISTS

Nevertheless, *The Structure of Scientific Revolutions* was a challenge to New Deferentialists both inductivist and deductivist to renew their efforts to articulate and defend the rationality of science. The commitment to a narrowly logical model was weakened, but not abandoned.

"Romantic excesses in reaction to excessive logical pedantry . . . are not unknown in the history of philosophy," Mary Hesse wrote in the introduction to *The Structure of Scientific Inference*; although, she continued, "their lifetime is usually brief" (which, in view of the steady rise of the New Cynicism over the quarter-century since her book was published, now seems somewhat over-optimistic). But most important for present purposes is Hesse's promise to map a "*via media* between the extremes of both formalism and historical relativism."[60] The description on the menu is appetizing; and Hesse's recipe includes some very promising-sounding ingredients.

Both Popper and Hempel had acknowledged that there is no sharp distinction between observational and theoretical statements; but neither was able comfortably to accommodate this acknowledgment in his official story. Hesse too repudiates the idea of sharply distinguished observational and theoretical languages; but she offers something to put in its place—a "network model," derived from Quine, of a mesh of interrelated predicates attached to observation by some of its knots.[61] All descriptive predicates are learned by means of direct empirical associations (i.e., by ostension), by means of other predicates so learned, or by some combination of the two; but no predicate can get its meaning from direct empirical associations alone.[62] Hesse also acknowledges the reliance of scientific inquiry on classifications of particular things, events, and phenomena into kinds. Rather than an absolute conception of universals, she adopts a generalized family resemblance approach according to which resemblances define significant classes when those predicates enter laws, supporting predictions about further properties of members of the class.

However, characterizing lawlike generalizations as those entailed or probabilified by already-accepted theories, Hesse construes the distinction between laws and accidental generalizations as epistemological rather than ontological[63]—a crypto-nominalism that thins her soup significantly. Moreover, as its title suggests, this book of Hesse's still belongs in the "logic of science" tradition; specifically, in the tradition of Carnapian inductive logic, approaching scientific inference via the theory of probability.

Like Carnap, Hesse thinks of probabilities as indicating betting ratios on propositions; but unlike Carnap she construes them, not objectively, but in a personalist sense, as degrees of rational belief—"rational" precluding degrees of belief such that a person betting on them would lose in all circumstances whatever. With unrestrictedly universal propositions a single piece of negative evidence could result in a loss, but no finite amount of positive evidence could result in a win; so such propositions have a probability of zero. Hesse avoids this objection by shifting towards instrumentalism, reconstruing unrestricted generalizations as rules for analogical inference from one kind of restricted generalization to another.[64] But she is more preoccupied with the objection that the probability approach is too weak, since *any* assignment of probabilities satisfying the axioms of the probability calculus is "rational" in the sense of coherent betting.

Bayes' theorem, a corollary of the multiplication axiom of the calculus of probabilities, states that (provided the probability of evidence E is not zero)

$$pr(H/E) = pr(H \& E)/pr(E).[65]$$

Bayesians suggest strengthening the personalist conception of rational belief by interpreting Bayes' theorem as a requirement on rational belief-change. In conformity with the Bayesian transformation, the posterior probability of hypothesis H—the degree of rational belief in H given new evidence E—should be the prior probability of H and E divided by the prior probability of E. Some even hope to account for convergence of opinion in the scientific community by appeal to the fact that (under certain conditions) different initial assignments of prior probabilities would eventually resolve to the same posterior probability. But like Carnap—who a quarter-century before had warned against putting too much epistemological weight on Bayes' theorem[66]—Hesse has reservations.

Posterior probabilities would have to be calculated taking all the available evidence into account; a fixed language would have to be assumed; and once the idea of a sharp distinction of observational and theoretical statements is given up, the corrigibility of evidence poses a tricky problem. Construing evidence statements as descriptions of reports of observation—her example, interestingly theory-laden, is, "it is reported that at time *t* NaCl dissolved in water"[67]—Hesse suggests the "correspondence postulate": observation statements contained in observation reports are true more often than not. But she acknowledges that what initial distributions of probabilities and transformation rules will work "depends crucially on how well adapted the learning organism is to actual contingent conditions in the world"; and so proceeds to investigate, not the necessity, but the sufficiency of Bayesian methods "as explications of certain local aspects of scientific induction."[68]

Hesse has identified two of the most important ingredients missing from Old

Deferentialist recipes: a robustly fallibilist account of observation and observation statements, and a serious appreciation of the importance of the relation of particulars and universals. Unfortunately, however, her most epistemologically nourishing ingredients just won't go through the Bayesian mouli-légumes. Perhaps that's why, even in 1974, Hesse had seemed a little reluctant to use the word "true" without benefit of precautionary scare quotes; and why, by 1995, we find her in post-modernist mood, expressing "sympathy with the explicit subordination of the philosophy of knowledge to value-systems of some kind," persuaded that "there is no universal . . . account of how we know," and seeing hope only in "patient commitment to continuing discourse"[69]—classic symptoms of epistemological malnutrition.

Imre Lakatos's much more influential "Falsification and the Methodology of Scientific Research Programmes" also attempted to acknowledge something of the richness of Kuhn's picture without sacrificing the rationality of the scientific enterprise; but, unlike Hesse's book, in a deductivist way. As Stove observes, in Lakatos the evasive scare quotes to which Popper sometimes resorts have become ubiquitous, a kind of literary tic. With Lakatos' "storms of neutralising quotation marks,"[70] it is a matter, not so much of the part where he says it and the part where he takes it back, as of a disturbing sensation of seeing double: the appealing-sounding sophisticated-critical-rationalist picture that appears if you ignore the scare quotes, and the startlingly irrationalist picture that appears when you take them seriously.

Following Popper, Lakatos describes himself as articulating the "logic of scientific discovery." Like Popper, Lakatos is preoccupied with the question of demarcation, though he presents himself as distinguishing not only science from non-science, but also honest inquiry from dishonest. But unlike Popper, who just vehemently protests what he perceives as Kuhn's relativism,[71] Lakatos works hard to acknowledge the complexities of scientific theories and their historical development, the interpenetration of theory, instrumentation, and observation; and to articulate the insight that "there is no instant rationality."[72]

The rationality of science can only be understood by taking a less atomistic view of theories, and a more diachronic view of scientific method. A scientific theory consists of a hard core surrounded by a protective belt of auxiliary assumptions which shield it from observational falsification.[73] The hard core of a scientific theory can't be broken by observation alone, only by an alternative theory, in what Lakatos calls a "problemshift." A problemshift is progressive only if it involves a modification predicting new facts—this is Lakatos's version of Popper's strictures against ad hoc modifications. A scientific research program is a series of successive theories with their associated auxiliary assumptions, instrumental procedures, etc. At any time, some research programs are progressive—good, honest, empirical, scientific—and others degenerating. The "firm rational strategy" that ensures the rationality of science is to sustain progressive research programs, programs which make consistently progressive, content-increasing, problemshifts[74]—of course, Lakatos adds, the new content should be, "at least now and then," corroborated—and to give up degenerating programs.

Sounds reassuring, doesn't it? But there are a couple of very large flies, including a by now all-too-familiar Popperian fly, in the soup. Like Popper in the part where he says it, Lakatos insists that basic statements can never be justified by observation—he even uses Popper's words: "no more than by thumping the table."[75] Like Popper in the part where he takes it back, Lakatos also argues that basic statements, which are theory-laden, are fallible; but offers no account of how observation contributes to justification. Instead, in line with the radical conclusion of the first argument, he resorts to writing of the role, not of observation, but of scare-quotes "observation."

The scare quotes appear first in Lakatos's description of the "sophisticated methodological falsificationist" position he attributes to Popper: "only those theories—that is, non-'observational' propositions—which forbid certain 'observable' states of affairs, and therefore may be 'falsified' and rejected, are 'scientific'."[76] But in his own behalf, also, he writes not of observational evidence but of "observational" evidence, not of falsification but of "falsification," not of the refutable auxiliary hypotheses in the protective belt of a theory but of "refutable" hypotheses, not of the empirical basis of science but of its "empirical basis," and so on. The plausible-sounding recommendation that scientists should pursue progressive research programs which consistently introduce new, refutable content which can be corroborated becomes the peculiar recommendation that scientists should pursue "progressive" research programs which consistently introduce new, "refutable" content which can be "corroborated."

And Lakatos is in trouble not only where he follows Popper, but also where he breaks away from him. Realizing that, if scientists followed Popper's simple falsificationist methodology, progress might be impeded by premature rejection of promising conjectures, Lakatos points out that it may advance science if a new theory which in its early stages is vague, muddled, or even incoherent, is protected from falsification—it may get better.[77] But, of course, it may also advance science if an old research program which has fallen on hard times and has been degenerating is protected from falsification—it, too, may get better (as, later, Lakatos admits). But this leaves us, after all, without *any* "firm rational strategy" for making scientific progress.

Figure 1. Betrayers of the truth? Left to right: Karl Popper, Imre Lakatos, Thomas Kuhn, and Paul Feyerabend. (These noted philosophers of science were featured in the October 15, 1987, issue of *Nature* as part of an article titled "Where Science Has Gone Wrong.")

So, when Paul Feyerabend dedicates *Against Method* "to Imre Lakatos, friend and fellow-anarchist" (Lakatos died before writing the reply that had been planned), he is not only being playfully infuriating as usual; he has noticed that Lakatos's philosophy of science—like the Popperian philosophy that was its point of departure[78]—is by no means so stably rationalist as advertised (see Figure 1).

THE WORST ENEMY OF SCIENCE, AND ITS BEST FRIEND

Feyerabend's penchant for the deliberately outrageous and the provocatively nutty is apt to induce a kind of intellectual vertigo—as if he had slipped something hallucinogenic in the soup! Certainly he often *sounds* like "the worst enemy of science."[79] But every now and then, e.g., when he recalls losing patience "when a debate about scientific achievements was interrupted by an attempt to 'clarify,' where clarification meant translation into some form of pidgin logic,"[80] he sounds surprisingly like a closet Critical Common-Sensist. "Ours is an age," writes Thomas Szasz, "in which partial truths are tirelessly transformed into total falsehoods, and then acclaimed as revolutionary revelations";[81] but disentangling Feyerabend's partial truths from the total falsehoods into which he tirelessly transforms them is not made easier by his avowed policy of writing so as to confuse the "ratiofascists" who try to bully us with talk of "truth," "objectivity," etc.

In his role as court jester of the Old Deferentialist establishment, poking fun at Popperian pidgin science and Positivist pidgin logic, Feyerabend poses as epistemological anarchist or Dadaist. (Tristan Tzara's *Dada Manifesto* opens: "In principle I am against manifestos, as I am against principles.") The only principle that doesn't inhibit progress is: anything goes. Oh, and "progress" can mean—whatever you like. There are no universal standards; talk of "truth," "objectivity," "rationality," "progress," etc., is propaganda serving the interests of the science establishment and its despicable hanger-on, the philosophy of science industry. Hey, *argument* is propaganda! Scientific theories aren't verified, confirmed, falsified, *or* corroborated by observation, which is theory-laden; they get "established" by rhetoric, politics, chicanery, opportunism. As one, sometimes incommensurable, theory replaces another, science may go forward, or backward, or sideways, or . . . ; oh, and "forward," etc., can mean—whatever you like.

Science is just one form of life among many; its supposed superiority to folk medicine, voodoo, astrology, etc., is an illusion fostered by the success of its imperialistic ambitions. Nor are the benefits of scientific civilization what they're cracked up to be; what we patronizingly describe as "primitive" cultures may do as well, or better, at enabling people to cope. In fact, science has become so dangerously powerful that, instead of trembling before the tired old ghost of Lysenko, we would do better to look to democratic politics to curb its excesses.

And yet, inside "the worst enemy of science," perhaps there is a friend—a friend, admittedly, given to the most unnerving extravagances, and who sometimes, as he thumbs his nose at the Old Deferentialism, puts me in mind of my grandmother's warning that, if the wind changes while you're making a face, you'll get stuck like it. (Not being a sociologist, my grandmother didn't think to tell me that the more bizarre the faces you make, the more prestigious and lavishly paid the academic jobs you will be offered.) At any rate, for all the wildness of his exaggerations, Feyerabend has some insight into the dangers of too narrowly logical an approach, a sense of the complexity and untidiness of science, of the role of luck, of intellectual opportunism, of good old-fashioned muddling-through, quite missing from Carnap's or Hempel's or Popper's or even Hesse's or Lakatos's philosophies.

Many, no doubt, think of W. V. Quine as among the best philosophical friends of science; as with Feyerabend, however, the situation is not as straightforward as it might seem. Certainly there is no doubting Quine's admiration for the achievements of science, especially of physics. And one finds in his writings, if not a systematic philosophy of science, valuable aperçus: his and Ullian's nice description of the mutual support between an explanation and what it explains,[82] for example, and his pregnant description of science as "solving one problem with the help of solutions to others."[83] (As, it goes without saying, is doing a crossword; a letter from Quine describing my crossword analogy as "a nice stick figure of scientific method" encouraged some of the thoughts I shall be developing later.) Nevertheless, a thesis of Quine's has been a powerful encouragement to the New Cynicism—as powerful, probably, as anything in Kuhn, whom, however, Quine coolly dismisses as an epistemological nihilist.[84]

The thesis in question is "underdetermination." For a thesis so influential, however, "the" underdetermination thesis is surprisingly hard to pin down. In "Two Dogmas" Quine writes that "any statement may be held true come what may, provided we make enough adjustments elsewhere in the system"; in *Word and Object*, that "we have no reason to suppose that man's surface irritations even unto eternity admit of any systematization that is scientifically better or simpler than all possible others. It seems likelier . . . that countless alternative theories would be tied for first place";[85] and in "On Empirically Equivalent Theories of the World," that "natural science is underdetermined . . . not just by past observation but by all observable events," or, in a more official form, that there are inevitably "theory formulations that are empirically equivalent, logically incompatible, and irreconcilable by reconstrual of predicates."[86] Moreover, as Quine's statements of underdetermination get more precise, they also get more hedged, until by the end of "Empirically Equivalent Theories," he is ready to commit himself only to what he calls a "last-ditch version": "our system of the world is bound to have empirically equivalent alternatives which, if we were to discover them, we could see no way of reconciling by reconstrual of predicates."

There is, Quine concludes, "no extra-theoretic truth"—a conclusion which, he admits, "has the ring of cultural relativism."[87] It begins to appear why he too has been called a Logical Negativist;[88] not, like Popper, because of a stress on the role of deduction in falsification, but because of his predilection for drawing strongly negative and even apparently skeptical or relativist conclusions from formal-logical results. Initial surprise that New Cynics should welcome a thesis of Quine's so warmly is soon dispelled; they hope that underdetermination will make room for social values to take up evidential "slack," a term of Quine's they adopt with enthusiasm. As intellectual vertigo threatens once again, the diagnosis is confirmed: the root of the trouble lies in the narrowly logical conception of rationality shared by the Old Deferentialists, both inductivist and deductivist, *and* by the New Cynics.

AND, IN CONCLUSION

By now, of course, the idea that narrowly logical, formal conceptions of science are inadequate is quite a familiar one; nor am I alone in aspiring to articulate an account which is (as I have put it) neither deferentialist nor cynical, and in which rational and social aspects of science intertwine. But my diagnosis of what is most conspicuously, and most consequentially, missing *both* from the narrowly logical approach of the Old Deferentialists, *and* from the historical-sociological-rhetorical approaches of the New Cynics—the world—will not be so familiar; nor my account of the nature and structure of evidence for scientific claims and theories, or my Critical Common-Sensist approach to questions about the methods of science.

In line with my diagnosis, the account developed in what follows will be, as I shall say, worldly.[89] It will also be eclectic, accommodating many insights from the older philosophy-of-science tradition I have sketched. Unlike Popper's "epistemology without a knowing subject," my account will begin with experiential evidence, with scientists' observations; but as the focus turns to evidence-sharing and division of scientific labor, it will be very much in the spirit of Popper's cathedral metaphor. It will acknowledge the importance of social aspects of science and of linguistic shifts and innovations, rightly emphasized by Kuhn, and the dependence of many of the special methods and techniques of the sciences on previous theory; but it will reveal the supposed paradigm-relativity of standards of evidence to be a kind of epistemological illusion. Like inductivism, it will allow that evidence may be supportive without being deductively conclusive; but, denying that supportiveness is a matter exclusively of form, it will acknowledge an element of truth in the deductivist's skepticism about "inductive logic." Like inductivism, it will acknowledge that there are objectively better and worse warranted claims; but as it reveals why basing a theory of warrant on the mathematical theory of probabilities won't do, it will share something of the spirit of Popper's critique of probabilism. It will accommodate Hesse's insight that we need a seriously fallibilist account of the role of observation and close attention to the place of universals, and Lakatos's shrewd observation that "rationality works more slowly than most people tend to think, and even then, fallibly."[90] And, recognizing the density of the world we investigate and our limitations as inquirers,[91] it will acknowledge that evidence is bound to be complicated, ambiguous, and often potentially misleading, inquiry difficult and demanding, and progress ragged and uneven.

I shan't attempt the overwhelming task of surveying all the many and various sociological, ethnomethodological, feminist, literary-theoretical, rhetorical, etc., etc., expressions of the New Cynicism; but some will come under scrutiny as my argument proceeds.[92] Nor shall I attempt the no less overwhelming task of sur-veying all the many and various Bayesian, decision-theoretic, neo-Popperian, error-theoretic, model-theoretic, neo-pragmatist, naturalized, neo-instrumentalist/constructive-empiricist, etc., etc., approaches developed in recent mainstream phi-losophy of science; but there will be opportunities to note some points of agree-ment, and of disagreement, as my argument proceeds.[93]

NOTES

1. Harris, "Rationality in Science," p. 46.

2. Though we would disagree on some of the details, my diagnosis has quite a lot in common with Laudan's in "The Sins of the Fathers . . . " (which I read after offering my diagnosis in the first draft of this chapter, but before writing the present version).

3. Quine's title, *From a Logical Point of View*—taken from the calypso to apply to essays on formal logic—is a play on the two uses.

4. Carnap, "The Old and the New Logic," p. 145. (Don't be confused by the fact that, earlier in the article, Carnap had described his conception of logic as "broad"; he means only that he includes *applied* narrow logic, which is what he takes epistemology to be.)

5. See Carnap, "Testability and Meaning."

6. But see chapter 5, pp. 135–39.

7. See especially Popper, "Philosophy of Science: A Personal Report," and "The Problem of Demarcation."

8. The criterion of demarcation is a convention: Popper, *The Logic of Scientific Discovery*, introduction to the English edition, p. 18; scientific knowledge is continuous with common-sense knowledge: *The Logic of Scientific Discovery*, p. 37.

9. Popper, *The Logic of Scientific Discovery*, p. 41.

10. Ibid., pp. 78 ff.

11. Ibid., pp. 42, 54.

12. Ibid., p. 31.

13. Ibid., p. 108.

14. Ibid., p. 97.

15. Ibid., p. 104.

16. Watkins, *Science and Scepticism*, p. 53.

17. Stove, *Popper and After*, chapters 1 and 2.

18. Olding, "Popper for Afters," p. 21 (for some reason Stove had let Popper, though not Lakatos, off this hook).

19. Popper, *The Logic of Scientific Discovery*, footnote 1*, pp. 251–52.

20. Popper, *Objective Knowledge*, pp. 21–22.

21. Popper, *The Logic of Scientific Discovery*, p. 111.

22. Popper, "The Verification of Basic Statements" and "Subjective Experience and Linguistic Formulation." See also Haack, *Evidence and Inquiry*, chapter 5.

23. In a lecture given to the Department of Physics, University of Miami, spring 1998. Though Bondi elsewhere (e.g., in *The Universe at Large*) refers admiringly to Popper, and discusses "Olbers' Paradox," I have not been able to find the argument he made in this lecture in a published source.

24. On *Daubert* see chapter 9, pp. 242–44 and 251–52, and my "Trial and Error: The Supreme Court's Philosophy of Science."

25. For example, by Hesse in "Positivism and the Logic of Scientific Theories," p. 97, and in *The Structure of Scientific Inference*, pp. 89 ff.

26. Popper, "Conjectural Knowledge," footnote 29, p. 20.

27. Hempel, "Studies in the Logic of Confirmation," pp. 4, 6.

28. Ibid., pp. 39 ff.

29. Ibid., pp. 22–24.

30. Ibid., pp. 10 ff; the quotation is from p. 19.

31. Ibid., pp. 35 ff.

32. Tarski, "The Semantic Conception of Truth."

33. Carnap, *Logical Foundations of Probability*, pp. 478–82.

34. Hempel, "Postscript (1964) on Confirmation," p. 50.

35. Goodman, "The New Riddle of Induction."

36. Hempel, "Postscript (1964) on Confirmation," p. 51.

37. Hempel, "The Irrelevance of the Concept of Truth for the Critical Appraisal of Scientific Theories," pp. 77, 78. As we shall see in chapter 5, p. 136, his argument is a non sequitur—or rather, a pair of non sequiturs.

38. Reichenbach, *The Theory of Probability*; *Experience and Prediction*; "On the Justification of Induction."

39. But like Frank Ramsey, whom Carnap acknowledges on p. 36 of *Logical Foundations of Probability*.

40. See chapter 3, pp. 75–76, for further disambiguation.

41. Carnap, *Logical Foundations of Probability*, p. 43.

42. Ibid., p. 20.

43. Ibid., pp. 204 ff.

44. Ibid., p. 20.

45. On the analogy with deductive logic, see Carnap, *Logical Foundations of Probability*, pp. 200–201 and 297–98.

46. Carnap, *Logical Foundations of Probability*, pp. 243, 229.

47. Hintikka, "Towards a Theory of Inductive Generalization"; see also Cohen, "Inductive Logic 1945–1977."

48. Carnap, *Logical Foundations of Probability*, p. 218.

49. Ibid., p. 218.

50. Kuhn, *The Structure of Scientific Revolutions*, p. 11.

51. Ibid., p. 171.

52. Ibid., p. 170.

53. Ibid., p. 42.

54. Ibid., p. 111.

55. Ibid., p. 118.

56. Ibid., p. 129.

57. Kuhn, "Logic of Discovery or Psychology of Research?" quotation from p. 16.

58. Kuhn, "Reflections on Receiving the John Desmond Bernal Award," pp. 28, 30.

59. Kuhn, "Afterwords," p. 336.

60. Hesse, *The Structure of Scientific Inference*, p. 4.

61. Ibid., p. 27.

62. Ibid., p. 11.

63. Ibid., p. 72.

64. As foreshadowed in Hesse's earlier book, *Models and Analogies in Science*.

65. Hesse, *The Structure of Scientific Inference*, p. 104 (Hesse points out, however, that the attribution to the Reverend Thomas Bayes is not historically accurate, since he actually proved a different corollary of the multiplication axiom).

66. Carnap, *Logical Foundations of Probability*, p. 332.

67. Hesse, *The Structure of Scientific Inference*, p. 127.

68. Ibid., pp. 124–25.

69. Hesse, "How to Be Postmodern without Being a Feminist," pp. 458, 459.

70. Stove, *Popper and After*, p. 10 (p. 34 in the new edition, *Anything Goes*).

71. Popper, "Normal Science and Its Dangers."

72. Lakatos, "Falsification and the Methodology of Scientific Research Programmes," p. 87.

73. Ibid., p. 48.

74. Ibid., p. 29.

75. Popper, *The Logic of Scientific Discovery*, p. 107; Lakatos, "Falsification and the Methodology of Scientific Research Programmes," p. 16.

76. Lakatos, "Falsification and the Methodology of Scientific Research Programmes," p. 25 (I shan't go into the intricacies of Lakatos's interpretation of Popper, though I think it incorrect—as did Popper).

77. Lakatos, introduction to *The Methodology of Scientific Research Programmes*, p. 6; "Falsification and the Methodology of Scientific Research Programmes," p. 65.

78. This is part of the point of Stove's classification of Popper, along with Lakatos, Kuhn, and Feyerabend, as the first of "four modern irrationalists."

79. Feyerabend, *Killing Time*, p. 146; the accusation, from Theocharis and Psimopoulos, "Where Science Has Gone Wrong," was already cited in chapter 1, p. 21.

80. Feyerabend, *Killing Time*, p. 142.

81. Szasz, *The Second Sin*, pp. 26–27.

82. Quine and Ullian, *The Web of Belief*, p. 79.

83. Quine, *From Stimulus to Science*, p. 16.

84. Quine, "Epistemology Naturalized," pp. 87–88.

85. Quine, *Word and Object*, p. 23.

86. Quine, "On Empirically Equivalent Theories of the World," pp. 313, 322.

87. Ibid., p. 327.

88. Wang, *Beyond Analytic Philosophy*, pp. 153, 174–76.

89. "Worldly" rather than "semantic" because the latter might be mistaken as an allusion to a Carnapian reliance on analytic meaning-relations among predicates or, in the context of more recent philosophy of science, to Patrick Suppes' approach in terms of formal, mathematical models of scientific theories.

90. Lakatos, "Falsification and the Methodology of Scientific Research Programmes," p. 87.

91. I have borrowed the term "density" from Norman Levitt's description of the world as "dense but not impenetrable" to human inquirers; see *Prometheus Bedeviled*, p. 37.

92. On sociological critiques of science, see chapter 7; on literary and rhetorical critiques, chapter 8; on feminist critiques, chapter 11, pp. 315–17, and chapter 12, p. 341; and on the New Cynicism generally, chapter 12, pp. 337–41.

93. For example, on Bayesianism see chapter 3, pp. 75–76; on constructive empiricism, chapter 5, pp. 137–39; on naturalism, chapter 11, pp. 306–10.

3

CLUES TO THE PUZZLE
OF SCIENTIFIC EVIDENCE

A More-So Story

> The liberty of choice [of scientific concepts and theories] is of a special kind; it is not in any way similar to the liberty of a writer of fiction. Rather, it is similar to that of a man engaged in solving a well-designed word puzzle. He may, it is true, propose any word as the solution; but, there is only *one* word which really solves the puzzle in all its parts. It is a matter of faith that nature—as she is perceptible to our five senses—takes the character of such a well-formulated puzzle. The successes reaped up to now by science . . . give a certain encouragement to this faith.
>
> —Albert Einstein, "Physics and Reality"[1]

What is scientific evidence, and how does it warrant scientific claims? That honorific usage in which "scientific evidence" is vaguely equivalent to "good evidence" is more trouble than it's worth. When I write of "scientific evidence" I mean, simply, *the evidence with respect to scientific claims and theories*. Scientific evidence, in this sense, is like the evidence with respect to empirical claims generally—only more so: more complex, and more dependent on instruments of observation and on the pooling of evidential resources.

The only way we can go about finding out what the world is like is to rely on our experience of particular things and events, and the hypotheses we devise about the kinds, structures, and laws of which those particular things and events are instances, checked against further experience and further hypotheses, and subjected to logical scrutiny. The evidence bearing on any empirical claim is the result of experience and reasoning so far, a mesh of many threads of varying strengths anchored more or less firmly in experience and woven more or less tightly into an explanatory picture. So I look at questions about evidence, warrant, etc., not in pristine logical isolation, but in the context of facts about the world and our place as inquirers in the world. And I deliberately eschew the familiar Old Deferentialist jargon of the confirmation of theories by data or by observation or basic statements, to signal that my conception of evidence, presupposing no distinction of observational and theoretical statements, is considerably ampler than "data"; that my conception of warrant is ineliminably temporal, personal, and social; and that my account of the determinants of evidential quality is not purely formal, but worldly, and not linear, but multi-dimensional.

Scientific evidence, like empirical evidence generally, normally includes both experiential evidence and reasons, and both positive evidence and negative. It is complex and ramifying, structured—to use the analogy I have long found helpful, but only recently found anticipated by Einstein—more like a crossword puzzle than a mathematical proof. A tightly interlocking mesh of reasons (entries) well anchored in experience (clues) can be a very strong indication of the truth of a claim or theory; that is partly why "scientific evidence" has acquired its honorific use. But where experiential anchoring is iffy, or where background beliefs are fragile or pull in different directions, there will be ambiguity and the potential to mislead.

Of course, the role of an analogy is only to suggest ideas, which then have to stand on their own feet; of course, the usefulness of one analogy by no means precludes the possibility that others will be fruitful too; and of course, an analogy is *only* an analogy. Scientific evidence isn't like a crossword puzzle in every respect: there will be nothing, for example, corresponding to the appearance of a solution in tomorrow's paper; nor (though a seventeenth-century philosopher thinking of scientists as deciphering God's Book of Nature would have thought otherwise) is there a person who designs it. Nor, unlike Kuhn's mildly denigratory talk of normal science as "puzzle-solving," is my use of the crossword analogy intended to convey any suggestion of lightness or of the merely routine. But the analogy will prove a useful guide to some central questions about what makes evidence better or worse.

All of us, in the most ordinary of everyday inquiry, depend on learned perceptual skills like reading, and many of us rely on glasses, contact lenses, or hearing aids; in the sciences, observation is often highly skilled, and often mediated by sophisticated instruments themselves dependent on theory. All of us, in the most ordinary of everyday inquiry, find ourselves reassessing the likely truth of this claim or that as new evidence comes in; scientists must revise their assessments over and over as members of the community make new experiments, conduct new tests, develop new instruments, etc. All of us, in the most ordinary of everyday inquiry, depend on what others tell us; a scientist virtually always relies on results achieved by others, from the sedimented work of earlier generations to the latest efforts of his contemporaries.

This, from a 1996 press report on that controversial Martian meteorite, conveys some idea of just how much "more so" scientific evidence can be:

> A recovery team found [a 4.3-pound meteorite, designated ALH84001] in 1984. . . . 4 billion years earlier, it was part of the crust of Mars. (Scientists know this because when the rock is heated, it still gives off a mix of gases unique to the Martian atmosphere.) . . . From this unprepossessing piece of rock scientists have teased out . . . evidence leading toward an astonishing conclusion. Team member Richard Zare, a chemist at Stanford, used lasers and an extremely sensitive detector called a mass spectrometer to spot molecules called polycyclic aromatic hydrocarbons. PAHs result from combustion; they are found in diesel exhaust and soot. . . . But they also come from the decomposition of living organisms. The residue in ALH84001, says Zare, "very much resembles what you have when organic matter decays." . . . Under another high-tech sensor, an ultra-high-resolution transmission microscope [scientists] found that the thin black-and-white bands at the edge of the carbonates were made of mineral crystals 10 to 100 nanometers across. . . . The crystals in the meteorite were shaped like cubes and teardrops, just like those formed by bacteria on earth. [David MacKay of the Johnson Space Center says] "We have these lines of evidence. None of them in itself is definitive, but taken together the simplest explanation is early Martian life." . . . Some scientists in the field express more optimism than others.[2]

Since then there has been heated controversy over whether or not this really was evidence of early Martian life. In 1998 new chemical studies comparing organic materials in the meteorite with those found in the surrounding Antarctic ice showed that significant amounts of the organic compounds in the meteorite are terrestrial contamination; but these studies didn't examine the crucial molecules, the PAHs. Controversy seems likely to continue at least until new samples of Martian rock and soil can be brought back by robotic spacecraft.[3]

As the example suggests, warrant comes in degrees, and is relative to a time;

a scientific idea, usually very speculative at first, tends either to get better warranted, or to be found untenable, as more evidence comes in or is flushed out. As the example also suggests, talk of the degree of warrant of a claim at a time, simpliciter, is shorthand for talk about how warranted the claim is at that time by the evidence possessed by some person or group of people.[4]

Since it is individuals who see, hear, etc., my account begins with the personal conception, the degree of warrant of a claim for a person at a time. The next step, distinguishing a person's experiential evidence and his reasons, and explaining how the two work together, is to articulate what makes a person's evidence with respect to a claim better or worse, and hence what makes the claim more or less warranted for him. Then, to articulate something of what is involved in evidence-sharing, I shall need to extrapolate from the degree of warrant of a claim for a person at a time to the degree of warrant of a claim for a group of people at a time; and then to suggest an account of the impersonal conception, of the degree of warrant of a claim at a time, simpliciter. Then I will be able to say something about how the concept of warrant relates to the concepts of justification and confirmation; to explain how degree of warrant ideally relates to degree of credence; and to discriminate what is objective, and what perspectival, in the concepts of warrant, justification, and reasonableness.

Because warranted scientific claims and theories are always warranted by somebody's, or somebodies', experience, and somebody's, or somebodies', reasoning, a theory of warrant must begin with the personal, and then move to the social, before it can get to grips with the impersonal sense in which we speak of a well-warranted theory or an ill-founded conjecture. This, obviously, is about as far as it is possible to be from Popper's ideal of an "epistemology without a knowing subject." Ironically enough, however, it is almost as congenial to his analogy of scientific knowledge as like a cathedral built over the centuries by generations of masons, carpenters, glaziers, gargoyle carvers, and so on, as to mine of scientific knowledge as like part of a vast crossword gradually filled in by generations of specialists in anagrams, puns, literary allusions, and so forth.

WARRANT—THE PERSONAL CONCEPTION

What determines the degree of warrant of a claim for a person at a time is the quality of his evidence with respect to that claim at that time. "His evidence" refers both to his experiential evidence (his seeing, hearing, etc., this or that, and his remembering having seen, heard, etc., this or that—his past experiential evidence), and to his reasons (other beliefs of his). There are significant asymmetries between experiential evidence and reasons, as between clues and crossword entries: most importantly, the question of warrant arises with respect to a person's reasons, as it arises with respect to crossword entries; but perceptual, etc., events and states, like clues to a crossword, neither have nor stand in need of warrant.[5]

Let me take experiential evidence first.

Both in the law and in everyday life, there is a usage in which "evidence" means "physical evidence," and refers to the actual fingerprints, bitemarks, documents, etc. We hear reports of new evidence about a plane crash brought up from the ocean floor, or of new evidence about a crime discovered in a suspect's apartment. My account will accommodate this usage, not directly, but in an oblique way, by taking for granted that in scientific observation, as in perception generally, we interact by means of our sensory organs with things around us— with the traces of the gases given off by that Martian meteorite when it is heated, with stuff on the slides under the microscope, with Rosalind Franklin's X-ray diffraction photographs of DNA, and so on. So in my account, the bits of airplane, the incriminating letter, etc., are the *objects* of experiential evidence, *what* is perceived. A person's experiential evidence is his perceptually interacting in one way or another—with the naked eye at a distance in poor light, by means of a powerful microscope in good light, etc.—with a thing or event.

Thinking of experiential evidence in science, it is natural to speak, not of perception, but of observation; and here—as when we speak of the observations made by a detective, or of a patient's being "under observation" in hospital—the word carries a connotation of deliberateness. Scientific observation is active, selective; it calls for talent, skill, and sometimes special training or background knowledge, as well as patience and sharp eyes. Very often it is mediated by instrumentation. Experiential evidence and reasons work together, as the reasonableness of a crossword entry depends in part on its fit with the clue and in part on its fit with intersecting entries. I don't assume a class of claims (the "observation statements" of some Old Deferentialist accounts) fully warranted by experience alone; rather, I see experiential evidence and reasons as carrying the burden in different proportions for different claims. But neither do I assume that each scientific claim has its own experiential evidence, as in a conventional crossword each entry does; often it is more like an unconventional crossword in which a clump of entries shares a clue, or a bunch of clues.

All this, obviously, takes the relevance of experience to warrant for granted. So what about Popper's argument for the *irrelevance* of experience—that, since there can be logical relations only among statements, not between statements and events, scientists' seeing, hearing, etc., this or that can have no bearing on the warrant of scientific claims and theories? It is true that logical relations hold only among statements (or whatever the truth-bearers are); but the conclusion Popper draws—that, e.g., someone's seeing a black swan is utterly irrelevant to the reasonableness or otherwise of his accepting the statement that there is a black swan at such-and-such a place at such-and-such a time, and hence to the reasonableness or otherwise of his rejecting the statement that all swans are white—is about as thoroughly implausible as a conclusion could be; so implausible that Popper himself elides it into the quite different thesis I have been defending, that experiential evidence is relevant but not sufficient.[6] This doesn't yet tell us *how* experience contributes to warrant; but it does tell us that the other assumption on which Popper's argument for the irrelevance of experience depends—that warrant is a matter exclusively of logical relations among statements or propositions—must be untrue.

So, how does experience contribute to warrant? A simple answer might rely on the old idea that, while the meanings of many words are learned by verbal definition in terms of other words, the meanings of observational words are learned by ostensive definition, as the language-learner hears the word used by someone pointing out something to which it applies.[7] So a person's seeing a dog warrants the truth of his belief that there's a dog present in virtue of the fact that "dog" is ostensively defined in such a way as to guarantee that it is appropriate to use it in just such observable circumstances as these. This picture, with its simple division of terms into observational and other, and of definitions into ostensive and verbal, won't do as it stands. Language is far subtler than that, the interconnections of words with observable circumstances and among themselves much more tangled—as the language-learner soon discovers as he masters "toy dog," "looks like a dog," etc., and learns more about what the truth of "it's a dog" requires and what it precludes. Nevertheless, the central idea seems right: our perceptual interactions with the world give some degree of warrant to claims about the world because of the connections of words with the world and with each other that we learn as we learn language.

Perhaps we can preserve this central idea while remedying the deficiencies of the simple dichotomy of ostensive versus verbal definitions. Even a very simple correction, replacing the dichotomy of observational versus theoretical predicates by a continuum of more and less observational, less and more theoretical, would be an improvement. But it would be better to make room for the possibility of different speakers learning a word in different ways, and of terms that can be learned *either* by a combination of ostension and verbal explanation *or* entirely by verbal explanation. Correcting the simple contrast of ostensive versus verbal definitions, allowing for the tangled mesh of extra- and intra-linguistic connections of words, we could explain both how experiential evidence can contribute to the warrant of a claim, and how the warrant given a claim by a person's experience may be enhanced, or diminished, by his reasons.

We nearly all encounter a sentence like "this is a glass of water,"[8] in the first instance, by hearing it used in normal circumstances in which a glass of water is visible to both teacher and learner. Subsequently, however, we learn a lot of caveats and complications: a glass of water looks, smells, tastes, etc., thus and so, *provided* the observer and the circumstances of observation are normal; if the stuff in the container is *really* water, it will give such and such results under chemical analysis; etc., etc. So seeing the thing can partially, though only partially, warrant the claim that there's a glass of water present; for a normal observer in normal circumstance can tell it's a glass of water by looking, even though there is room for mistake.

A molecular biologist has to learn to read an X-ray diffraction photograph, as all of us had to learn to read. Someone who had learned the predicate "helix," ostensively or otherwise, by reference to simple examples like a telephone cord, but who had no experience of X-ray diffraction photographs, wouldn't be able to make much of Rosalind Franklin's photograph of the B form of DNA. As soon as James Watson saw it, however, he was firmly convinced that the DNA molecule is helical. And his seeing the photograph partially, but only partially, warranted this claim. For a trained observer in appropriate circumstances can tell it's a helix by looking at a (good enough) X-ray diffraction photograph, even though there is room for mistake.

In sum: a person's seeing, etc., this or that can contribute to the warrant of a claim when key terms are learned by association with these observable circumstances—the more (the less) so, the more (the less) the meaning of those terms is exhausted by that association. Experiential evidence consists, not of propositions, but of perceptual interactions; and it contributes to warrant, not in virtue of logical relations among propositions, but in virtue of connections between words and world set up in language-learning.

Now let me turn to reasons.

When, earlier, I rather casually referred to a person's reasons as other beliefs of his, I hadn't forgotten that according to some philosophers, among them both Peirce and Popper, belief has no place in science. I agree that faith, in the religious sense, does not belong in science; though in their professional capacity scientists accept various claims as true, this usually is, or should be, tentative, and always in principle revisable in the light of new evidence. By my lights, however, to believe something *is* to accept it as true, in just this fallibilist sense;[9] that's why I shall sometimes write of the "degree of credence" a person places in a claim or theory.

Unfortunately, it won't quite do simply to construe a person's reasons as those propositions in which he places some degree of credence, ignoring the fact that some of his beliefs are strongly held and others weakly—any more than it would do, in judging the plausibility of a crossword entry, to ignore the fact that one intersecting entry is written firmly in ink, another only faintly in pencil. If a crossword entry intersecting the entry at issue is only lightly pencilled in, it counts for less, positively or negatively, than if it is indelibly inked in; similarly, if a person gives a reason for or against a claim only a modest degree of credence, it should count for less, positively or negatively, than if he holds it very firmly. One way of handling this might be to treat a person who places some but less than complete credence in a proposition as giving full credence to a hedged version of the same proposition, including among his reasons "there is a good chance that p," "it is likely that p," or "it is possible that p" (in which case we will need to find a way to accommodate such hedged propositions into our account of supportiveness of evidence). Another way, which I shall explore in more detail below, would be to include the propositions without the hedges, and compensate by adjusting the degree of warrant of the claim for, or against, which they are reasons (in which case we will need to avoid introducing inconsistencies by misrepresenting someone who has no idea whether or not p as giving both p and its negation some degree of credence.)

Unlike his experiential evidence, a person's reasons are propositional; and so it might seem that here at least we must be squarely in the domain of logic. Not so, however. Reasons ramify, more like the entries in a crossword puzzle than the steps in a mathematical proof. The plausibility of a crossword entry depends not only on how well it fits with the clues and any intersecting entries, but also on how plausible those other entries are, independent of the entry in question, and on how much of the crossword has been completed. Similarly, the quality of a person's evidence with respect to a claim depends not only on how supportive his reasons are of that claim, but also on how warranted those reasons are, independent of the claim in question, and on how much of the relevant evidence his evidence includes. Moreover, as it turns out not even supportiveness—not even conclusiveness, the limit case of supportiveness—is quite simply a matter of logic.

For reasons to be conclusive with respect to a claim—i.e., to support it to the highest possible degree—it is not sufficient that they deductively imply the claim. For inconsistent propositions deductively imply any proposition whatever (from p and not-p, q follows, whatever q may be);[10] but inconsistent reasons aren't conclusive evidence for anything, let alone for everything (p and not-p isn't conclusive evidence for any q, let alone for every q).[11] For example, suppose the evidence is: that the murderer is either Smith or Jones; that whoever committed the murder is left-handed; that Smith is right-handed; and that Jones is right-handed. This deductively implies that Jones did it; *and* that Smith did it; *and* that aliens did it. But it is certainly not conclusive evidence for any of these claims, let alone for all of them. However, if the evidence were: that the murderer is either Smith or Jones, that whoever committed the murder is left-handed, that Smith is right-handed, and that Jones is left-handed, it would be conclusive with respect to the claim that Jones did it. So conclusiveness requires that the evidence deductively imply the claim in question, *but not also its negation*; i.e., that it deductively imply that claim differentially, and not just in virtue of the fact that, being inconsistent, it implies every proposition whatsoever.[12]

The principle that everything follows deductively from a contradiction is a principle of classical logic. So non-classical logicians may object that while the inference from "*p* and not-*p*" to an arbitrary "*q*" is valid in classical logic, there is a whole range of non-classical systems—paraconsistent logics, relevance logics, connexivist logics, etc., etc.—in which this inference is not valid; and propose that we close the gap between conclusiveness of evidence and deductive implication by resorting to such a logic. I suspect that the motivation for such non-standard systems derives at least in part from a confusion of logical with epistemological issues; but I don't rule out the possibility that they might shed some light on how inconsistent evidence could, in some circumstances, be better than simply indifferent with respect to supportiveness.[13]

Again, lawyers might object that inconsistent testimony can be extremely informative. Indeed it can; but that witness A says that *p*, while witness B says that not-*p*, does not constitute inconsistent evidence in the sense at issue here (i.e., evidence of the form "*p* and not-*p*"). Granted, a person who is aware of an inconsistency in his evidence with respect to some claim is in something like the position of a lawyer faced with inconsistent testimony; and if he is sensible he will try to identify the background beliefs responsible for the inconsistency, and assess which are better warranted. Witness A saw the murder from close by, a juror might reason, B only from a distance, so A's testimony is likelier to be right; or: A is the defendant's brother-in-law, while B is a stranger to him, so B has less reason to lie. A scientist who realizes that there is an inconsistency in his evidence may reason in a similar way: "my confidence that DNA is composed of the four nucleotides in regular order is less well warranted than my confidence that bacterial virulence is contained in nucleic acid, not protein; so of my evidence that DNA is the genetic material, and my evidence that it isn't, the former is likelier to be right." But this is quite compatible with my point, which is only that inconsistent evidence is not conclusive evidence.

*   *   *

Against the background of the familiar quarrels between the inductivist and deductivist wings of the Old Deferentialism, it may seem that to acknowledge that there is such a thing as supportive-but-not-conclusive evidence must be to declare allegiance to the inductivist party. Not so, however. There is supportive-but-not-conclusive evidence; but there is no syntactically characterizable inductive logic, for supportiveness is not a purely formal matter.

David Mackay observes that, though the evidence derived from that meteorite is not definitive, "the simplest explanation is early Martian life." He takes for granted that the evidence so far supports the idea of early Martian life because there having been bacterial life on Mars long ago would explain how things come to be as the evidence says. And, whether or not he is right about bacterial life on Mars, he is right to assume a connection between supportiveness and explanation.

The connection is not, however, simply that evidence supports a claim in virtue of the claim's being the best explanation of the evidence. Supportiveness of evidence is not categorical, but a matter of degree. That there is a significantly greater incidence of lung cancer in smokers than in non-smokers, for example, supports the claim that smoking causes lung cancer; but the degree of support is very significantly enhanced by additional evidence of specific genetic damage connected to lung cancer and caused by smoking. Moreover, there is that "mutual reinforcement between an explanation and what it explains."[14] In the example just given, the evidence supports the claim in virtue of the claim's potential to explain the evidence. But the explanatory connection may go either way; in other cases it is a matter, rather, of the evidence potentially explaining the claim. That there is a trough of low pressure moving in a southeasterly direction, for example, supports the claim that Hurricane Floyd will turn north before it reaches the South Florida coast, because there being such a trough of low pressure would explain the hurricane's turning north. So "inference to the best explanation" is too one-directional, and captures only a small part of a larger picture in which degree of supportiveness of evidence is tied to degree of explanatory integration of the evidence with the claim in question.

Explanatory integration is a pretty concept, but not easy to spell out. But it is clear, at any rate, that neither explanation nor, a fortiori, explanatory integration or supportiveness of evidence, can be narrowly logical concepts. For explanation, like prediction, requires the classification of things into real kinds. Knowing that geese migrate south as the weather cools, we predict that when the weather gets cooler this goose will fly south, and explain that this goose flew south because the weather got cooler—which is only possible because classifying something as a goose identifies it as of a kind members of which behave thus and so. There is the same covert generality in the previous examples: e.g., if "trough of low pressure" and "hurricane" didn't pick out real meteorological phenomena connected by real laws, the prediction would be unjustified and the appearance of explanatoriness bogus. Explanatoriness is not a purely logical, but a worldly, concept.

So if we think of supportiveness as a relation among sentences, it will be a vocabulary-sensitive relation, requiring kind-identifying predicates; in other words, it will not be syntactic, a matter of form alone, but broadly semantic, depending on the extensions of the predicates involved. (The point is masked, but not obviated, if we think of supportiveness as, rather, a relation among propositions.) This suggests why scientists so often find themselves obliged to modify the vocabulary of their field, shifting the use of old terms or introducing new ones: a vocabulary can not only be more or less convenient or more or less transparent in meaning, but also—most importantly—more or less successful at identifying kinds of thing, stuff, or phenomenon.

How plausible a crossword entry is depends not only on how well it fits with the clue and any already-completed intersecting entries, but also on how plausible those other entries are, independent of the entry in question, and how much of the crossword has been completed. Analogously, the degree of warrant of a claim for a person at a time depends not only on how supportive his evidence is, but also on how comprehensive it is, and on how secure his reasons are, independent of the claim itself.

A person's evidence is better evidence with respect to a claim, the more (less) warranted his reasons for (against) the claim in question are, independently of any support given them by that claim itself. So (in line with the second possible way of handling weakly believed reasons) we can include a proposition among a person's reasons if he gives it any degree of credence, without giving partially believed reasons more weight than we should; for the unhedged "p" or "q" included as proxy will be less independently secure than the hedged "possibly p" or "maybe q" that would more accurately represent the person's low degree of credence. A weakly believed reason for a claim will contribute less to its warrant.

Although the independent security clause mentions warrant, there is no vicious circularity. In a crossword, the reasonableness of an entry depends in part on its fit with other entries, and hence on how reasonable they are, independent of the entry in question. Similarly, the warrant of a claim depends in part on the warrant of other claims that support it, independent of any support given to them by the claim itself. This interlocking of mutually supportive claims and theories no more conceals a vicious circle than the interlocking of mutually supportive crossword entries does. Nor does it threaten us with an infinite regress, or leave the whole mesh floating in mid-air; for experiential evidence, which stands in no need of warrant, serves as anchor for scientific claims, as clues do for crossword entries.

The quality of a person's evidence, and hence the degree of warrant of a claim for him, also depends on how much of the relevant evidence his evidence includes. Comprehensiveness is one of the determinants of evidential quality, not an afterthought to be relegated to methodology. (Stating the comprehensiveness requirement precisely, however, would call for an extension of the conception of evidence on which I have relied thus far; for in this context "all the relevant evidence" has to mean something like "answers to all the relevant questions.") Even if it strongly supports the claim in question, even if it is very secure itself, evidence is poorer insofar as relevant information is missing. Weakness on the dimension of comprehensiveness is apt to make evidence misleading, i.e., supportive of a false conclusion; and since the evidence with respect to a scientific claim is never absolutely comprehensive, there is always the possibility that, as new evidence comes in, the evidence so far will turn out to have been misleading.

Because the determinants of evidential quality are multi-dimensional, and one claim may do well on one dimension and another on another, there is no guarantee of a linear ordering of rival claims with respect to degrees of warrant. Moreover, the three dimensions interact. Evidence which is poor on the dimension of comprehensiveness is often also poor on the dimension of supportiveness; while evidence which is highly supportive of a claim is often rather lacking in independent security.

This sheds some light on an old disagreement about the status of negative evidence. Popper's thesis that scientific claims can be falsified, but not verified or confirmed, derives in part from his criterion of demarcation (equating "scientific" with "falsifiable"), and in part from the assumption that a single negative instance falsifies a generalization. The position associated with Quine and Duhem, by contrast, is that scientific laws and generalizations are no more decisively falsifiable than they are decisively verifiable. The demarcation issue aside, there is disagreement about the evidential relation of negative instances to generalizations. Popper is right, of course, that negative evidence needs to be taken into account; and right that, for instance, that there is a black swan at Perth airport at such-and-such a time is a conclusive reason against the generalization that all swans are white. But his critics are right to insist that this doesn't settle the matter. Conclusiveness is the highest degree of supportiveness; but supportiveness is only one dimension of evidential quality. It does not guarantee decisiveness, which would require in addition that the conclusive reason be perfectly independently secure, and that it be all the relevant evidence.[15]

WARRANT—THE SOCIAL CONCEPTION

Now let me turn to the warrant of a claim for a group of people.[16]

In 1954 George Gamow set up the RNA Tie Club, a group of 20 people—one for each amino acid—devoted to figuring out the structure of RNA and the way it builds proteins. Each member was to have a black RNA tie embroidered with green sugar-phosphate chain and yellow purines and pyrimidines, and a club tie-pin carrying the three-letter abbreviation of his assigned amino acid; later there was even RNA Tie Club stationery, with a list of officers ("Geo Gamow, Synthesizer, Jim Watson, Optimist, Francis Crick, Pessimist, . . .").[17] Very few scientific communities, however, are as definitely identifiable as this; the notion of a scientific community is notoriously vague, and specifying criteria for what is to count as a *scientific* community, let alone for what is to count as *one* such community, is a formidably difficult task.

In fact, "the" scientific community to which philosophers of science sometimes optimistically refer is probably more mythical than real; the reality is a constantly shifting congeries of sub-communities, some tightly interconnected and some loosely, some nested and some overlapping, some short-lived and some persisting through several generations of workers. So it is just as well that I can sidestep the awkward problems about the individuation of scientific communities and sub-communities, because my present task is to specify on what the degree of warrant of a claim depends for *any* collection of scientists, whether that collection is a close-knit sub-community or a scattered or gerrymandered group.

"One man's experience is nothing if it stands alone," wrote C. S. Peirce,[18] meaning that its engagement of many people, within and across generations, is one of the great strengths of the scientific enterprise. He was right; and not least because this enables the sciences to extend their evidential reach far beyond that of any individual. But it is not an unmixed blessing. For in any group of scientists there will likely be disagreements both about the claim the warrant of which is at issue, and about the reasons for or against that claim; experiential evidence, furthermore, is always some individual's experiential evidence; and in even the most close-knit group of scientists there will be failures of communication, with each member having only imperfect access to others' evidence.

Given that different individuals within a group of scientists may disagree not only in the degree of credence they give the claim in question, but also in their background beliefs, we can't construe the group's evidence as a simple sum of all the members' evidence. But the crossword analogy suggests a way to overcome this first difficulty. Think of several people working on the same crossword, agreeing that 2 down is "egregious," and 3 across "gigantic," but disagreeing about 4 across, which some think is "intent," and others think is "intern." What would determine how reasonable, given the evidence possessed by this group of people, an entry which depends on 4 across is? Presumably, how reasonable it is if the disputed entry is *either* "intent" *or* "intern" (or equivalently, since the rival entries agree in their first letters, if the last letters are either "nt" or "rn.") Similarly, where there is disagreement in background beliefs within a scientific community, the best approach may be to construe the group's evidence as including not the conjunction of the rival background beliefs, but their disjunction. However, this one-size-fits-all solution will need considerable adjustment to accommodate disagreements of different shapes and sizes: the community may, for example, be more or less evenly divided, or there may be just one dissenter.

It is always an individual person who sees, hears, remembers, etc. In scientific work, however, many people may make observations of the same thing or event; of an eclipse from observatories in the northern and in the southern hemispheres, for example. By observing the same thing or event from different places, scientists have access to more of the information the thing or event affords. And by having several people make the same observation, they can discriminate the eccentricities of a particular individual's perceptions from what can be perceived by all normal observers. Sometimes one person claims to be able to see what no one else can: all the observations supposedly confirming that a homeopathic dilution of bee-venom degranulates blood cells, apparently, were made by one observer, Elisabeth Davenas. In such circumstances, either the person involved is an especially talented observer (as Jacques Benveniste maintains Mlle Davenas is), or else he or she is, as we say, "seeing things" (as John Maddox and the team he sent from *Nature* to investigate the work of Benveniste's lab maintain Mlle Davenas must be).

In relying on others' observations, scientists depend on those others' perceptual competence, on the working of the instruments on which they rely, and on the honesty and accuracy of their reports. It is a matter, not simply of mutual trust, but of justified mutual confidence (usually grounded implicitly in the observer's, or the instrument's, credentials). Scientists will reasonably take into account that an observer's commitment to this or that theory may make him readier to notice some aspects of what he or she sees than others; and if they have grounds for suspecting the observer of perceptual defect, instrumental failure, dishonesty, or self-deception—whether directly or, as with those homeopathy experiments, because the supposed results are so extraordinary—they may reasonably doubt the reliability of his or her observational reports. In a group of scientists, even if each has his own experiential evidence, most depend at second hand on others'. So the warrant of a claim for the group will depend in part on how reasonable each member's confidence is in others' reports of their observations; and in part (now I turn to the third difficulty mentioned earlier) on how good communication is within the group.

It hardly seems appropriate to allow that a claim is warranted for a group in which evidence is not shared, but merely scattered: as with two scientists centuries apart, the later quite unaware of the work of the earlier, or with rival research teams neither of which has ever seen the other's reports. We would not count a claim as well warranted for a group of people, even if between them they possess strong evidence for it, unless that evidence is communicated among the members of the group. Only when their evidence is shared—as when the several people working on the same part of the crossword puzzle are all able to look over the others' shoulders—can their joint evidence warrant a claim. "Efficiency of communication" covers a whole range of issues: how effectively refereeing and publishing processes ensure that good work is published quickly, and not drowned in a sea of worthless busywork; how good the means are of finding relevant material; how far conferences manage to be occasions for genuine communication and mutual education rather than mere self-promotion and networking; how cogently and clearly work is presented.

So we could think of the degree of warrant of a claim for a group of scientists as the degree of warrant of that claim for a hypothetical individual whose evidence is the joint evidence of all the members of the group, only construed as including not the conjunctions but the disjunctions of disputed reasons, and discounted by some measure of the degree to which each member is justified in believing that others are reliable and trustworthy observers, and of the efficiency or inefficiency of communication within the group.

WARRANT—THE IMPERSONAL CONCEPTION

Now I can say something about the impersonal conception, of the degree of warrant of a claim at a time, simpliciter.

When, looking at science from the outside, you wonder which claims and theories are well and which poorly warranted, it is this impersonal conception which is most salient. But to say that a claim or theory is well or poorly warranted at a time must be understood as an elliptical way of saying that it is well or poorly warranted by the evidence possessed by some person or some group of people at that time. And since a claim may be well warranted for this group or person, but poorly warranted for that group or person, the question is on *whose* evidence "impersonal" warrant is appropriately taken implicitly to depend.

I shall construe it as depending on the evidence of the person or group of people whose evidence is, in a certain sense, the best. "In a certain sense," because in this context, "best evidence" means "best indicator of the likely truth of the claim or theory in question." This is adapted from a legal conception of "best evidence,"[19] and should not be confused with "best evidence" in the sense of "evidence which gives the highest degree of warrant to the claim or theory in question." The difference is that in the sense relevant here what matters is how secure and how comprehensive the evidence is, *whether it is supportive or undermining, favorable or unfavorable* to the truth of the claim in question.

When there is efficient communication within a group, the group's shared evidence may be better than that of any individual member; but when communication within the group is poor, an individual's evidence may be the best. For example, after trying unsuccessfully to interest Karl Nägeli in his work, Gregor Mendel published his "Experiments in Plant Hybridization" in the journal of the Society of Natural Science in Brunn, Moravia, where it languished unread for decades; so for a time his evidence with respect to the particulate theory of inheritance was much better than anyone else's—in both senses. The theory was more warranted for him than for anyone else, but his was also the best evidence in the sense at issue here: i.e., the best indicator of the likely truth of the theory. It was also, in the relevant sense, the best evidence with respect to the rival, blending theory; to which, however, it was unfavorable.

Again: in 1944 Oswald Avery wasn't ready to say in his scientific publications, as he suggested in a letter to his brother, that his experiments indicated that DNA, not protein, is the genetic material. So for a time his evidence with respect to the protein-as-genetic-material hypothesis was, in the sense presently at issue, better than anyone else's: i.e., the best indicator of the likely truth of the hypothesis—but unfavorable. His was also, in the relevant sense, the best evidence with respect to the rival DNA-as-genetic-material hypothesis; to which, of course, it was favorable. Much later, however, in his textbook on molecular biology, Watson would refer to "Avery's bombshell," the "first real proof of the genetic role for nucleic acids";[20] and in the Rockefeller University calendar for the academic year 2001, beside a rather charming photograph of Avery in a funny hat at Rockefeller's 1940 Christmas party, the text informs us that it was Avery, with his colleagues Colin MacLeod and Maclyn McCarty, who "showed for the first time that genes are made of DNA."

WARRANT, JUSTIFICATION, AND CONFIRMATION

Now I can tackle the question of the relation of warrant to such other concepts as justification and confirmation.

For a claim to be warranted to some degree, I shall require (not that the evidence indicate that the claim is more likely than not, but) only that the evidence indicate that the claim is non-negligibly likely. The claim that p is well warranted for an individual if his evidence strongly indicates that p; the claim is fairly warranted for him if his evidence fairly strongly indicates that p; it is weakly warranted for him if his evidence weakly indicates that p; and it is unwarranted for him if his evidence does not indicate that p—whether because it indicates that not-p, or because it is too impoverished even weakly to indicate either p or not-p.

That a claim is highly warranted for a person doesn't guarantee that he is in good epistemic shape with respect to that claim. A scientist may accept a claim, with greater or lesser confidence, as true; or accept its negation, with greater or lesser confidence, as true; or give no credence either to the claim or to its negation. Ideally, he would give p the degree of credence it deserves. But he may fall short of this ideal either because p is well warranted for him, but he gives it too little credence, or because p is poorly warranted for him, but he gives it too much credence. These failings may be described, respectively, as underbelief and overbelief.

Moreover, it may not be the evidence a scientist possesses that moves him to give a claim whatever degree of credence he does. He may give some degree of credence to a claim because he is impressed by the fact that an influential figure in his profession has endorsed it, or because he very much wants things to be as the claim says, or, etc. In such a case I shall say that, even if the claim is warranted for him, he is not justified in giving it the degree of credence he does.[21] (Note to Karl Popper: justification, in the sense just explained, is a partly causal notion; and experiential evidence can contribute to the justification of a person's belief precisely by contributing causally to his accepting it.)

At any time, some scientific claims and theories are well warranted; others are warranted poorly, if at all; and many lie somewhere in between. Sometimes several competing claims may all be warranted to some degree. When no one has good enough evidence either way, a claim and its negation may be both unwarranted; in which case, the best option is—admitting that at the moment we just don't know—to seek out more evidence, and to rack our brains for other candidate hypotheses.

Most scientific claims and theories start out as informed but highly speculative conjectures; some seem for a while to be close to certain, and then turn out to have been wrong after all; a few seem for a while to be out of the running, and then turn out to have been right after all. Many, eventually, are seen to have been right in part, but also wrong in part. Some mutate, shifting in content to stand up to new evidence in an adapted form. Ideally, the degree of credence given a claim by the relevant scientific sub-community at a time—assuming we can give some sense to this not entirely straightforward idea—would be appropriately correlated with the degree of warrant of the claim at that time. The processes by which a scientific community collects, sifts, and weighs evidence are fallible and imperfect, so the ideal is by no means always achieved; but they are good enough that it is a reasonable bet that much of the science in the textbooks is right, while only a fraction of today's frontier science will survive, and most will eventually turn out to have been mistaken. Only a reasonable bet, however; all the stuff in the textbooks was once speculative frontier science, and textbook science sometimes turns out to be embarrassingly wrong.

I shall say that a claim is confirmed when additional evidence raises its degree of warrant, the degree of confirmation depending on the increment of warrant. Thus construed, the concept of confirmation is not only distinct from, but presupposes, the concepts of warrant and supportiveness. Some Old Deferentialists, however, used "confirm" indifferently for supportiveness, warrant, *and* confirmation. The confusions such ambiguities generate linger on in, for example, the still-common idea that evidence already possessed at the time a theory was proposed cannot support it; which seems plausible only if supportiveness and confirmation are run together.[22]

In my usage, we can describe evidence as confirming a claim (1) when new evidence, i.e., evidence not previously possessed by anyone, raises its degree of warrant; (2) retrospectively, when a claim previously already warranted to some degree became more warranted when such-and-such then new, but now familiar, evidence came in; or (3) to assess the degree of warrant of a claim first relative to such-and-such evidence, and then relative to that evidence plus additional evidence not previously included in the reckoning.

This suggests a way of approaching an old controversy about whether true predictions are especially confirmatory. On the one hand, it is certainly impressive when astronomers predict that Halley's comet will reappear or that the sun will be eclipsed at such-and-such a future time, and turn out to be right. On the other hand, it is certainly puzzling how the fact that a statement is about the future, in and of itself, could endow it with any special epistemological importance. The explanation is that, though the intuition that successful prediction can be strongly confirmatory is correct, the reason is not simply that it is a true prediction. Verification of a prediction derived from a claim is always new evidence, in the sense required by (1) or (retrospectively) by (2). However, new evidence may concern past events, and not only future ones; e.g., if an astronomical calculation has the consequence that there was a solar eclipse at such-and-such a time in ancient history, and then new evidence is found that in fact there was, this true "postdiction" confirms the theory no less than a true prediction would do. Moreover, additional evidence in the sense of (3), even if it is not new evidence in the sense of (1) or (2), may also be confirmatory.

Thus far, I have said only that confirmatory evidence raises the degree of warrant of a claim. In ordinary usage, however, "confirm" is quite often used comparatively, to indicate that additional evidence warrants p over some rival q. We might say that additional evidence which raises the degree of warrant of p but lowers the degree of warrant of q "confirms p over q." In ordinary usage, again, "confirm" also often carries a suggestion that the claim confirmed is now not merely *more* warranted, but *firmly* warranted. We might say that additional evidence that raises the degree of warrant of p beyond some specified cut-off point is "strongly confirmatory."

You may have noticed that though I have talked in terms of degrees of credence, degrees of warrant, and degrees of confirmation, and occasionally of likelihoods, I have thus far rather pointedly avoided "probable." By now, the reason should be pretty obvious: the classical calculus of probabilities, originally devised to represent the mathematics of games of chance, looks like a poor match for degrees of warrant. It could hardly constitute a theory of warrant, if this concept is as subtle and complex as it seems to be. Nor could it constitute a calculus of degrees of warrant; for the probability of p and the probability of not-p must add up to 1, but if there is insufficient evidence either way, neither a claim nor its negation may be warranted to any degree. For example, scientists now believe that mad-cow disease is caused by prions, protein molecules abnormally folded up in the cell;[23] but neither this claim nor its negation was even intelligible until the concept of macromolecule was developed, and neither was warranted to any degree until the significance of the folding of macromolecules began to be understood, and mad-cow disease was identified.

Naturally, given my reservations about probabilism generally, I am disinclined towards Bayesianism specifically (nor have I forgotten that even so determined a probabilist as Carnap warns of the dangers of putting too much epistemological weight on Bayes' theorem). Of course, there's nothing wrong with the theorem itself, *qua* theorem of the calculus of probabilities; and presumably, when they engage in statistical reasoning, scientists sometimes calculate probabilities in a Bayesian way. However, as even the most enthusiastic Bayesians acknowledge, degrees of credence, construed purely descriptively, need not satisfy the axioms of the calculus of probabilities; they may not be coherent. And if, as I argued above, degrees of warrant need not satisfy the axioms of that calculus either, then there is good reason (over and above familiar worries about where the priors come from) for denying that Bayes' theorem could be an adequate model of scientists' readjustments of degrees of warrant in the light of new evidence.

Complex and diffuse as it is, evidence is a real constraint on science. And though the degree of warrant of a claim at a time depends on the quality of some person's or some group's evidence at that time, the quality of evidence is not subjective or community-relative, but objective.

However, it doesn't follow from the objectivity of evidential quality that it is transparent to us. In fact, judgments of the quality of evidence depend on the background beliefs of the person making the judgment; they are perspectival. If you and I are working on the same crossword, but have filled in the much-intersected 4 down differently, we will disagree about whether the fact that an entry to 12 across ends in an *F*, or the fact that it ends in a *T*, makes it plausible. If you and I are on the same hiring committee, but you believe that handwriting is an indication of character while I think that's all nonsense, we will disagree about whether the fact that a candidate loops his *f*s is relevant to whether he should be hired—though whether it *is* relevant depends on whether it is *true* that handwriting is an indication of character.

Quite generally, a person's judgments of the relevance of evidence, and hence of how comprehensive this evidence is, or of how well this claim explains those phenomena, and hence of how supportive it is, are bound to depend on his background assumptions. If he thinks fur color is likely to vary depending on climatic conditions, he will think it relevant to a generalization about the varieties of bear whether the evidence includes observations from the Arctic and the Antarctic; if he thinks the structure as well as the composition of a molecule determines how it functions, he will insist on asking, as Roger Kornberg reports that with the advent of structural chemistry molecular biologists began to do, "[h]ow do you do it with nuts and bolts; how do you do it with squares and blocks and the sorts of thing that we know molecules are made of?"[24] And so on.

When there are serious differences in background beliefs between one group of scientists and another, there will be disagreement even about what evidence is relevant to what, and about what constitutes an explanation—disagreements that will be resolved only if and when the underlying questions are resolved (or which may, as Max Planck famously observed, just fade away as the supporters of one side to the dispute retire or die off).[25] What has been taken for paradigm-relativity of evidential quality is a kind of epistemological illusion; again as in the graphology example, whether evidence is relevant, whether this is a good explanation of that, how strong or weak this evidence really is, how well or poorly warranted this claim actually is, is an objective matter.

Sometimes scientists know that they don't have all the evidence relevant to a question; and sometimes they have a pretty shrewd idea what the evidence is that they need but don't have. But sometimes, given the evidence they have, they may be unable to judge, or may misjudge, whether or what additional evidence is needed. They can't always know what it is that they don't know; they may not, at a given time, even have the vocabulary to ask the questions answers to which would be relevant evidence. Nor can they always envision alternative hypotheses which, if they did occur to them, would prompt them to revise their estimates of the supportiveness of their evidence. And so on. Since evidential quality is not transparent, and scientists can only do the best they can do, a scientist may be reasonable in giving a claim a degree of credence which is disproportionate to the real, objective quality of his evidence, if that real quality is inaccessible to him. Reasonableness, so understood, is perspectival.

TO ILLUSTRATE: THE EVIDENCE FOR THE DOUBLE HELIX

Thus far, this has all been quite austerely, not to say agonizingly, abstract. To make the picture more concretely vivid, I shall look at Watson and Crick's evidence for their model of the structure of DNA; but first let me give a brief sketch of some landmarks in the history of genetics and molecular biology.

When, a century before Watson and Crick's breakthrough, Darwin proposed the theory of evolution, he implicitly accepted a blending theory of inheritance (which in fact posed difficulties for evolution). Unknown to Darwin, Mendel was already working out the particulate theory; but Mendelian genetics were not integrated with the theory of evolution until the 1930s.

The fourth edition of a standard text on the mechanisms of heredity, published in 1951, shortly before Watson and Crick's breakthrough, summarizes what was then known: that genes are carried in sperm or eggs or both, since only these bridge the gap between generations; that generally, within a species, sperm and egg contribute equally to inheritance of genes; that since, though the egg has quite a lot of cytoplasm, the sperm is almost all nucleus, the nucleus must be the essential part of the gamete for transmission of genes; that of the constituents of the nucleus, only chromatin material is accurately divided at mitosis and segregated during maturation; that there are striking parallels between the behavior of genes as seen in the results of breeding and the behavior of chromosomes as seen under the microscope; so that "it would appear to be an inescapable conclusion . . . that Mendelian genes are carried in the chromosomes."[26]

The stuff we now know as DNA was discovered in 1869 by Friedrich Meischer, who called it "nuclein" because it was a component of the cell nucleus distinct from the proteins; he thought its chief function was to store phosphorous. By 1889 Richard Altmann had succeeded in obtaining nuclein free of protein, and had suggested the name "nucleic acid."[27]

In the early twentieth century it was assumed that most polymeric molecules were aggregates of much smaller molecules. The idea of a macromolecule, the kind to which we now know DNA belongs—very long molecules held together by covalent bonds and compactly folded in the cell—was first introduced by Hermann Staudinger in 1922. It was so controversial that in 1926, when Staudinger presented the idea at the Zurich Chemical Society, several distinguished members of the audience tried to dissuade him, and by the end of the meeting he was reduced to "shouting '*Hier stehe ich, ich kann nicht anders*' in defiance of his critics."[28]

For a good while protein was thought to be the genetic material. According to the tetranucleotide hypothesis, usually attributed to Phoebus Levene,[29] DNA was built up of the four nucleotides following each other in a fixed order; and so was too simple a molecule to carry the genetic information (which was why, in 1944, Avery was reluctant publicly to draw what now appears to be the obvious conclusion of his work). But by 1950 the tetranucleotide hypothesis had been ruled out by Erwin Chargaff's evidence that the four bases in DNA occur in widely varying proportions in yeast, bacteria, oxen, sheep, pigs, and humans. The specificity that could be carried by different sequences of nucleotides, Chargaff realized, "is truly enormous." And yet there is a remarkable uniformity among this diversity—an almost exact equivalence in the ratio of purines to pyrimidines; "whether this is accidental," Chargaff continued, "cannot yet be said."[30]

By the time Watson and Crick got interested in the composition and structure of genes and the transmission of inherited characteristics, the conjecture that protein is the genetic material had been ruled out not only by Avery's work (which showed that bacterial virulence was contained in nucleic acid rather than protein), but also by Hershey and Chase's radioactive tracing (which showed that it is not the protein but the DNA of a bacteriophage that enters the bacterium and multiplies).[31] In *The Double Helix* Watson describes those who still regarded the evidence for DNA over proteins as inconclusive as "cantankerous fools who unfailingly backed the wrong horses."[32] Hence the laconic opening paragraph of the paper on which I shall rely (not the very short paper in which Watson and Crick first announced their discovery, but the longer and more detailed piece, published the same year, in which they suggested a mechanism for DNA replication),[33] observing that "it would be superfluous to discuss the importance of DNA."

Perhaps it isn't quite superfluous, however, to slip in a few words to give you an idea of the scale of the thing. The measurements on which Watson and Crick rely are in angstrom units, an angstrom being one ten-billionth of a meter. Or if you prefer, here is John Kendrew writing in imperial feet and imperial inches: "All the DNA in a single human being would reach right across the solar system," and yet "[s]omehow three feet of it must be wrapped in a single cell perhaps a thousandth of an inch across."[34]

Taking for granted that DNA carries the genetic specificity of viruses and must therefore be capable of exact self-replication, Watson and Crick present chemical and physical-chemical evidence indicating that DNA is a long fibrous molecule folded up on itself in the cell; evidence that the fiber consists of two chains; evidence that its structure is a double helix of two complementary chains; and a possible mechanism by which such a structure could undergo the exact self-replication required to carry genetic information.

They begin with what is known about the chemical formula of DNA: a very long chain, the backbone of which is made up of alternate sugar and phosphate groups, joined together in regular 3' 5' phosphate di-ester linkages, with a nitrogenous base attached to each sugar, usually one of four kinds (the purines, adenine and guanine; and the pyrimidines, thymine and cytosine). The structure is regular in one respect (the internucleotide linkages in the backbone), but irregular in another (the sequence of the different nucleotides in the bases stacked inside). Physico-chemical analysis involving sedimentation, diffusion, light scattering and viscosity measurements indicate that DNA is a very asymmetrical structure approximately 20 Å wide and many thousands of angstroms long, and relatively rigid. These results are confirmed by electron microscopy, revealing very long thin fibers about 15–20 Å wide.

The evidence for two chains comes mainly from X-ray diffraction work using the sodium salt of DNA extracted from calf thymus, purified and drawn out into fibers. These indicate that there are two forms of DNA: the crystalline A form, and the less ordered paracrystalline B form, with a higher water content. X-ray photography indicates how far apart the nucleotides are spaced; and the measured density of the A form, together with the cell dimensions, show that there must be two nucleotides in each such group, so it is very likely that the crystallographic unit consists of two distinct polynucleotide chains. Correspondence of measurements indicates that the crystallographic unit and the fiber studied by electron microscopy are the same.

All this suggests that DNA must be regular enough to form a three-dimensional crystal, despite the fact that its component chains may have an irregular sequence of purine and pyrimidine nucleotides; and, as it contains two chains, these must be regularly related to each other. "To account for these findings" Watson and Crick propose

> a structure in which the two chains are coiled around a common axis and joined together by hydrogen bonds between the nucleotide bases. Both chains follow right-handed helices, but the sequences of the atoms in the phosphate-sugar backbones run in opposite directions and so are related by a dyad perpendicular to the helix axis. The phosphates and sugar groups are on the outside of the helix while the bases are on the inside. [To fit the model to observations of B-type DNA] our structure has a nucleotide on each chain every 3.4 Å in the fiber direction, and makes one complete turn after 10 such intervals, i.e., after 34 Å. Our structure is a well-defined one and all bond distances and angles, including van der Waal distances, are stereochemically acceptable.
>
> . . . The bases are perpendicular to the fiber axis and joined together in pairs . . . only certain pairs will fit into the structure [since] we have assumed that the backbone of each polynucleotide chain is in the form of a regular helix. Thus, irrespective of which bases are present, the glucosidic bonds (which join sugar and base) are arranged in a regular manner in space. . . . The result is that one member of a pair of bases must always be a purine and the other a pyrimidine, in order to bridge between the two chains. (p. 125)

Acknowledging that this model has not yet been proven correct, they observe that three kinds of evidence support it: X-ray evidence of the B form strongly suggests a basically helical structure, with a high concentration of atoms on the circumference of the helix in accordance with a backbone-out model, and indicates that the two polynucleotide chains are not spaced equally along the axis but are displaced from each other by about three-eighths of the fiber axis (X-ray evidence of the A form is more ambiguous); the anomalous titration curves of undegraded DNA with acids and bases strongly suggests that hydrogen bond formation is characteristic of the structure; the analytical data show that, though the ratio of adenine to cytosine can vary, the amount of adenine is close to that of thymine, and the amount of guanine is close to that of cytosine plus 5-methyl cytosine—Chargaff's rules—a "very striking result" suggesting a structure involving paired bases.

"We thus believe that the present experimental evidence justifies the

working hypothesis that the essential features of our model are correct. . . ,"
Watson and Crick conclude; and go on to suggest that, on this assumption, each
of the complementary DNA chains might serve as a template for the formation
onto itself of a new companion chain.

Just about all the essential ingredients of my analysis of the concepts of evidence
and warrant are found in this example: degrees of warrant, shifting over time;
confirmation, increment of warrant, as new evidence comes in; the sharing of
evidential resources; positive evidence and negative; observational evidence and
reasons working together; the role of special instruments and techniques of
observation; the ramifying structure of evidence; supportiveness, independent
security, and comprehensiveness as determinants of evidential quality; the intimate connection of supportiveness with explanatory integration, and hence its
sensitivity to the identification of kinds.

By the time Watson and Crick started their work, they could take for granted
that they were dealing with a *kind of molecule* (of which there turned out to be
two, and by 1980, when DNA with a left-handed twist—Z-DNA—was found,
three, forms);[35] of a type, the macromolecules, the structures of which were gradually being solved; and which conformed to these and those known chemical categories and laws.

Watson and Crick's evidence includes both experimental/observational
results, and other biological, chemical, etc., presumed knowledge. Their observational evidence relies on all kinds of complicated techniques and equipment
(electron microscopy, X-ray crystallography, procedures for extracting and puri-
fying the material under investigation, titration, sedimentation, etc., etc.). The
reliability of these depends, in turn, on other background theory and other observational evidence.

Much of their evidence is drawn from the work of others—Franklin's X-ray
photography, Chargaff's rules, etc., etc. In the notes to the paper on which I am
focusing, twenty-three other papers are cited; but this is only the tip of an enormous iceberg, for Watson and Crick also depend implicitly on a vast body of
what could by that time be simply taken for granted as background knowledge.

Because its presentation in speech or writing is of necessity linear, it can
seem that the structure of evidence must likewise be chain-like. But the case at
hand shows how far this is from the truth. The evidence Watson and Crick
present, and the background knowledge they take for granted, cannot plausibly
be construed in any simple linear form. It is a matter, rather, of ramifying clusters of evidence: the crossword analogy comes not only to my mind, but also,
independently, to Paul Meehl's[36]—indeed, given Watson and Crick's puzzling

over A (adenine), G (guanine), T (thymine), and C (cytosine), the analogy is just about irresistible.

There is mutual support: of the double-helical model by its ability to explain self-replication of the gene, of the explanation of self-replication by the model; of the model by its consonance with Chargaff's rules, of the biological significance of those rules by their consonance with the model; of the interpretation of those X-ray photographs as suggesting a double helix by theoretical crystallographic considerations, of the significance of those techniques by the explanatory power of the double-helical model; and so forth. There is a pervasive interdependence of perceptual evidence and background knowledge: of patterns observed in X-ray photographs and theoretical considerations about X-ray crystallography, and so on.

Watson and Crick rule out this or that hypothesis as inconsistent with what is (they think, reliably enough) known; as Crick was later to observe, solving a problem of any complexity requires a whole sequence of steps, and since any false move may put one on entirely the wrong track, "it is extremely important not to be trapped by one's mistaken ideas."[37] And as new evidence comes in they assess and reassess the likelihood that their model is correct, much as one might assess the reasonableness of a crossword entry: *given* these and those completed entries, *given* that they seem to be sufficiently warranted, this conjecture about the solution to 5 down can be ruled out, that one looks likelier.

They argue for the consistency of their hypothesis with what is already known/observed; and for its ability to suggest the mechanism whereby a phenomenon known to occur could take place. They allude to grounds for believing in the reliability of the techniques on which they depend; and they point to additional evidence which, were it obtained, would raise or lower the likelihood that their hypotheses are true.

Watson and Crick express considerable but not perfect confidence in their structure for DNA, less in their proposed duplication scheme. Already by the time of the longer paper cited here they are more confident of their double-helical structure than they had been in the briefer papers earlier the same year— more evidence had already come in. The relatively lower degree of warrant of their account of the duplication of DNA is due to difficulties which they can only hope are "not insuperable." In fact, it was not until the early 1980s, when the rival hypothesis of side-by-side chains, which made the problem of how the two chains separated seem easier, was ruled out, that the double-helical structure of DNA was, as Crick writes, "finally confirmed."[38]

BLACK RAVENS, RED HERRINGS, GRUE EMERALDS, AND ALL THAT

Compared with real-life evidence for real-life scientific claims, the old preoccupation with "this is a raven and this is black," and its relation to "all ravens are black," looks astonishingly simple and one-dimensional. Still, you may wonder if I have anything helpful to say about black ravens, red herrings, grue emeralds, and all that; and you deserve an answer. In brief: such puzzles are artifacts of the narrowly logical conceptions of evidence, warrant, and confirmation that I have been contesting; they evanesce with the recognition that supportiveness of evidence is not a purely formal matter, but depends on the substantial content of predicates, their place in a mesh of background beliefs, and their relation to the world.

The "raven paradox," remember, is that it follows from apparently obvious assumptions that "y is non-black and y is a non-raven," like "x is a raven and x is black," confirms "all ravens are black." Hempel replies that the "paradoxical" conclusion is true; it only seems counter-intuitive because we smuggle in the background information that there are many more non-black things than non-ravens. Goodman observes that, if it were all the evidence, "this is non-black and a non-raven" would confirm "all non-black things are non-ravens," and even "nothing is either black or a raven," as well as "all ravens are black." So "it is no longer surprising that under the artificial restrictions of the example, the hypothesis that all ravens are black is also confirmed."[39] But the example is far more artificial even than Goodman and Hempel acknowledge.

By "confirmation," Hempel and Goodman mean relative confirmation, conceived as a relation among propositions analogous to, but weaker than, deductive implication. But supportiveness of evidence is much less like deductive implication than they take "confirmation" to be: it depends on explanatory integration, which it is not straightforwardly atomistic, but a matter of the intermeshing of evidence, and not purely syntactic, but sensitive to the content and extension of predicates. Moreover, it is not clear it even makes sense to imagine someone's having *just* the evidence that this is non-black and a non-raven, or that black and a raven, without knowing anything about what a raven is, what a bird is, etc.

Actually, *Webster's* (ninth edition, 1991) defines a raven as a glossy black corvine bird, "corvine" meaning "of the crow family"; and we routinely identify birds as ravens in part on the basis of their color—though my bird book describes the Chihuahuan raven as having a thin band of white feathers at its neck, and ornithologists might yet discover a related variety just like familiar ravens except for its color. But the important point here is that "raven" is no simple observational term, but a kind predicate; and that even in the simplest cases of support of a scientific generalization by observed instances, laws and kinds are implicitly involved.

We know that in some species of bird many color-schemes are found, that in some there are differences of color between males and females, or differently colored varieties in different climatic zones, or seasonal changes of plumage; and that there are occasional albinos. The class of all ravens being, though finite, unsurveyable, we have to discover—checking males and females, temperate and non-temperate varieties, summer and winter plumage, etc.—what knot of properties holds of all things of the kind, and what properties hold of only some, and which, and why. That's why (as with Chargaff's work on DNA) we are concerned with variation of instances, more interested in a black female raven than yet another black male, in a black raven from the Arctic than yet another black raven from a temperate zone, and in a black swan or a blue parrot than a white shoe or a red herring.

Goodman's puzzle is called the "new riddle of induction" because it relocates Hume's old problem: since not all, but only some, predicates are inductively projectible, we should ask, not how we can know that unobserved instances will be like observed instances, but which predicates are projectible, and why—why, in particular, "green" but not "grue." "Grue," as Goodman characterizes it, "applies to all things examined before *t* just in case they are green but to other things just in case they are blue."[40] The paradox is that at *t* the hypothesis that all emeralds are grue, which predicts that emeralds subsequently examined will be blue, is as well "confirmed" as the hypothesis that all emeralds are green, which predicts that emeralds subsequently examined will be green. From a narrowly logical point of view, the relation of "all so far observed emeralds have been green" to "all emeralds are green" is just the same as the relation of "all so far observed emeralds have been grue" to "all emeralds are grue." But, though the raven paradox gave him no qualms about the formal conception of "confirmation," Goodman acknowledges that the difference between "green" and "grue" is not narrowly logical.

According to Goodman, the relevant difference between "green" and "grue" must be socio-historical: only predicates entrenched in scientific usage are projectible. This sounds as if it would impose a crippling conceptual conservatism on the scientific enterprise. According to Quine, the relevant difference is in a broad sense semantic, to do with the extension of the term: only natural-kind predicates are projectible.[41] This sounds as if it relies on the strange idea that green things—grass, zucchini, kiwi fruit, certain Chrysler vehicles, many lawn-mowers, a few of the shirts in my wardrobe, etc., etc.—constitute a natural kind.

But there is an element of truth in Goodman's answer which, properly understood, can be combined with an element of truth in Quine's to yield a more plausible solution: entrenchment in the language of science is not a simple historico-sociological accident, but an indication (albeit, of course, a fallible indication) of the intermeshing of evidence and of the intimate relation to the world to which scientific language aspires.

As earlier I set aside the problem that "black" occurs in *Webster's* definition of "raven," so now I will set aside the problem that *Webster's* definition of "emerald" is "a rich green variety of beryl"[42] (and that "emerald," like "raven," is sometimes used as an adjective of color). The grue hypothesis doesn't entail that any emerald will change color; but it does entail that any emerald first examined after *t* will be blue and not green. On the conveyor belt coming out of the deep, dark mine, the emeralds that emerge before *t* are green, while those emerging after that time will be blue; the new blades of grass first visible in the lawn the day after *t*, unlike the old, green blades still there from the day before, will be blue;[43] and so on.

The evidence we have about the color of emeralds is not simply that all so far observed emeralds have been green and that all so far observed emeralds have been grue (which is as implausible as the idea that "this is black and this is a raven" might be all the evidence we have relevant to the color of ravens); it is a whole mesh of evidence about the composition of gemstones, the optics of color perception, etc. None of this evidence offers any encouragement to the idea that mineral composition, or optical laws, etc., might be different at different times; nor, therefore, to the idea that emeralds first examined after some future time *t* will be blue. And all this is itself part of a much larger clump of crossword entries; for the new riddle of induction isn't just about grue emeralds, but also about blite ravens, whack polar bears, etc.

But couldn't there be a language in which "grue" and "bleen" were primitive, and "green" and "blue" defined? And if we used such a language, wouldn't all our evidence speak the other way? If you think in narrowly logical terms, the possibility of such a language seems undeniable. But how could such a language be learned? We understand "grue" because we understand "green" and "first examined before *t*"; and we understand "green" because our visual apparatus is sensitive to light of such-and-such wavelength.[44] So we can be trained to use "green" in the presence of emeralds, grass, zucchini, kiwi fruit, etc. It is not clear to me, however, that we *could* learn "grue" and "bleen" from scratch.

Still, suppose for the sake of argument that we could; and that a grulor-vocabulary, with "grue" and "bleen" primitive and "green" and "blue" defined by reference to a time *t* before now—say, the beginning of 2000—had been entrenched. By now, presumably, scientists would have noticed that new blades of grass are coming up bleen, not grue, and that the sapphires now coming out of the mine are grue, not bleen; and so, presumably, would have begun to suspect that something was badly wrong with their physics of color (or grulor) and their optics of color- (or grulor-) perception.

To be sure, it is possible that there are singularities of which we are not yet, and perhaps even of which we could never, be aware. If there were, we might never get things right. But this is only to recognize the imperfection of our epistemic condition.

Grue emeralds and red herrings haven't played a significant role in encouraging the New Cynics' disenchantment with the ideas of evidence, warrant, etc.; but Quine's underdetermination thesis has not only encouraged constructive empiricism and other kinds of non-realism in mainstream philosophy of science, but also played a very significant role in the New Cynicism. It is cited by Feyerabend on the left wing of philosophy of science, by Longino and Nelson in feminist science criticism, by Bloor and Collins in sociology of science—all of whom, with many others, take it as a radical threat to the objectivity of evidential quality, warrant, etc.

Sometimes, it seems, cynics simply invoke Quine's authority in support of the idea that when the evidence available at a given time is insufficient to decide between rival theories, it is okay to decide on political grounds which to accept. But since Quine's thesis is presumably focused on the strictly theoretical, political considerations seem simply out of place. Is quark theory or kwark theory politically more progressive?—the question makes no sense. Beyond this, however, there are difficulties about just what thesis, or theses, "underdetermination" refers to, and what the grounds are for believing it, or them, to be true.

Distinguishing several variants, Larry Laudan argues in effect that "underdetermination" runs together claims which are true but not radical (e.g., "theories are not logically implied by their positive instances"), with claims which are radical but not true (e.g., "every theory is as well supported by the evidence as any of its rivals").[45] Though I would take issue on many details, and prefer to his replace his "confirmed by positive instances," etc., with a different vocabulary, Laudan's strategy has merit. The formula "theories are underdetermined by the data" gestures vaguely towards a whole unruly family of theses: theories are never conclusively verified or falsified, or, theories are never better or worse warranted; for any theory and any evidence, there is always an equally warranted rival theory, or, all theories are equally warranted; theories are underdetermined by all available data, or, by all possible data; by observation statements, or, by evidence in a broader sense; etc. Perhaps underdetermination is true but not radical when "data" and "underdetermination" are interpreted narrowly, radical but not true when they are interpreted broadly.

Taking all the possible permutations and combinations in turn would be a Herculean task; I shall focus here on the interpretation which equates underdetermination with "empirical equivalence": the thesis that, for any scientific theory, there is another which is empirically equivalent to, but incompatible with, the first. Two theories are empirically equivalent, Quine tells us, just in case they entail the same set of "observation conditionals," i.e., statements the antecedents of which specify spatio-temporal co-ordinates and the consequents of which apply some observational predicate. Two theories are incompatible just in case, for some statement which follows from one, either its negation or some statement which translates into its negation follows from the other. So for the empirical equivalence thesis even to be statable requires a way of distinguishing incompatible theories with the same empirical consequences from notational variants of one and the same theory, and of identifying the class of observation statements constituting the empirical consequences of a theory. The first of these presupposes robust notions of meaning and translation (for, as Quine acknowledges, what look like empirically equivalent but incompatible theories may really be only verbal variants of one another); the second presupposes a clean distinction of observational and theoretical predicates.

But Quine is officially committed to denying both presuppositions! (So perhaps it's no wonder that the more precisely he formulates the empirical equivalence thesis, the more he hedges his commitment to it.) Quine's skepticism about the intensional and his thesis of the indeterminacy of translation undermine the first presupposition; and his commitment to the idea that observationality is a matter of degree undermines the second. If Quine is right to reject either or both of those presuppositions, the empirical equivalence thesis is in trouble.

Quite apart from Quine's views about meaning and translation, it is doubtful that clear criteria for identifying and individuating scientific theories are feasible. Rather, under a single rubric such as "theory of evolution," familiar terms may take on additional layers of meaning and shed some old connotations, new claims may be added and old claims dropped. But I shall concentrate here on the second presupposition, the observational/theoretical distinction.

When he is giving the empirical equivalence thesis its most explicit articulation, in "Empirically Equivalent Theories of the World," Quine suggests that those predicates are observational that can be learned ostensively. But this is seriously to backslide from the far more sophisticated account he had given in *Word and Object*, where he had treated observationality as a matter of degree, and ostensive and verbal learning as intertwined. And this conception of language-learning (to which the account I spelled out earlier in explaining the relevance of experience to warrant is closely akin)[46] is far superior to the crudely dichotomous account to which Quine resorts when his back is against the wall. It plainly implies, however, that there is no cleanly identifiable class of observation statements or observational predicates. And unless there is such a class, the empirical equivalence thesis is not even statable, and a fortiori not true (though not false, either).

As that parenthetical caveat indicates, I am not proposing an "empirical inequivalence" thesis, but repudiating the terms in which Quine's thesis is stated. At any time, there will be many questions which the available evidence is quite inadequate to settle. Some, at least, will get settled when new evidence—the nature of which we may not yet even be able to imagine—comes in. For example, not long after Henri Poincaré had written that, while Fresnel believed that vibration is perpendicular to the plane of polarization, and Neumann believed it was parallel, "we have looked for a long time for a crucial experiment that would decide between these two theories, and have not been able to find it,"[47] the question was settled by a experiment of Heinrich Hertz. Nevertheless, as in a real crossword there might be different words fitting the clue and all the letters determined by intersecting entries, so there may be scientific questions which could never be settled however long scientific inquiry continued. There is no guarantee that even the best evidence we could have would leave no gaps or irresoluble ambiguities. In that case, the most that we could find out by inquiry is that *either* p *or* q. But this, again, is only to recognize the imperfection of our epistemic condition.

AND, IN CONCLUSION

This is still quite far from a fully detailed account of the nature and structure of scientific evidence and of what makes it stronger or weaker—more like a preliminary analysis of the chemical composition of those concepts than the detailed account of their molecular structure I would ideally like. How, more specifically, do observation and background beliefs work together? What, more specifically, is involved in such concepts as explanatory integration and kinds, to which I have thus far simply helped myself? What does my long story about evidence and warrant have to do with the truth of scientific claims, or with progress in science? And—the question I shall tackle next—what does any of this have to do with the traditional preoccupation with "scientific method"?

NOTES

1. Einstein, "Physics and Reality," p. 295.
2. Rogers, "Come In, Mars," pp. 56–57.
3. Wilford, "2 New Chemical Studies Find Meteorite Samples Show No Traces of Past Life on Mars," p. A22.
4. Compare Russell: "individual percepts are the basis of all or knowledge, and no method exists by which we can begin with data . . . public to many observers" (*Human Knowledge*, p. 8).
5. This sentence must be read in the spirit of the foundherentist account offered in *Evidence and Inquiry*, chapter 4; *not* as saying, as a foundationalist would, that experiential *beliefs* support, but are not supported by, other beliefs.
6. See chapter 2, pp. 35–36.
7. Something like this is suggested by Russell in *Human Knowledge*, pp. 4, 63 ff., 501, 502. Only "something like this," however; Russell is at pains to point out that ostensive definition leaves room for differences in the meaning attached to a word by one individual and by another.
8. The example comes from Popper, from a passage where, rather than denying the relevance of experience, he is insisting on the fallibility of "basic statements."
9. Both Popper and Van Fraassen, in different ways, distinguish between belief and acceptance; I do not.
10. Proof: From "p & not-p" it follows that p. From "p" it follows that p or q. From "p & not-p" it also follows that not-p. From "not-p" and "p or q" it follows that q. QED.
11. Hempel notices the problem; he suggests that contradictory observation reports could be excluded "by a slight restriction of the definition of 'observation report,'" but that "there is no important reason to do so" ("Studies in the Logic of Confirmation," p. 31, note 48). Carnap discusses the problem in a note entitled "Remarks on the Exclusion of L-false Evidence," *Logical Foundations of Probability*, pp. 295–96.
12. Perhaps it would be desirable to add, as a precaution against a potential parallel difficulty in case the claim in question is necessarily true, that conclusiveness requires that this evidence, *but not any evidence whatsoever*, deductively imply the claim in question. But I shall set these complications aside.
13. See Routley et al., *Relevant Logics and Their Rivals* (on relevance, paraconsistent, etc., logics); Haack, *Philosophy of Logics*, pp. 197–203; and *Evidence and Inquiry*, pp. 83–84. I was glad to find Thagard independently taking a line quite like mine in his *Conceptual Revolutions*.
14. Quine and Ullian, *The Web of Belief*, p. 79, already cited in chapter 2, p. 50.
15. Though I arrived at it independently, this is, I now realize, in essence the diagnosis Hempel gave (in his rather confusing vocabulary of "absolute" versus "relative" verification) in 1945, in section 10 of "Studies in the Logic of Confirmation." See also Hesse, *The Structure of Scientific Inference*, pp. 130–31.

16. Like Kitcher (*The Advancement of Science*, chapter 8), I take the epistemology of science to be social in a relatively conservative sense, as involving interactions among individuals.

17. My source is Judson, *The Eighth Day of Creation*, pp. 264–65.

18. Peirce, *Collected Papers*, 5:402, second footnote. References to this work will be by volume and paragraph numbers.

19. "[T]he best evidence principle . . . expresses the obligation of litigants to provide evidence that will best facilitate this central task of accurately resolving disputed issues of fact": Nance, "The Best Evidence Principle," p. 233.

20. Watson, *Molecular Biology*, p. 52.

21. In ordinary English, "warrant" and "justification" are more or less interchangeable; but I am deliberately exploiting the availability of the two words to make a needed distinction. (In *Evidence and Inquiry*, as here, I took justification to be a partly causal concept; but I did not, as here, also employ the purely evidential notion of warrant.)

22. See, for example, Laudan, "A Critique of Underdetermination," p. 91. See Mayo, *Error and the Growth of Experimental Knowledge*, pp. 206ff., for a discussion of this idea in the context of Popper's conception of severity of tests.

23. Cowley, "Cannibals to Cows," p. 54.

24. Judson, *The Eighth Day of Creation*, p. 495.

25. Planck, *Scientific Autobiography*, pp. 33–34.

26. My source is Meehl, "Corroboration and Verisimilitude," pp. 25–26, 54, citing Snyder, *Principles of Heredity*, p. 301.

27. Portugal and Cohen, *A Century of DNA*, chapter 1.

28. Olby, *The Path to the Double Helix*, pp. 6–10, quoting (p. 7) Frey-Wyssling, "Frühgeschichte und Ergebnisse der submikroskopischen Morphologie," p. 5; Staudinger's response to his critics may be translated as, "Here I stand; there is nothing else I can do."

29. But see Olby, *The Path to the Double Helix*, pp. 89ff. on the accuracy of the attribution.

30. Chargaff, "Chemical Specificity of Nucleic Acids and Mechanism of Their Enzymatic Degradation." Chargaff's table of his data is reproduced in Bauer, *Scientific Literacy and the Myth of Scientific Method*, p. 22.

31. Hershey and Chase, "Independent Functions of Viral Protein and Nucleic Acid in Growth of Bacteriophage."

32. Watson, *The Double Helix*, p. 14.

33. References in what follows are to Watson and Crick, "The Structure of DNA." The two shorter papers published the same year are "Molecular Structure of Nucleic Acids," and "Genetical Implications of the Structure of Deoxyribonucleic Acid."

34. Kendrew, *The Thread of Life*, p. 63.

35. Crick, *What Mad Pursuit*, p. 73.

36. Meehl, "Corroboration and Verisimilitude," p. 27. Portugal and Cohen, *A Century of DNA*, p. 3, suggest the analogy of a jigsaw puzzle.

37. Crick, *What Mad Pursuit*, p. 70.

38. Ibid., pp. 71 ff.

39. Goodman, "The New Riddle of Induction," pp. 70–71; the quotation is from p. 71.

40. Ibid., p. 74.

41. Quine, "Natural Kinds."

42. A point due to Judith Thomson, "Grue."

43. It was inevitable, I suppose: shortly after I finished this chapter I read (in Read, "For Parched Lawns, A Patch of Blue") that recently introduced lawn-patching seed mixes, supposed to camouflage dead spots while new grass takes hold, produce patches of azure-blue grass!

44. While the human sensory system perceives intensity of light continuously, it breaks continuously varying wavelengths of light into the more or less discrete units of the color spectrum; and though color vocabularies differ from language to language, they do so, not at random, but in up to eleven basic color units in a particular order. I rely here on Wilson, *Consilience*, pp. 161–65. He refers to papers by Denis Baylor, John Gage, John Lyons, and John Mollon in Lamb and Bourriau, *Colour: Art and Science*, and Lumsdem and Wilson, *Promethean Fire*.

45. See Laudan, "Demystifying Underdetermination," and Laudan and Leplin, "Empirical Equivalence and Underdetermination."

46. And quite akin, also, to Hesse's account (see chapter 2, pp. 45–47). Hesse also notes Quine's shifts on the question of the observational/theoretical distinction (see *The Structure of Scientific Inference*, p. 27).

47. Poincaré, *Electricité et optique*, p. vi.

4

THE LONG ARM
OF COMMON SENSE

Instead of a Theory of Scientific Method

> The scientific method, as far as it is a method, is doing one's
> damnedest with one's mind, no holds barred.
>
> —Percy Bridgman, "The Prospect for Intelligence"[1]

Picture a scientist as working on part of an enormous crossword puzzle: making an informed guess about some entry, checking and double-checking its fit with the clue and already-completed intersecting entries, of those with their clues and yet other entries, weighing the likelihood that some of them might be mistaken, trying new entries in the light of this one, and so on. Much of the crossword is blank, but many entries are already completed, some in almost-indelible ink, some in regular ink, some in pencil, some heavily, some faintly. Some are in English, some in Swahili, some in Flemish, some in Esperanto, etc. In some areas many long entries are firmly inked in, in others few or none. Some entries were completed hundreds of years ago by scientists long dead, some only last week. At some times and places, on pain of firing or worse, only words from the Newspeak dictionary may be used; at others there is pressure to fill in certain entries this way rather than that, or to get going on this completely blank part of the puzzle rather than working on easier, partially filled-in parts—or not to work on certain parts of the puzzle at all. Rival teams squabble over some entries, pencilled or even inked in and then rubbed out, perhaps in a dozen languages and a score of times. Other teams cooperate to devise a procedure to churn out all the anagrams of this chapter-long clue or a device to magnify that unreadably tiny one, or call to teams working on other parts of the puzzle to see if they already have something that could be adapted, or to ask how sure they are that it really must be an *S* here. Someone claims to notice a detail in this or that clue that no one else has seen; others devise tests to check whether he is an especially talented observer or is seeing things, and yet others work on instruments for looking more closely. From time to time accusations are heard of altered clues or blacked-out spaces. Sometimes there are complaints from those working on one part of the puzzle that their view of what's going on in some other part is blocked. Now and then a long entry, intersecting with numerous others which intersect with numerous others, gets erased by a gang of young turks insisting that the whole of this area of the puzzle must be re-worked, this time, naturally, in Turkish—while others try, letter by letter, to see if most of the original Welsh couldn't be kept . . . . I don't mean to fob you off with a metaphor instead of an argument. But I do mean my word-picture to suggest, what I believe is true, that scientific inquiry is far messier, far less tidy, than the Old Deferentialists imagined; and yet far more constrained by the demands of evidence than the New Cynics dream.

The crossword analogy proved a helpful guide to questions about evidence; but some may feel that, where questions of method are concerned, the analogy looks conspicuously *un*helpful. After all, what is there to say about the "method" of solving crossword puzzles, except that you should make a guess at some entry in the light of its clue, then try other entries in the light of their clues and any already-completed entries; and that when an otherwise plausible-seeming entry turns out to be incompatible with others, you should neither give it up too easily, nor hang on to it too obstinately? And what would that tell us about the "scientific method," except that you should make an informed conjecture about the explanation of the phenomenon that concerns you, see how well it stands up to the evidence you have and any further evidence you can lay hands on; and that when your otherwise plausible-seeming conjecture turns out to be incompatible with some of this evidence, you should neither give it up too easily, nor hang on to it too obstinately?

Indeed; but to my way of thinking this *is* helpful. It nudges us towards the correct conclusion: there is less to "scientific method" than meets the eye. Is scientific inquiry categorically different from other kinds? No. Scientific inquiry is continuous with everyday empirical inquiry—only more so. Is there a mode of inference or procedure of inquiry used by all and only scientists? No. There are only, on the one hand, modes of inference and procedures of inquiry used by all inquirers, and, on the other, special mathematical, statistical, or inferential techniques, and special instruments, models, etc., local to this or that area of science. Does this undermine the epistemological pretensions of science? No! The natural sciences are epistemologically distinguished, have achieved their remarkable successes, in part precisely because of the special devices and techniques by means of which they have amplified the methods of everyday empirical inquiry.

That annoying honorific use of "science" and its cognates notwithstanding, not all and not only scientists are good inquirers. And there is no distinctive procedure or mode of inference used by all and only practitioners of science, and guaranteeing, if not true, approximately true, or probably true, or more nearly true, or more empirically adequate results—no "scientific method," as that phrase has often been understood. Inquiry in the sciences is continuous with other kinds of empirical inquiry. But scientists have devised many and various ways to extend and refine the resources on which we all rely in the most ordinary of everyday empirical inquiry. Controlled experiments, for example—sometimes thought of as distinctive of the sciences—aren't used by all scientists, or only by scientists; astronomers and evolutionary theorists don't use them, but auto mechanics, plumbers, and cooks do. In many areas of science, however, techniques of experimental control have been developed to a fine art.

Already in *Evidence and Inquiry* I had suggested that the sciences, though distinguished epistemologically, are not privileged; and that "there is no reason to think that [science] is in possession of a special method of inquiry unavailable to historians or detectives or the rest of us";[2] but these brief first efforts to articulate the place of the natural sciences within inquiry generally were politely ignored by the epistemological community. Later, I recalled John Dewey's observation that "[s]cientific subject-matter and procedures grow out of the direct problems and methods of common sense,"[3] and James B. Conant's *Science and Common Sense*; and then, to my surprise and pleasure, found Thomas Huxley observing that "[t]he man of science simply uses with scrupulous exactness the methods which we all, habitually and at every minute, use carelessly,"[4] Albert Einstein that "the whole of science is nothing more than a refinement of everyday thinking,"[5] Percy Bridgman that "there is no scientific method as such, . . . the most vital feature of the scientist's procedure has been merely to do his utmost with his mind"[6]—and Gustav Bergmann describing the sciences, in a marvelously resonant phrase, as the "long arm" of common sense.[7] This is the spirit in which I offer here, not a new theory of scientific method, but an exploration of the constraints on everyday empirical inquiry, and the resources it requires; and of the astonishingly various ways and means the sciences have found to make themselves "more so."

"NOTHING MORE THAN A REFINEMENT OF EVERYDAY THINKING"

There is "a good deal of ballyhoo about scientific method," Bridgman writes; but, he continues, it is really much ado about not very much.[8] Scientists are in the business of seeking true answers to their questions; so, like all serious empirical inquirers, they should check the evidence as exhaustively as possible, and try to curb any tendency to wishful thinking. Beyond that, they should use whatever specific methods or techniques can be devised, given the subject-matter and in light of what is already known, that may help deal with the questions that concern them.[9] This is right, so far as it goes. But what are the standards and resources of well-conducted empirical inquiry generally? What does well-conducted empirical inquiry demand of inquirers? And how, exactly, is scientific inquiry more so?

Inquiry, unlike composing a symphony, cooking dinner, writing a novel, or pleading a case before the Supreme Court, is an attempt to discover the truth of some question or questions; though sometimes the upshot is not an answer, but a realization that the question was in some way misconceived, and quite often, once you've answered one question, you find yourself facing a whole slew of new ones. In the academy, in politics—everywhere, in fact—pseudo-inquiry is ubiquitous: sham reasoning, making a case for some proposition to the truth of which you are already unbudgeably committed; and fake reasoning, making a case for some proposition to the truth of which you are indifferent, but advancing which you believe will benefit yourself. But genuine inquiry is a good-faith effort to arrive at the truth of the matter in question, whatever the color of that truth may be.[10]

There is empirical inquiry and (at least apparently; but now is not the time to get into questions about the epistemology of mathematics or logic)[11] non-empirical inquiry. Within the category of empirical inquiry there is the natural-scientific, the social-scientific, the historical, the forensic, and so on, as well as everyday inquiry into when your plane leaves, where to buy chappatie flour, how to make the printer do italics, what it was you ate that upset you, and so forth. Some inquiry is better conducted—more scrupulous, more thorough, more imaginative, etc.—and some worse. This goes for inquiry of every kind, scientific inquiry included.

Like all empirical inquirers, scientists aim to give a true account of how some part or aspect of the world is; not just any old true account, though—"either the universe originated in a big bang, or not" won't do—but a substantial, significant, explanatory account. This is not easily achieved. All we have to go on, after all, is what we see, etc., and what we make of what we see, etc. Coming up with plausible explanatory conjectures can stretch the human mind to its limits; evidence is always incomplete and ramifying, often potentially misleading, and frequently ambiguous; getting more evidence is hard work and may require considerable ingenuity in experimental contrivance. And there is no guarantee of success, so the temptation is ever-present to skimp the work and believe what we hope, or fear, or think our sponsors would like to hear is how things are.

Though scientific theories are sometimes startlingly at odds with common-sense beliefs, scientific inquiry is recognizably continuous with everyday empirical inquiry of the most familiar kind. Henry Harris imagines prehistoric people trying to find out whether the river that runs through (what we now call) Oxford is the same river that runs through Henley, by floating logs daubed in a particular color into the river at Oxford and asking contacts at Henley whether they saw them. Then he describes physiologists' efforts to find out what happens to the large numbers of lymphocytes that enter the blood from the lymph: until eventually, by tagging them with radioactive isotopes, J. L. Gowans discovered that the lymphocytes pass from the blood into the lymph and then back into the blood.[12]

All empirical inquirers—molecular biologists and musicologists, entomologists and etymologists, sociologists and string theorists, investigative journalists and immunologists—make informed conjectures about the possible explanation of the phenomena that concern them, check out how well these conjectures stand up to the evidence they already have and any further evidence they can lay hands on, and then use their judgment whether to stick with their conjecture, drop it, modify it, or what. They need imagination, to think up plausible potential explanations of problematic phenomena, to devise ways to get the evidence they need, and to figure out potential sources of error; care, skill, and persistence, to seek out any relevant evidence no one yet has, as well as relevant evidence others have; intellectual honesty, the moral fiber to resist the temptation to stay out of the way of evidence that might undermine their conjectures, or to manipulate unfavorable evidence they can't avoid; rigorous reasoning, to figure out the consequences of their conjectures; and good judgment in assessing the weight of evidence, unclouded by wishes or fears or hopes of getting tenure or resolving a case quickly or pleasing a patron or mentor or becoming rich and famous.

Making an informed conjecture ("informed" being the operative word) requires inference: that this conjecture implies, that another is consistent with, that yet another is inconsistent with, your background information. And checking how a conjecture stands up to evidence requires inference; that, if the hypothesis is true, this consequence follows, that the hypothesis is confirmed to some degree when that consequence is found to obtain, that it is likely false if the consequence is found not to obtain, and so on. The point isn't that scientists don't make such inferences, nor that logic has nothing to tell us about them. It is only that detectives, investigative journalists, historians, and the rest of us make such inferences too; so that logic cannot by itself explain how the sciences have achieved their successes (let alone why they often fail).

The Old Deferentialists' search for "scientific method"—an inductive logic of discovery or confirmation, the conjecture and refutation-by-*modus tollens* of Popperian deductivism, repeated applications of Bayes' theorem, or whatever—focused on just one part of a whole complicated story. What we need instead is a multi-dimensional explanation of the successes, and the failures, of the sciences: an explanation which will not ignore the nature and structure of evidence, but which will also appeal in part to historical contingencies, in part to the subject-matter of the natural sciences, and in part to the ways in which scientists have managed to extend, deepen, and strengthen ordinary everyday inquiry.

Among the reasons for the successes of natural-scientific inquiry, one would have to be—not chance, exactly, but the fact that there happened to be a few remarkable individuals with the temperament and the talent for speculation about how the world works, and enough others to take up intersecting parts of the crossword, at a time and in a place where the social and intellectual climate allowed them to pursue their inquiries and communicate their results. This suggests where we might start looking for an explanation of why modern science grew up when and where it did, rather than earlier or elsewhere; the phrase that comes to mind is "critical mass."

Another reason lies in the subject-matter of the natural sciences, in the deep interconnectedness of natural phenomena. This may be what Wilson has in mind when he suggests that the method of science is "reductionism," a systematic investigation of smaller and smaller component parts of things. But such analysis, important as it is in its place, is only one of many scientific approaches. By my lights, the epistemological significance of the interconnectedness of natural phenomena lies rather in the way each new step in understanding potentially enables others.

Now the phrase that comes to mind is "nothing succeeds like success." This draws our attention to the many and various helps to inquiry that scientists have devised—my chief focus in what follows. For expository purposes I shall rely on a rough and ready division of helps to the imagination, helps to the senses, helps to reasoning, and helps to evidence-sharing and intellectual honesty—although, because they are all tied to the goals of inquiry and to human inquirers' cognitive capacities and limitations, they are really quite intimately intertwined.

SCIENTIFIC HELPS TO INQUIRY

The word "helps" is borrowed from Francis Bacon, who—for all his misguided hopes of a mechanical inductive logic of discovery, and for all that (as William Harvey unkindly said of him) "he wrote about science like a Chancellor"—had something important exactly right when he urged the dangers of a factitious despair of the possibility of successful investigation, and the necessity of devising ways of overcoming our sensory and cognitive limitations and the fragility of our commitment to finding out.[13] As Bacon well understood, we humans are fallible creatures, our imagination, our senses, and our cognitive capacities limited, and our intellectual integrity fragile; haste, sloppiness, busy-work, and wishful thinking come more easily to us than the difficult and demanding business of thorough, honest, creative inquiry. Nevertheless, we *are* capable of well-conducted inquiry; and of devising means, albeit themselves imperfect and fallible means, of overcoming our natural limitations and weaknesses; even of figuring out when these imperfect and fallible means of overcoming those limitations and weaknesses are likeliest to fail, and devising ways to avoid the pitfalls.

The underlying demands and resources common to all empirical inquiry are constant; but scientific helps are constantly evolving. Though some extend across the sciences and even beyond, many are local to specific areas of science. They generally rely on previous scientific work, and when the earlier work on which they rely is solid, they enable science to build on its successes; of course, when the previous work on which a scientific help relies is not solid, scientists may be led astray—just like doing a crossword.

The models, metaphors, and analogies which aid scientists' imagination have encouraged some New Cynics to assimilate science to imaginative literature, and some to complain that the metaphors and analogies reflected in the content of scientific theories are socially regressive; the instruments of observation which extend scientists' perceptual powers have encouraged the idea that observation is too theory-dependent to constitute a genuinely objective evidential check; the artificial laboratory situations which are sometimes needed to test theoretical claims have encouraged the notion that scientific theories describe, not the natural world, but only a "reality" created by scientists themselves; and the social character of scientific inquiry has encouraged a conception of scientific knowledge as nothing but a social construction serving the interests of the powerful. And some New Cynics maintain that the local, evolving character of scientific helps reveals that standards of well-conducted inquiry are relative to context or paradigm.

But all this is an overreaction. True, because there is always the risk that the earlier work on which this or that instrument or technique relies will turn out to have been mistaken, there is always a risk of failure. True, in judging how thoroughly or carefully work has been conducted, scientists must rely on what they think they know about what evidence is relevant, what the potential sources of experimental error are, and so on. Scientific inquiry is fallible, in other words; and judgments of better- and worse-conducted inquiry, like judgments of the worth of evidence, are perspectival, dependent on background beliefs. But the relativist conclusions to which Kuhn and others seem to be drawn obviously don't follow.

All inquiry, scientific inquiry included, requires imagination. As C. S. Peirce put it, "[w]hen a man desires ardently to know the truth, his first effort will be to imagine what that truth can be. . . . [T]here is, after all, nothing but imagination that can ever supply him an inkling of the truth. . . . For thousands of men a falling apple was nothing but a falling apple, and to compare it to the moon would by them be deemed 'fanciful'." But unlike an artist or a writer, a scientist "dreams of explanations and laws"[14]—explanations and laws which, when he is successful, are not imaginary, like fictional entities, but real.

Our imagination, like our other powers, is limited; and so, among the helps on which scientists rely are models, analogies, and metaphors. In *La théorie physique*, published in 1914, Pierre Duhem contrasted the abstract, logical, systematizing, geometric mind-set characteristic, as he thought, of Continental physicists, with the visualizing, imaginative mind-set characteristic of the English, which he saw as a distraction from the mathematical abstraction at the real heart of physics. In Oliver Lodge's book on electrostatics, he complained, "[w]e thought we were entering the tranquil and neatly ordered abode of reason, but we find ourselves in a factory"; for there is nothing in the book but "strings which move around pulleys, which roll around drums, which go through pearl beads [and] toothed wheels which are geared to one another and engage hooks."[15] I can just see Duhem tearing out his hair as John Kendrew invites his readers to imagine "a man . . . magnified to the size . . . of the United Kingdom," and explaining that then "a single cell might be as big, perhaps, as one factory building. . . . [O]n this scale a molecule of nucleic acid . . . would be thinner than a single piece of electric light flex in our factory."[16] These days, biology students are sometimes taught to think of the cell as a complex city, with mitochondria as power plants, Golgi apparatuses as post offices, and so on. Unlike Duhem, however, I see mathematics versus model-building, analysis versus analogy, systematization versus simile, as false dichotomies; models, analogies, and metaphors play an important role not only pedagogically but also in theory-construction, as helps to the imagination.

"Models, analogies, and metaphors" are something of a mixed bag, including everything from actual physical arrangements of templates or poppet beads, etc., to such master metaphors as the "Invisible Hand" of Adam Smith's theory of markets; all, however, liken a less to a more familiar or accessible phenomenon. One role of physical models, such as Watson and Crick's series of mock-ups of the DNA molecule, is as a visual aid to the imagination, enabling scientists to envisage the molecule, like the model, in three dimensions. And while some scientific metaphors are more decorative than functional, others suggest questions to investigate, directions in which to look. Of course, a metaphor may prompt different scientists to look in different directions; and sometimes may draw inquiry in what turns out to be the wrong direction.

As science critics like to point out, sometimes the metaphors and analogies scientists use call upon familiar social phenomena: such as the "master molecule" metaphor, of which Evelyn Fox Keller makes a big feminist deal (so far as I can determine, however, it was never, as she suggests, an influential idea in molecular biology). Scientific metaphors *can* be cognitively important, and *may* mislead the imagination as well as guide it in fruitful directions. But the fruitfulness or otherwise of a metaphor does not depend on the desirability or otherwise of the social phenomenon on which it calls; such considerations simply have no bearing on whether thinking in terms of molecular chaperones (to use a genuinely influential example),[17] or parental investment, or whatever, will lead to what turns out to be fruitful territory.

All empirical inquiry depends on experience; but among the ways in which inquiry in the sciences is often "more so" is that the experience on which it depends is not unassisted, but enabled and mediated by instruments of every kind; is not unchecked, but open to scrutiny by others in the community; and is not left to chance or merely serendipitous, but is deliberate, contrived, controlled.

Let me begin with instruments of observation, from the familiar microscope or telescope to the much fancier and more esoteric instruments that now extend scientists' sensory powers. In the history of such instruments there is perhaps no more striking case of "nothing succeeds like success" than medical imaging. Wilhelm Roentgen took the first X-ray photograph in 1895 (and in 1901 received the first-ever Nobel Prize). The first dental X-rays were made in 1896; the first diagnostic X-rays were made on the battlefield in Abyssinia in 1897, the same year X-ray photographs were first used in court. Over the next couple of decades the theory of X-rays—electromagnetic waves of very short wavelength, between

0.01 and 10 nanometers—was developed. Between 1919 and 1927, contrast media, first air, then lipoidal, then sodium iodine, then thorium dioxide, began to make X-ray photographs more informative. The tomograph, a picture of an internal slice of the patient, was introduced by Jean Keiffer in the 1930s; X-ray crystallography derives from the same period. By the 1970s, much faster imaging machines were being developed, and now, with sophisticated computer technology, we have the CAT scan; MRI, in which bones disappear and tissue shows up; the PET scan, using radioactive tracing materials, and instruments to detect those traces and create images; and the EBT scan.[18]

The theory on which an instrument relies might turn out to be mistaken, and perhaps in ways that will undermine scientists' confidence in its workings; but, if the theory is well warranted, this upshot, though possible, is not likely. Instrumentation is theory-dependent; but what explains the workings of an instrument is seldom if ever the very theory some consequence of which it is being used to test, and which the observations mediated by the instrument are therefore bound to confirm. It is a matter, rather, of scientists' looking to optics to explain the workings of the microscope by means of which they study the constitution of these cells, or of the telescope by means of which they study the movements of that star. True, scientific theories interlock, as crossword entries do, so the remote possibility of a damaging mutual dependence cannot quite be ruled out in principle. True, especially in the early days, scientists sometimes needed to go to elaborate lengths to persuade others of the reliability of their instruments. True, sometimes scientists mistake artifacts of their instruments for bona fide evidence (as some believe is the case with the supposed evidence of bacterial life on Mars). But none of this means that instruments aren't, usually, real helps.

Serious inquirers of every kind are active in seeking out evidence: historians hunt down documents, interview survivors, and so forth; detectives follow and observe suspects, and so on. But natural-scientific inquiry is often "more so," insofar as it involves contriving the circumstances in which this or that evidence will be available. Scientific inquirers are put on their mettle to design experiments which will be as informative as possible, isolating just the variable in which they are interested—like working on the intersecting entries that cross this one precisely at the letters which differentiate rival solutions from one another. Oswald Avery's experiments to identify the stuff responsible for "bacterial transformation" provides a vivid example of the finesse, and the factual knowledge, that good experimental design requires.

In mice injected with a live but non-virulent R type pneumococcal preparation and a dead but virulent S type, Frederick Griffith had found living, virulent pneumococci of the S form. Later, associates of Avery's found that the same bacterial transformations could be produced in vitro. To discover what was responsible, Avery first devised an elaborate process to extract the "transforming principle," whatever it was, obtaining just under a hundredth of an ounce from twenty gallons of culture. Then he subjected this extract to standard tests for protein, with negative results; and to standard tests for DNA, with strongly positive results. Under chemical analysis, he found the extract to have the 1.67:1 ratio of nitrogen to phosphorous that would be expected with DNA, but not with protein. Then he found that enzymes known to degrade proteins or RNA left the extract intact, but those known to degrade DNA destroyed it. Then, using immunological tests, he found that neither pneumococcal protein nor capsular polysaccharide was present. He spun a sample of the extract on the ultra-high-speed centrifuge and found a sedimentation pattern matching DNA from calf thymus. Finally, he found that under electrophoresis the molecules of the extract stayed together and moved relatively fast, as nucleic acids do; and that the extract absorbed ultraviolet light to yield the same profile as DNA.

In a letter to his brother, Avery wrote: "[T]he substance is highly reactive & . . . conforms *very* closely to the theoretical values of pure *desoxyribose nucleic acid.* . . . Who could have guessed it?"[19] (Because of the influence of the tetranucleotide hypothesis, however, he was careful to point out in the conclusion of his published paper that he had not yet excluded the possibility that the transforming principle was not the DNA itself, but minute amounts of something else adsorbed to the DNA.)

Now think of the multiplicity of ways of ensuring that experiments and observations are not contaminated, physically or otherwise: from the mundane, such as prohibitions on bringing food into the lab where biochemical experiments are going on, checklists for systematically ensuring that complicated equipment is functioning properly, standardized forms on which to mark observational findings so that relevant details don't get forgotten; through the more sophisticated, such as double-blinding; to hard issues about, e.g., what sampling procedures are most appropriate given the question at issue, and which might skew the results.

The appropriateness of taking these precautions rather than those—prohibiting food or pets in the lab, say, but not shoes or ballpoint pens—depends on assumptions about what kind of thing might interfere. These assumptions could conceivably be mistaken, and are sometimes very tricky to figure out; what aspects of how the experimenter conducts this psychological test, or what facts about the experimenter himself, might affect the response? When I criticize the design of a psychological study in which subjects were told ahead of time that the purpose of the interviews in which they were participating was to identify their "women's ways of knowing," I take for granted that telling them this introduced the likelihood that the results would be biased by suggestion.[20] Conceivably (just), I might be mistaken. But that precautions are not infallible doesn't mean that they are not, usually, real helps.

When we criticize poorly conducted inquiry, we may complain of insufficient effort to get hold of relevant evidence; of a lazy detective: "he didn't even bother to track down the maid and ask her what she saw." Or we may complain of insufficient care in assessing the worth of evidence; of a sloppy historian—a real example, this, from the brouhaha about a letter supposedly showing that Marilyn Monroe had blackmailed President Kennedy:[21] "he jumped to the conclusion that the letter was genuine, ignoring the fact that the address includes a zip code, and that correction ribbon was used, when neither existed when the letter was supposedly written." And so on.

These kinds of complaint are equally relevant to scientific inquiry; but there we also justifiably complain when specific precautions appropriate in a given area are not taken. As Bridgman observes, "[w]hen the scientist ventures to criticize the work of his fellow scientist, as is not uncommon, he does not base his criticism on such glittering generalities as failure to follow the 'scientific method,' but his criticism is specific, based on some feature of the particular situation."[22] Critics of a study supposedly establishing the medical efficacy of prayer, for example, object that it was not completely double-blinded;[23] the same objection that Maddox and his team make to the work of Benveniste's lab supposedly establishing the effectiveness of high-dilution homeopathic remedies: "no substantial effort has been made to exclude systematic error, including observer bias" and "the phenomenon described is not reproducible in the ordinary meaning of the word." Interestingly enough, however, a very Humean thought underlies both these criticisms: the idea that prayers made on his behalf but unknown to the patient might aid recovery is alien to medical science; the idea that a "solution" so dilute that it contains not a single molecule of the supposed "solute" works because it leaves traces in the water's "memory," is egregiously at odds with all accepted chemical theory, so much so that it is much more likely that the alleged findings are the result of experimental error or self-deception on the part of the experimenter, than that they are genuine.[24]

Investigating our susceptibility to error in this or that respect improves our understanding of which results are likely due to experimental or other kinds of error. For example, the detection of fraud in scientific inquiry is often triggered by someone's noticing that these or those alleged results are "too good to be true"—confirming a hypothesis more neatly than what we know about human, or instrumental, fallibility, suggests is likely.

* * *

Already, considering the design of experiments and precautions against experimental error, I began to venture into the realm of helps to the intellect; but as yet I have barely touched upon the many refinements and amplifications of reasoning on which the sciences have relied.

One of the most striking features of much natural-scientific inquiry is its mathematical character. Indeed, modern science might be seen as born of the union of mathematics and physics (anticipated by Archimedes, begun in full earnest by Galileo, whom Gillespie calls Archimedes' best student, and perfected in Newton).[25] You need only recall that, not so long ago, the word "computer" referred, not to the machines which were then only a gleam in Alan Turing's eye, but to the women employed to calculate the trajectories of artillery (and before that to women employed to calculate the positions of stars by hand),[26] to appreciate the role of mathematics as part of the evolution of scientific helps: from counting by means of notches in sticks or knots in a piece of rope, to Roman, and then Arabic, numerals, to the calculus, and now the computer.

Rather than being devised specifically for scientific applications, the mathematics that proves vital to the sciences is often developed for quite other reasons. Non-Euclidean geometries, for example, were developed in response to questions about the (in)dependence of Euclid's parallel postulate, without any idea of physical applications; but Riemann's general theory of curved spaces of two, three, or more dimensions was precisely the formalism Einstein needed to articulate the mathematics of general relativity. It was serendipity, too, when in 1960 Murray Gell-Mann and Yuval Ne'eman found that the already-existing mathematics of Lie groups could characterize the mixing of electrons and neutrinos in the electroweak theory.[27]

Already at the time of Watson and Crick's discovery of the structure of DNA, biological work was becoming almost as mathematical as physics or chemistry. By now, sophisticated calculations are made much faster and more accurately by computers than by humans; and we read in the press of the month-long computational marathon in which physicists at Brookhaven National Laboratory calculated the anomalous magnetic momentum of the muon, and of the vast super-computing operation by means of which Celera Genomics produced its map of the human genome. To their familiar repertoire of experiments in vivo (with a living creature) and in vitro (in a glass tube or dish), biologists have now added experiments "in silica"—i.e., simulated on a computer.[28]

But the computer is only the most obvious example from the large category of "helps to reasoning." Another would be the various still-controversial techniques of meta-analysis, combining a whole bunch of epidemiological or other studies to squeeze out more information than any of them can provide alone. In

2001, for example, two Danish researchers re-analyzed evidence for the "placebo effect," the existence of which had been taken for granted since 1955, when Henry Beecher reported that about 35 percent of patients are helped by placebos. Some physicians now use placebos as treatment; and some medical researchers devote their time to figuring out how placebos might work. But Drs. Hrobjartsson and Gotzsche suggest that the supposed effect is largely mythical. Of 727 potentially eligible trials, they reanalyzed 114 studies (involving altogether about 7,500 patients with 40 different conditions) which they judged "well designed"; these divided patients into (1) those receiving real medical treatment, (2) those receiving a placebo, and (3) those receiving nothing at all. Except for a small effect in pain reduction, they conclude, there is little evidence that placebos have significant clinical effects.[29]

Hrobjartsson and Gotzsche criticize Beecher's research methods as inadequate to distinguish the effects of placebos from the natural course of diseases, regression to the mean, and other factors (e.g., self-medication by patients, the possibility that some "placebos" are not entirely inert after all). But they also acknowledge that their own methods may need further refinement, observing that various types of bias may have affected their findings too, especially where subjectively reported outcomes are concerned. And indeed, during the year following the publication of their paper, the authors of several other papers suggested that Hrobjartsson and Gotzsche's conclusion was exaggerated.[30] Whatever the truth turns out to be, this work, and other scientists' reactions to it, vividly illustrates the finesse required by such meta-analytic refinements of reasoning, and their dependence on factual knowledge.

The social character of scientific inquiry, sometimes seen as a threat to its epistemic pretensions, is better conceived as another kind of fallible help. There are lots of jobs that get done better if many people are involved; but scientific work isn't much like shelling a hundredweight of peas, which can be done quicker the more people are involved, nor much like carrying a heavy log, which can be lifted by several people but not by one. It is complex, intricate, multi-faceted— yes!—like working on a vast crossword puzzle.

Warrant is a matter of degree; degrees of warrant of rival theories, furthermore, won't necessarily be linearly ordered. In the large grey area where either a more or a less optimistic attitude towards a somewhat-but-not-overwhelmingly warranted claim is reasonable, there is no easy step from degrees of warrant to "rules of acceptance and rejection." An as-yet half-baked conjecture may be worth exploring despite being, thus far, unwarranted; most importantly, "the" best way to proceed is often for different members of the community to proceed differently.

As Duhem observed, the sciences often find themselves in a state of indecision between bolder and more cautious approaches to an as-yet unresolved issue.[31] Some of us are quick to rub out a crossword entry when the constraints it imposes on other entries begin to cramp our style, others are slower to change; if we worked together, I might sometimes prevent you from hanging on to an idée fixe too long, and you might prevent me from giving up a promising idea too easily. Scientific inquiry, this suggests, is apt to go better when the community includes, as it usually will, some who are quick to start speculating about possible new theories when the evidence begins to disfavor the presently dominant view, and others who are more inclined patiently to try modifying the old.[32]

Michael Polanyi has already made the important point that, though the best way to organize an army of pea-shuckers might be to have one person direct, this is as inappropriate to scientific inquiry, because of its articulated complexity, as it would be to organizing a team of people working on a giant jigsaw[33]—or, as he didn't say, but I will, on a vast crossword. As Polanyi saw, it is better for different people, with their more and less conservative temperaments, their different strengths and weaknesses, each to do what he does best. Among the many and various talents useful in science, the extraordinary intellectual creativity which enabled those heroes of the history of science to make their startling theoretical leaps comes first to mind. But the list is long and diverse, including inter alia that special gift for spotting patterns in what you see with which some scientists seem to be gifted as other people are with musical talent; ingenuity in experimental design or devising instruments or tests or mathematical models; and sheer patience and painstaking in checking and double-checking.

Now issues about communication, expertise, and authority start to come into focus. If effort is not to be wasted by reduplication, the work of each needs to be freely and conveniently available to others. And scientists need, not only to be able to look over one another's shoulders as they work, but also to stand on the shoulders of those who have gone before; for each would have to start from scratch if he couldn't take earlier results for granted. Issues of communication, of transmission of results, intersect with issues about expertise and authority. Journals clogged with papers by cranks and crackpots will make the work of inquiry harder, not easier. Hence the need for ways of discriminating the nut and the incompetent from the competent inquirer—credentials, peer review—so as to ensure that what the journals make available is not rubbish but worthwhile work.

Like the precautions taken against contamination in the laboratory and so forth, these mechanisms are fallible. Imperfectly honest reviewers may be tempted to try to prevent the publication of rivals' work; imperfectly qualified reviewers may lack the background knowledge necessary for a fair assessment;

unimaginative reviewers may fail to appreciate truly innovative ideas. There is, furthermore, a broad spectrum from the creatively unorthodox scientific conjecture to the pathologically nutty; the crackpot idea does not necessarily come marked on its face as hopeless, nor the creatively unorthodox as promising. What work is judged worthwhile and what worthless has to depend on what is taken to be known fact. Thus Martin Gardner writes, dismissing Immanuel Velikovsky as beyond the scientific pale, that his views, "if correct, would require rewriting physics, astronomy, geology, and ancient history."[34] What is taken to be known fact is sometimes false—Darwin's work, after all, required rewriting much that was accepted as known fact about the fossil record—so these judgments may be mistaken; but, as with the precautions against experimental contamination discussed earlier, that these precautions are fallible doesn't mean that they are not, usually, real helps.

We need to go beyond thinking in terms of "many hands make light work" to an understanding that what is involved is the subtlest kind, or kinds, of division of labor; and to go beyond thinking of science as depending on "trust" to an articulation of the delicate balance of institutionalized mutual criticism and checking and the institutionalized authority of well-warranted results that natural-scientific inquiry has achieved at its best. Scientists rely on each others' competence and expertise; but, as those words were chosen to indicate, the trust and the authority involved are not the privilege of certain persons or even of certain positions, but distinctions that have to be earned.

To my picture of scientists looking over one another's shoulders, or standing on the shoulders of those who have gone before, I need to add, since inquiry in the sciences can be competitive as well as cooperative, that scientists are sometimes found elbowing each other out of the way (I do my limited best, but painting the whole picture really calls for the talents of a Brueghel!). Though it is not invariably or necessarily so, competition between rival individuals or teams can be productive: the hope of beating the other guys out for the Nobel Prize can spur miracles of intellectual effort; proponents of one approach or theory are motivated to seek out evidence disfavoring its rival, which its proponents are motivated to neglect or play down. Competition can be another of those helps; not, like the microscope, an aid to human beings' limited senses, nor, like a well-chosen analogy, an aid to our limited imaginations, but an aid to our limited energy and fragile intellectual integrity.

Social helps, like the others, are fallible. As the hearsay quality of the literature on the placebo effect suggests, with its layer upon layer of cross-references in scores of scientific papers all leading back to that one questionable study of Beecher's, scientists' confidence in others' work is sometimes misplaced. Competition can lead to wasted resources, even to the suppression of worthwhile work; cooperation may prolong a wild goose chase. But "doing one's damnedest with one's mind" also includes trying to figure out in what circumstances instruments or citation practices are likely to mislead, in what circumstances competition is likely to turn sour and counter-productive, in what circumstances cooperation is likely to degenerate into bureaucratic expansionism, busy-work, or mutual promotion. And if we can figure this out, we can devise safeguards—imperfect and fallible, as always, but useful nonetheless—against the failure of precautions.

Scientific helps are often local. (A physicist doesn't have to worry, like a biochemist, about pets in the lab—or not for the same reasons, anyway!) Nevertheless, scientific inquiry and other kinds are by now interwoven in all sorts of ways, and "scientific" helps are used by others too. Detectives rely on techniques of forensic science such as blood-typing, genetic fingerprinting and the rest. Ancient historians recently began to use a technique devised for the detection of breast cancer to decipher the lead "postcards" on which Roman soldiers wrote home.[35] Other historians relied on neutron analysis to show that pieces of jasper found in a settlement in northern Newfoundland contain trace elements present only in jasper from Greenland and Iceland, confirming the conjecture that the Vikings reached North America half a millennium before Columbus.[36] General Motors uses statistical techniques developed by the Centers for Disease Control to detect potential "epidemics" of defects in their cars and trucks.[37] And so on. Occasionally, though less often, scientists borrow from other inquirers; when, for example, a family tree compiled by the father of a sufferer proved a vital clue to the identification of the genetic defect responsible for hereditary pancreatitis.[38]

TO ILLUSTRATE: THE SEARCH FOR THE DOUBLE HELIX

Since I used their scientific papers to illustrate some themes about scientific evidence, it seems appropriate to turn to Watson's account of how he and Crick discovered the structure of DNA to illustrate some themes about scientific inquiry. Watson recounts events as they seemed to him at the time; others involved, we know—Crick included—didn't always see things the same way.[39] Still, *The Double Helix*, for all its brashness, its personalities, its tasteless comments about the "popsies" at Cambridge parties and Rosalind Franklin's dress sense, precisely because of its artlessness is revealing about how science has sometimes managed to make the most of human strengths, and of how ego can be put in the service of creativity and respect for evidence. It is a book that calls poignantly to mind that aphorism of Diderot's: "to say that man is made up of strength and weakness, of insight and blindness, of pettiness and grandeur, is not to criticize him; it is to define him."[40]

In his preface, Watson writes that he hopes his book will show that "science seldom proceeds in the straightforward logical manner imagined by outsiders. Instead, its steps forward (and sometimes backward) are often very human events." So far, so good. But he continues, "human events in which personalities and cultural traditions play major roles"; which some cynics might take as endorsing the idea of science as "social negotiation." It soon transpires, however, that he is doing nothing so radical, but only musing on the quirks of the British, especially their reluctance to muscle in on a problem they see as someone else's turf.

But I anticipate. Within a few days of arriving in Cambridge and meeting Francis Crick, Watson tells us, "we knew what to do; imitate Linus Pauling and beat him at his own game." If they could do that, they could solve "a smashingly important problem."[41] Already you hear the intimate combination of ambition and respect for evidence that characterizes the whole story.

Competing with Pauling required cooperating with Maurice Wilkins and Rosalind Franklin in London, however much Crick would prefer to avoid the personal awkwardness of jumping in on a problem Wilkins had worked on for years (those "cultural factors" at work)—because the London team had potentially crucial X-ray diffraction photographs of DNA, which might save months of work by ruling out theoretically possible configurations.

Before long Watson and Crick thought the solution was almost in their grasp. Crick and Cambridge crystallographer Bill Cochran both arrived, by different routes, at a plausible theory of the diffraction of X-rays by helical molecules, confirmed by visual inspection of Max Perutz's X-ray diagrams. And from Watson's report of a lecture of Franklin's, it seemed that "only a small number of formal solutions were compatible both with the Cochran-Crick theory and with Rosy's experimental data."[42]

A day's fiddling with their tinker-toy models, and they had something promising—a three-chain model compatible with the Cochran-Crick theory and with the locations of the X-ray reflections in Franklin's photographs, but needing to be checked against their relative intensities. But when Wilkins and Franklin arrived from London to look at it, the result was not a triumph but an embarrassing debacle. Franklin had in any case always disapproved of the tinkering with models, and was not yet convinced that DNA was helical. More importantly, "her objections were not mere perversity"; Watson had misremembered the water content of her DNA sample—the correct model had to contain at least ten times more water than was found in his and Crick's. Though it might be possible to fudge the extra water molecules into vacant regions on the periphery of the helix, Watson continues, with the extra water involved, "the number of potential DNA models alarmingly increased."[43] The early optimism about a quick solution was gone.

Watson learned X-ray diffraction techniques, but then got caught up in problems about the sex-life of bacteria. From time to time he and Crick fiddled some more with their DNA models, but "almost immediately Francis saw that the reasoning which had momentarily given us hope led nowhere." And then a letter from Pauling to his son, at that time a research student in Cambridge, conveyed "the long-feared news": Pauling had a structure for DNA.[44] Perhaps, they thought, if they could reconstruct what Pauling had done before they saw his paper, they could get equal credit. And then, when Pauling's paper arrived, and his model sounded suspiciously like their aborted effort of a year before, perhaps, they thought, they might already have got the credit if they'd not held back. And then, on studying the illustrations, they realized that Pauling had either made a revolutionary breakthrough in the underlying chemical theory, or else had made an elementary mistake. But there would be no reason to keep such a theoretical breakthrough secret, so it had to be—an "unbelievable blooper."

"Though the odds still appeared against us, Linus had not yet won his Nobel";[45] but it might be fatal to waste time gloating over Pauling's mistake . . .

A few days later, Watson saw Franklin's photographs of the B form of DNA; the pattern, "unbelievably simpler" than in the earlier photographs, was overwhelming evidence that the structure was helical. Since "the meridional reflection at 3.4 Å was much stronger than any other reflection," Watson figured, "this could only mean that the 3.4 Å-thick purine and pyrimidine bases were stacked on top of each other in a direction perpendicular to the helical axis," and both electron-microscope and X-ray evidence suggested that the helix diameter was about 20 Å. Crick was not yet convinced that they should be looking for a two-chain rather than three-chain model; but Watson was already persuaded, on the grounds that the biological likes to come in pairs. Watson was not yet convinced by Franklin's hunch that the backbone must be outside and not inside the structure; but he began to warm to the idea after taking apart "a particularly repulsive backbone-centered model."[46]

Spending his evenings at the cinema, disappointed that Hedy Lamarr's nude scenes had been cut from *Ecstasy*, and vaguely dreaming that at any moment the answer would suddenly hit him, Watson came up with a solution so apparently promising that he mentioned it in a letter to Max Delbrück—only to have it torn to shreds within the hour when crystallographer Jerry Donohue "happily let out that for years organic chemists had been favoring particular tautomeric forms over their alternatives on only the flimsiest of grounds. In fact, organic-chemistry texts were littered with pictures of highly improbable tautomeric forms. The guanine picture I was thrusting toward his face was almost certainly bogus."[47] No gimmick emerged to salvage the idea.

But very soon Watson came up with another, compatible with Donohue's reservations, and startling in its explanatory power:

> Suddenly I became aware that an adenine-thymine pair held together by two hydrogen bonds was identical in shape to a guanine-cytosine pair held together by at least two hydrogen bonds. . . . [N]o fudging was required. . . . I suspected we now had the answer to the riddle of why the number of purine residues exactly equaled the number of pyrimidine residues. Two irregular sequences of bases could be regularly packed in the center of a helix if a purine always hydrogen-bonded to a pyrimidine. Furthermore, the hydrogen-bonding requirement meant that adenine would always pair with thymine, while guanine could pair only with cytosine. Chargaff's rules then suddenly stood out as a consequence of a double-helical structure for DNA. Even more exciting, this type of double helix suggested a replication scheme much more satisfactory than my briefly considered like-with-like pairing.[48]

At first Crick had taken scant interest when Watson told him of Chargaff's discovery of regularities in the relative proportions of purine and pyrimidine bases in DNA, but he soon became convinced of their importance. Checking, he found no other way to satisfy Chargaff's rules, and was ready to tell everyone that they had found the secret of life. But more model-building, careful measurements to ensure that they hadn't fudged a series of atomic contacts each separately almost acceptable but jointly impossible, further consultations with Wilkins and Franklin comparing the X-ray data with the predictions of the model, were still needed.

The same week, letters from Pasadena revealed that Pauling "was still way off base." Though Watson preferred that Pauling not be told yet, Delbrück, who "hated any form of secrecy in scientific matters," told him anyway; and "the overwhelming biological merits of a self-complementary DNA molecule made him effectively concede the race." By then it hardly mattered that Pauling knew, since "the evidence favoring the base pairs was steadily mounting.[49]

Even potted, this marvelous scientific detective story illustrates what I said earlier, that *"scientific inquiry is far messier, far less tidy than the Old Deferentialists imagined; and yet far more constrained by the demands of evidence than the New Cynics dream."*

Determined cynics aren't deterred. Of late, students of the rhetoric of science have turned their attention to Watson's story, and even to Watson and Crick's first paper in *Nature*, with its supposedly parallel rhetorical structure: S. Michael Halloran suggests that Watson and Crick's work began a revolution in biology because of their rhetorical daring in using "we" instead of the passive voice; Alan Gross, that "the sense that a molecule of this structure exists at all . . . is an effect only of words, numbers and pictures judiciously used with persuasive effect."[50] But to the contrary—and despite the fact that he seems to have exaggerated the competition with Pauling, to have misrepresented Franklin's position at King's College and her attitude to helices, etc., etc.[51]—Watson's account illustrates the intimate interweaving of evidential, personal, and social aspects of science, and the role of scientific helps to inquiry.

An informed guess is checked out with reference to the evidence they have and the additional evidence they can get, is modified, but founders because of the mistake about water content; further, better-informed guesses are tried, checked, modified, discarded, until one finally stands up and is inked in—sorry, is submitted to *Nature*, and eventually earns its authors the Nobel Prize. The interplay of inference and judgment, the dependence on background beliefs themselves fallible, is ever-present. The conjecture relying on the mistake about the water content *might* have been right, nevertheless—but, no, it was too much of a stretch. Pauling's "long-feared" structure *might* have been right, too, *if* he had evidence for the revolutionary changes that would be required in the underlying chemical theory—but, no, he hadn't. The backbone-in models *might* have been right—but, no, those awkward efforts were going nowhere. The double-helical, backbone-out structure with like-with-like base pairs *might* have been right, had the organic-chemistry texts not been arbitrarily favoring certain tautomeric forms over others—but they had.

We hear, in Watson's story, snatches of familiar themes from the Old Deferentialist debates: an early conjecture is falsified when the mistake about water content is discovered, and Pauling's structure is falsified when the chemical blooper is noticed; the backbone-in models start to look like a degenerating research program; and, of course, eventually the double-helical, backbone-out, etc., solution is confirmed (warranted) to such a degree that it would be unreasonable not to accept it. But these inferences are all made against the background of a mesh of other beliefs, much more like doing a crossword puzzle than constructing a mathematical proof.

And then there are the models and metaphors, the instruments, and the cooperation and competition with other scientists that helped Watson and Crick to envisage potential structures for DNA, to obtain needed evidence, and to stay motivated and honest. Those tinker-toy models, aided and abetted by metaphors about "which molecules like to sit next to each other," serve both as aids to spatial imagination, and, made to scale, as evidential checks on what structures are stereo-chemically possible. The observational evidence is mediated by X-ray diffraction techniques—the theory of which is being worked out as they go along; but as soon as Watson sees Franklin's photographs of the B form of DNA, his "mouth falls open and [his] pulse begins to race," and when Crick sees them it takes him just "ten seconds" to realize their significance.[52] And this is a tale of competition and cooperation from the beginning; but, far from turning the quest for the structure of DNA into mere social negotiation, competition is a spur to intellectual effort, cooperation a way of dividing labor and extending evidential reach.

More than thirty years later, musing on how he and Watson did it, Crick writes that "we passionately wanted to know the details of the structure. . . . If we deserve any credit at all, it is for persistence and the willingness to discard ideas when they became untenable. . . . We considered [the problem] so important that we were determined to think about it long and hard, from any relevant point of view. . . . [Our discussions were] very demanding and sometimes intellectually exhausting."[53] I am reminded of Newton's reply to an admirer who wanted to know how he made his discoveries: "By always thinking unto them";[54] and of Santiago Ramon y Cajal's *Advice for a Young Investigator*, that the crucial thing is sustained concentration, the "steady orientation of all our faculties toward a single object of study . . . which the French call *esprit de suite*."[55]

THE OLD DEFERENTIALISM
AND THE NEW CYNICISM REVISITED

And looking back at the Old Deferentialism and the New Cynicism, I am irre-sistibly reminded of that marvelous children's poem, "The Blind Men and the Elephant." One blind Hindu, groping along the elephant's side, decides that an elephant "is very like a wall," another, grasping its tail, that an elephant "is very like a rope," etc.:

> And so these men of Hindostan
> Disputed loud and long,
> Each in his own opinion
> Exceeding stiff and strong,
> Though each was *partly* in the right
> *And all were in the wrong.*[56]

Old Deferentialists rightly took for granted that the natural sciences have had striking successes. But they were wrong insofar as they supposed that the explanation of these successes must be a method of inquiry of a narrowly logical or quasi-logical character, exclusive to science, and guaranteeing, if not success, at least progress. New Cynics are rightly skeptical that, in the hoped-for sense, there is such a thing as "the scientific method"; and see that concentrating too exclusively on narrowly logical aspects of science disguises the significance of the fact that, whatever else it is, science is a social institution. But they are wrong when they conclude that the whole idea that natural-scientific inquiry is in some way epistemologically distinguished must be an illusion. Rather than labor the general point further, let me apply it, in the light of the account articulated here, to some old debates about the problem of demarcation and the discovery/justification distinction.

For Old Deferentialists—for the Popperians especially, but in an oblique way for some Positivists, too—distinguishing science from non-science was an important preoccupation, both encouraging and encouraged by the honorific use of "science" and the myth of a uniquely rational scientific method.[57] But if the Old Deferentialists were too preoccupied with demarcation, the New Cynics (whether thinking of science as just another large, powerful social institution, or as just another imaginative creation, not essentially different from fiction) are too dismissive of the epistemological pretensions of the sciences. Regarding the honorific usage of "scientific" as a nuisance, and skeptical of "scientific method" in the Deferentialist sense, I am disinclined to give the "problem of demarcation" the high priority some Old Deferentialists did. But, taking "science" to pick out a federation of epistemologically distinguished kinds of inquiry, I give epistemological concerns a much more central place than the New Cynics.

The first thing to say is that "non-science" is an ample and diverse category, including the many human activities other than inquiry, the various forms of pseudo-inquiry, inquiry of a non-empirical character, and empirical inquiry of other kinds than the scientific; and that, to make matters even more complicated, there are plenty of borderline and mixed cases. The use of "scientific" and its cognates as all-purpose terms of epistemic praise tempts scientists as well as laypersons (and judges—but I'll get to that later) to criticize poorly conducted science as not really science at all; but "not scientific" is as unhelpful as generic epistemic criticism as "scientific" is as generic epistemic praise.

The phrase "pseudo-science," which presumably refers to activities which purport to be science but aren't really, deserves special attention. Its pejorative tone derives in part from its imputation of false pretenses, but also in part from the honorific connotations of "scientific." Here is Bridgman again: "[t]he working scientist is always too much concerned with getting down to brass tacks to be willing to spend his time on generalities."[58] By my lights too, rather than criticizing work as "pseudo-scientific," it is always better to specify what, exactly, is wrong with it: that it is not honest or serious inquiry; that it rests on assumptions for which there is no good evidence, or which are too vague to be susceptible to evidential check; that it uses mathematical symbolism, or perhaps elaborate-looking apparatus, purely decoratively; etc.

Lakatos worries that "if Kuhn is right, there is no explicit demarcation between science and pseudoscience, no distinction between scientific progress and intellectual decay, there is no objective standard of intellectual honesty. . . . [W]hat criteria can he then offer to demarcate scientific progress from intellectual degeneration?"[59] In just a couple of sentences, Lakatos has run together science, scientific progress, intellectual honesty, and the good intellectual health of a culture—all quite distinct concepts, though interrelated in complex and subtle ways. In particular, willingness to acknowledge negative evidence is not a criterion of genuine science, but a condition of intellectual honesty—for scientists as for inquirers of every kind. Darwin writes that he always followed "a golden rule, namely, that whenever a published fact, a new observation or thought came across me, which was opposed to my general results, to make a memorandum of it without fail and at once; for I had found by experience that such facts and thoughts were far more likely to escape from the memory than the favourable ones":[60] an experience familiar to anyone—scientist, historian, journalist, or even philosopher—who has ever engaged in serious inquiry.

Of course, for some purposes it is necessary to draw a rough and ready line between science and other things. One way might be to think of the sciences as differing from such other activities as clog-dancing or advocacy in being kinds of inquiry; as differing from other kinds of empirical inquiry such as history or legal or literary scholarship in virtue of their subject-matter; and, perhaps, as differing from natural theology in virtue of the kinds of explanation they acknowledge. Rough as it is, this isn't a bad starting point for an explanation of, say, how science differs from literature or the entertainment industry. But if we want to understand the historical aspects of social science or evolutionary biology or cosmology, but to avoid assimilating science and history, we shall need something subtler. And if we want to understand how creationism differs epistemologically from physical cosmology or evolutionary biology, we will do better to focus directly on questions of evidence and warrant, instead of fussing over whether creationism is bad science, or not science at all.

Now for the discovery/justification distinction. Deferentialists Old and New hoped to confine the social and psychological aspects of science to the context of discovery, and concentrate on the nice, tidy, logical, context of justification.[61] But New Cynics, naturally, dispute the legitimacy of the distinction. This kerfuffle has encouraged both the deferentialist tendency to over-simplify the process of inquiry, and a cynical readiness to ignore the role of evidence.

The problem is less that there is *no* distinction for the contrast of discovery versus justification to pick out, than that there are too many. There is certainly a difference between the question of how a theory was arrived at, and the question of how good the evidence with respect to it is. There are certainly different stages of inquiry: a theory is conceived, developed, tested, refined, modified, presented in the journals, etc. And there is certainly a difference between psychological and sociological questions about scientific inquiry, and epistemological ones. But to identify the context of justification with the presentation stage of inquiry, as Reichenbach does, is to risk running together the question of evidential quality and the question of what a scientist does to persuade his colleagues of the truth of his theory. And to identify the context of discovery as the sphere of the psychological or sociological and the context of justification as the sphere of the logical, as Popper does, is to risk disguising the fact that arriving at a hypothesis is not usually blind, but involves inference, and that collecting, sharing, and appraising the evidence with respect to a hypothesis is generally a community undertaking.

As with the problem of demarcation, what I'd recommend is a shift of emphasis: this time, to the role of inference in the process by which scientists arrive at theories, and of interactions among scientists as they appraise the worth of evidence. No rule of inference guarantees a good conjecture; but a good conjecture must be compatible with (if possible, entail) what is already known. Think how strong the constraints were on possible solutions to the structure of DNA, given the vast mesh of background beliefs on which Watson and Crick relied, and Franklin's photographs of the B form. Watson arrived at the exact nature of the base pairs "by serendipity," Crick recalls, but it could have been done by elimination, systematically testing the pairs suggested by Chargaff's rules.[62] And, as something inferential is involved in coming up with a plausible conjecture, something social—the interaction of more and less conservative members of the scientific community as the evidence needed to decide between a conjecture and its rivals is sought and sifted—plays a role in the process by which it eventually comes to be put in the textbooks, or discarded.

AND, IN CONCLUSION

Commenting on that recent article in the *New England Journal*, Dr. Clement McDonald, who eighteen years before had published a paper questioning the genuineness of the placebo effect, observes that "the good thing about science is that sooner or later the truth comes out."[63] No doubt this sounds a little naive; nonetheless, in a crude way it captures something important. Progress has been ragged; nevertheless, thanks to the natural sciences, we now know far more about the world than we did, say, 400 years ago.

If the DNA story illustrates the glory, the saga of that Martian meteorite illustrates the raggedness. 1996: scientists suggest that the carbonates the meteorite gives off constitute evidence of early bacterial life on Mars; 1997: other scientists suggest rival explanations of the carbonates; 1998: others suggest that the bacterial traces are terrestrial contamination; 1999: satellite photographs indicate there may be water beneath the Martian permafrost, and the Polar Lander spacecraft is sent off to search for further evidence; January 2000: the Lander is thought to have been lost as it crashed into a canyon on Mars; February 2000: it is conjectured that mysterious signals coming through may be being sent by the Lander after all; April 2000: NASA investigators think the Lander crashed into the frozen Martian surface and broke up; April 2001: NASA's Odyssey probe is launched; June 2002: it is announced that data from the new probe suggests that oceans of ice still cover Mars, raising the likelihood that there was once life there.[64] I haven't much doubt that the truth will come out eventually.

Let me suggest a friendly re-interpretation of Lakatos's observation that "there is no instant rationality," and of Kuhn's conception of the history of science as "written by the winning side." Progress may be small and incremental, or large and revolutionary, or anywhere in between. It may be a happy accident or even a serendipitous mistake, as when in ignorance or confusion a scientist comes up with a conjecture incompatible with what is taken for known fact, but, as it happens, is no fact at all. At some times and in some areas, scientific inquiry stagnates, or even regresses; and it may be only with hindsight that it becomes clear this or that shift was progressive. Nonetheless, if my account is right, it is no mystery why, by and large and on the whole and in the long run, natural-scientific inquiry *has* progressed. For it relies on helps which, though fallible and imperfect, generally tend to aid the imagination, extend evidential reach, and stiffen respect for evidence. By no means every step will be in the right direction; but to the extent that the helps succeed, the general tendency will be in the direction of stronger experiential anchoring and improved explanatory integration.

Still, many readers, the cautiously sophisticated as well as the brashly cynical, are likely to suspect that my modest-sounding but reassuring conclusion may covertly rely on an uncritical and indefensibly ambitious realism. To speak to such concerns, I shall have to probe more deeply into issues about observation and theory, generals and explanation, truth and progress.

NOTES

1. Bridgman, "The Prospect for Intelligence," p. 535.

2. Haack, *Evidence and Inquiry* (1993), p. 137. Shortly thereafter, I developed these ideas in "Puzzling Out Science" (1995) and "Science as Social?—Yes and No" (1996).

3. Dewey, *Logic, The Theory of Inquiry*, p. 66.

4. My source is Grove, *In Defence of Science*, p. 13; he doesn't give an exact reference, and I have been unable to locate the passage in question.

5. Einstein, "Physics and Reality," p. 290.

6. Bridgman, "New Vistas for Intelligence," p. 554.

7. Bergmann, *Philosophy of Science*, p. 20.

8. Bridgman, "On 'Scientific Method'," p. 81.

9. Ibid., pp. 81–83.

10. The term "sham reasoning" comes from Peirce, *Collected Papers*, 1:57; see also Haack, "Confessions of an Old-Fashioned Prig," and "Preposterism and Its Consequences."

11. See Haack, *Deviant Logic, Fuzzy Logic: Beyond the Formalism*.

12. Harris, "Rationality in Science," pp. 40–44.

13. Bacon, *The New Organon*, aphorism 2.

14. Peirce, *Collected Papers*, 1:46–48.

15. Duhem, *The Aim and Structure of Physical Theory*, chapter 5, quoted in Hesse, *Models and Analogies in Science*, pp. 1–2.

16. Kendrew, *The Thread of Life*, p. 20.

17. See Craig, "Chaperones, Molecular," in Meyers, *Molecular Biology and Biotechnology*, pp. 162–65: "Molecular chaperones are ubiquitous proteins that play a critical role in the cellular processes of protein folding and translocation of proteins across membranes into organelles."

18. "CAT" stands for "computer-assisted [or "axial"] tomograph"; "MRI" for "magnetic resonance imaging" (magnetism is to MRI what radiation is to X-rays); "PET" for "positron emission tomograph"; and "EBT" for "electron beam tomography." My source is Kevles' history of medical imaging technology, *Naked to the Bone*, and, on EBT, Parker-Pope, "New Tests Go beyond Cholesterol to Find Heart-Disease Risks."

19. My source is Judson, *The Eighth Day of Creation*, pp. 35–39.

20. Belenky et al., *Women's Ways of Knowing*; Haack, "Knowledge and Propaganda," p. 125.

21. Thomas, Hosenball, and Isikoff, "The JFK-Marilyn Hoax."

22. Bridgman, *Reflections of a Physicist*, p. 81.

23. Tessman and Tessman, "Efficacy of Prayer," criticizing Byrd, "Positive Therapeutic Effects of Intercessory Prayer in a Coronary Care Unit."

24. Benveniste et al., "Human Basophil Degranulation Triggered by Very Dilute Antiserum Against IgE"; Maddox et al., "'High-Dilution' Experiments a Delusion" (quotations from p. 287).

25. Gillespie, *The Edge of Objectivity*, pp. 16, 144.

26. See Bird, *Enterprising Women*, p. 117.

27. Weinberg, *Dreams of a Final Theory*, pp. 153 ff.

28. Johnson, "In Silica Fertilization."

29. Hrobjartsson and Gotzsche, "Is the Placebo Powerless?" See also Bailar, "The Powerful Placebo and the Wizard of Oz"; Kolata, "Placebo Effect Is More Myth than Science, Study Says," and "Putting Your Faith in Science?"

30. Begley, "In the Placebo Debate, New Support for the Role of the Brain in Healing."

31. Duhem, *The Aim and Structure of Physical Theory*, p. 217.

32. The desirable kind of pluralism sketched here should not be confused with the politicized race and gender pluralisms criticized in chapter 11, pp. 313 ff.

33. Polanyi, "The Republic of Science."

34. Gardner, *Science: Good, Bad and Bogus*, p. 235. Gardner's reference to ancient history confirms my earlier remarks about the overlap of scientific and other inquiries.

35. "Wish You Were Here," *Oxford Today* 10, no. 3, Trinity 1998, p. 40.

36. Begley, "The Ancient Mariners," p. 54.

37. White, "GM Takes Advice from Disease Sleuths to Debug Cars."

38. Bounds, "One Family's Search for a Faulty Gene."

39. Reviews of the book—not, however, Chargaff's, permission to reprint which was denied—are conveniently collected in Gunther Stent's edition of *The Double Helix*. See also Olby, *The Path to the Double Helix*, and Judson, *The Eighth Day of Creation*.

40. Diderot, *Addition aux pensées philosophiques*.

41. Watson, *The Double Helix*, p. 32.

42. Ibid., p. 49.

43. Ibid., p. 59.

44. Ibid., p. 91.

45. Ibid., p. 95.

46. Ibid., p. 103.

47. Ibid., p. 110.

48. Ibid., p. 114.

49. Ibid., p. 128.

50. Gross, *The Rhetoric of Science*, chapter 4; the quotation is from p. 54. I shall deal with Halloran and Gross in more detail in chapter 8, pp. 218–22.

51. Sayre, *Rosalind Franklin and DNA*, especially pp. 17 ff.; and Judson, *The Eighth Day of Creation*, pp. 147 ff.

52. Watson, *The Double Helix*, p. 98.

53. Crick, *What Mad Pursuit*, pp. 70, 74, 75.

54. Gillespie, *The Edge of Objectivity*, p. 117.

55. Ramon y Cajal, *Advice for a Young Investigator*, p. 32; *esprit de suite* might be roughly translated as "spirit of persistence, following through."

56. Saxe, "The Blind Men and the Elephant."

57. Of course, the problem of demarcation goes back, in a somewhat different form, to Plato and Aristotle. Laudan, "The Demise of the Demarcation Problem," is useful both on the history, and on the problems, of the preoccupation with this question.

58. Bridgman, "On 'Scientific Method'," p. 81.

59. Lakatos, *The Methodology of Scientific Research Programmes*, p. 4.

60. Darwin, *Autobiography*, p. 45.

61. Popper, *The Logic of Scientific Discovery*, chapter 1, section 2; Reichenbach, *Experience and Prediction*, pp. 6–7. Hoyningen-Huene, "Context of Discovery and Context of Justification," gives a history of the distinction (which in some form goes back at least to Herschel's *Preliminary Discourse on the Study of Natural Philosophy* of 1830/31, p. 164), and a decomposition into the several distinct distinctions he sees as having been run together.

62. Crick, *What Mad Pursuit*, pp. 65–66; cf. Meehl, "Corroboration and Verisimilitude," p. 27.

63. Kolata, "Putting Your Faith in Science?"

64. Rogers, "Come in, Mars" (1996); Begley and Rogers, "War of the Worlds" (1997); "Meteorite—or Wrong?" (1998); Wilford, "2 New Chemical Studies Show No Traces of Life on Mars" (1998); Hayden, "A Message, but Still No Answer" (1999); "NASA Scientists Seem Close to Confirming . . ." (1999); "Did the Mars Lander Crash in a Grand Canyon?" (January 2000); Broad, "Evidence Builds That Mars Lander is Source of Mystery Signal" (February 2000); Murr, "Final Answer: It Crashed" (April 2000); and Guterl, "Water, Water Everywhere" (June 2002).

5

REALISTICALLY SPEAKING

How Science Fumbles,
and Sometimes Forges, Ahead

Let us not pretend to doubt in philosophy what we do not
doubt in our hearts.
—C. S. Peirce, "Some Consequences of Four Incapacities"[1]

I have tried to make my story realistic, in the most ordinary, everyday sense
of the word: neither too optimistic nor too pessimistic, fully acknowledging
the achievements of science, but also the pervasive fallibility, the imperfections and flaws, the sheer untidiness, of this remarkable but thoroughly human
enterprise. Now it is time to acknowledge explicitly that, implicit in my insistence that understanding the complexities of scientific evidence and scientific
inquiry requires attention not only to matters of logic, but also to the world and
our place as inquirers in the world, are other more technically philosophical
kinds of realism.

There is one, real world; and the sciences aim to discover something of how
this world is. Of course, human beings intervene in the world, and we, and our
physical and mental activities, are part of the world. The world we humans
inhabit is not brute nature, but nature modified by our physical activities and
overlaid by our semiotic webs, including the imaginative constructions of writers
and artists, and the explanations, descriptions, and theories of detectives, historians, theologians, etc.—and of scientists. The imaginative constructions of nov-

elists and artists, their fictional characters and events, are both imaginative and imaginary. But, when they are successful, the imaginative constructions of inquirers, their theoretical entities and categories, are not imaginary but real, and their explanations true.

Successful scientific inquiry, like successful empirical inquiry of any kind, is possible only because we, and the world, are a certain way. Even the most routine empirical inquiry would be impossible if we didn't have sense organs competent to detect information about particular things and events around us, and the intellectual capacity to make generalized conjectures and devise ways to check those conjectures against further evidence; or if the particular things and events in the world of which we can be perceptually aware were not of kinds and subject to laws. If this weren't so, we could not categorize things or discover useful generalizations about them; nor could the natural sciences—deeper and more detailed than everyday empirical inquiry, far better unified, more accurate, yet still thoroughly fallible, imperfect, and incomplete—gradually have managed to identify real kinds of thing or stuff, discern their inner constitution, and discover laws of nature.

My approach is realistic about perception, about kinds and laws, about the world, about truth. But though my realism is extensive, it is not extreme; in fact, it is very modest. Our sensory organs put us in touch with things and events in the world, but our senses are limited, imperfect, and sometimes distorted by our expectations; and there is no cleanly identifiable class of purely observational statements, or of observable things. There are real kinds; but this is only to say that some knots of properties are held together by laws. There are objective truths, and the sciences sometimes succeed in discovering some of them; but truth is not transparent, and progress is not guaranteed.

These days, to acknowledge even such an Innocent Realism as this is to invite criticism on the one hand that you have naively failed to grasp the conceptually or linguistically or textually constructed character of the world, or the relativity of truth or evidence to community or paradigm; and on the other that you must be committed to indefensibly ambitious metaphysical or epistemological ideas. But my Innocent Realism is, I hope, modest enough to escape the charge of indefensible ambition, and subtle enough to accommodate the complexities that have tempted some to succumb to indefensibly ambitious forms of linguistic or conceptual idealism.[2]

I don't aspire to offer a decisive argument that there is a real world, etc.; I am somewhat at a loss to know what less controversial propositions could possibly serve as premises for such an argument. What moves me, and I hope will move you too, is a rather different kind of consideration: scientific inquiry is a highly sophisticated, complex, subtle, and socially organized extension of our everyday reliance on experience and reasoning. Like everyday empirical inquiry, it can succeed only if there is a real world knowable to some extent by creatures with sensory and intellectual powers such as ours. Peirce comments that "a man must be downright crazy to doubt that science has made many true discoveries";[3] and I have no doubt that scientific inquiry *has* often succeeded. But to detach the Innocent Realist conclusion that there is a real world knowable to some extent by creatures like us, it is not strictly necessary to establish that scientific inquiry sometimes succeeds; it is sufficient that everyday empirical inquiry does— which, in his heart, no one really doubts.

So the task before me is to stretch out the compactly folded conceptual structure in need of articulation here into a long, thin, verbal fiber. Perhaps it's not too fanciful to think of the twin themes of Innocent Realism, in metaphysics, and Critical Common-Sensism, in epistemology, as forming the double-helical backbone of my model, intertwined around the series of conceptual base pairs on which I shall focus in what follows: observation and theory, generals and explanation, truth and progress.

OBSERVATION AND THEORY

The method of everyday empirical inquiry, and the method of science ("insofar as it is a method," I will repeat, echoing Bridgman once more), is the method of experience and reasoning; or, since the two work so intimately together, perhaps "the method of experience-and-reasoning" is a better phrase. It is sometimes assumed that to acknowledge the inter-dependence of perception and background belief, of observation and theory, is eo ipso to deny that experience can be a real evidential constraint. But this is a mistake. Is perception dependent on background beliefs, observation on theory? Yes, in more ways than one. Is there a privileged category of infallible observation statements, or of observable things? No, and no again. And yet the evidence of the senses ultimately anchors our theories in the world; and it is a real constraint.

Interacting with things and events around us, we pick up some of the information they afford—though only some. That is why we are vulnerable to certain kinds of illusion (and can predict, for example, that in such-and-such circumstances the road ahead will look wet even if it isn't, and explain why); and why we are susceptible to ambiguous puzzle pictures designed to invite us to fill in gaps this way—and that. We all know to turn a thing around, or go behind it, to check the back; to smell or taste or poke as well as look; to get someone else to come and take a look too. For all the limitations of our perceptual apparatus, for all our susceptibility to misperception and to the distorting influence of our expectations, there is something stubborn about how things look, something receptive about the perceptual event.

Scientists contrive to interact with things and events around them in subtler, more indirect, and more complex ways than the rest of us, and are even more skeptical about supposed phenomena that only one observer can see. To convey the appropriate sense of deliberateness or contrivance, in the context of the sciences we usually speak of "observation" rather than simply of perception; for scientific observations, as Kuhn remarks, are better described as "the collected with difficulty" than as "the given."[4] To be sure, some scientific observations are more serendipitous than contrived. A recent press report tells of a graduate student searching for specimens on a beach and noticing a peculiar-looking centipede that turned out to have an even number of pairs of legs—a startling finding, since until now the received wisdom had been that all centipedes have an odd number of leg pairs. But even here it was the student's background knowledge that led him to think of counting, and thus to make a "marvelous discovery which adds a good deal to our knowledge of centipedes."[5] But very often, what observations scientists choose to make—to follow around this female monkey and her offspring rather than that male and his buddies, to watch the eclipse from this place as well as that, to check these and those aspects of the chemical reaction—depends on the background beliefs that determine what experiential evidence they take to be relevant to the claim in question.

Very often, also, scientific observation relies on instrumentation, and hence on theoretical presuppositions built into the design of the instruments. These presuppositions are rarely from the same domain as the claims they help scientists check; and they need not be known to the scientist making the observations, who more than likely just relies on the integrity of this manufacturer's products, and hence indirectly on other scientists' competence.

In a deliberately paradoxical inversion, Sir Arthur Eddington once advised that it is "a good rule not to put too much confidence in the observational results . . . *until they are confirmed by theory.*"[6] This draws to our attention that, as observation and theory work together, things can go wrong: the presuppositions which lead a scientist to make these observations rather than those, or which underpin the workings of the instruments of observation on which he relies, may be false; and if they are, he may make the wrong observations, or fail to make the right ones, or be misled by what is really only an artifact of his instrument. But when things go right, and the presuppositions which lead scientists to make these observations rather than those, or which underpin the workings of the instruments of observation on which they rely, are true, they can build on those earlier successes and move scientific inquiry forward.

How things look to an observer depends on his eyesight, on lighting conditions, distance, angle, etc., and on any instrument of observation he uses. But what judgment an observation triggers depends not only on how things look to him, but also on his background beliefs. Robert Root-Bernstein, who reports that about 10 percent of his students believe on Biblical grounds that men have fewer ribs than women, had them examine skeletons and X-rays for themselves; but even afterwards some still claimed that the male skeletons had one pair of ribs fewer, and he had to stand by them as they re-counted two or three times before they acknowledged that the number is the same.[7]

Rosalind Franklin's X-ray diffraction photographs of DNA look the same to an expert observer and to a layperson; but the expert notices what the layperson does not, and—or rather, "that is," for noticing *is* a matter of being moved by what you see to judge that the thing is thus and so—his judgment about what he sees is different. Similarly with Russell Hanson's point that, unlike an Eskimo baby, a scientist will see an X-ray machine *as* an X-ray machine, or see *that* it is an X-ray machine.[8] This is true enough; but "seeing as" and "seeing that" are locutions that refer, not simply to the perceptual event, but also, less overtly, to the judgment it triggers.

The more you know, the more you notice. In 1973, at the big accelerator at CERN, Gargamelle—a huge bubble-chamber containing twelve cubic meters of liquid propane, named for the mother of Gargantua—took many thousands of photographs. A scanner recognized one as strange and new; the important thing is a limp little curlicue, "the signature of an electron" as Gerald Holton puts it, an indication of a rare but revealing event. The scanner took the photograph to a research student, who carried it up to the deputy group leader, who showed it to the institute director, who acknowledged it as precisely what they had been hoping for: an indication of neutrino-electron scattering, confirmation of the theory of the unification of the electromagnetic and the weak forces, the "electroweak" force.[9]

Kuhn refers to an experiment in which subjects presented with anomalous playing cards such as a red six of spades first took what they saw for normal cards, then began to hedge ("it's the six of spades, but there's something wrong with it—the black has a red border"), and later expressed severe discomfort and confusion.[10] Perhaps forgetting that it is a presupposition of the experiment, itself grounded in observation, that the anomalous cards feature red spades, Kuhn is tempted to draw the radical conclusion that how things look is too theory-dependent for experiential evidence to serve as anchor for scientific claims. But this is quite unlike an experiment conducted under non-standard lighting conditions, or with the subjects wearing tinted contact lenses; and one possible interpretation is quite conservative: the subjects register the available visual input normally, but their judgment of what they see is skewed by their expectation that spades must be black. Another, however, is more radical: the visual input registered is distorted by the subjects' expectations, as if a mental template were somehow superimposed.[11]

In any case, failure to notice some aspect of what you see, mistaken judgment of what you see, and even the kind of misregistration of visual input which, on the more radical interpretation, explains the subjects' responses in the playing-card experiment, are all potentially curable by giving the observer more information. With a little time and patience, the experimenter in the playing-card experiment could show the subjects how they had been fooled, by correcting the false presupposition that these are regular playing cards; with a lot more time and patience, an X-ray crystallographer could show the layperson what he had failed to notice, by imparting the missing knowledge about what pattern is to be expected from a helical molecule. And, of course, that Eskimo baby might grow up to be a radiologist. To be sure, as Kuhn points out, a scientist "can have no recourse above and beyond what he sees with his eyes and instruments."[12] This doesn't mean, however, that scientists' misperceptions are incurable, but that it may be only with hindsight, after further scientific developments, that the cure becomes available.

In every perceptual event, as I said, there is something received, something resistant to one's will and independent of one's expectations. In every judgment, however, even in those judgments most immediately triggered by observation, there is something in some degree interpretive, and hence fallible. There is no sharply distinguishable category of statements the meaning of which is exhausted by experience. This isn't to say that observations are somehow propositional—they aren't; they are events. It is only to say that the meaning of any statement depends on the links between its words and others as well as on the links with the world learned directly by ostension. Take, for example, a scientist looking at a gauge, and judging that the needle on the dial points to 7. Not even such a judgment is entirely free of interpretation; it is triggered by the scientist's seeing the dial, but only against the background of his beliefs about meters, dials, and the conventions about arrows and numerals.

To deny that there is a sharply distinguishable category of interpretation-free observation statements is not to say that "it's a glass of water" or "it's gone green" or "lo, a raven" are really theoretical statements after all; it is only to insist that even sentences learned initially by ostension have meaning-conferring links with other sentences as well. Neither is it to deny that, at any time, some statements attribute properties which are in no way observable—"purely theoretical" statements, as we might say. Surely there are such statements; but the boundary of the "purely theoretical" constantly shifts with advances in instruments of observation.

What is observable to us depends in part on our human perceptual apparatus (dogs can hear sounds we cannot, bats can navigate by sonar, and so on). Does this mean that, though there is no sharply delimited class of observational *statements*, there is a sharply delimited class of observable *things* (including elephants, dahlias, and so forth, but excluding electrons, DNA, and so on)? No. What is observable depends not only on our perceptual capacities, but also on our ingenuity in devising instruments to extend and improve our powers of detection; the boundary of the observable, in other words, like the boundary of the "purely theoretical," constantly shifts with advances in instruments of observation.[13]

GENERALS AND EXPLANATION

What scientists observe is particular things and events. And quite often, too, they make claims about particulars: about the movements of this or that planet, the significance of this or that fossil, the train of events in the first few minutes after the Big Bang. Nevertheless, generalization is crucial to the sciences, for supportiveness of evidence is tied to explanatory integration, and explanation to kinds and laws. In fact, as I said earlier, the very possibility of scientific inquiry requires that there be, as Peirce would have put it, "real generals"; otherwise, though we could describe particular things and events, we could neither explain nor predict.

In their earliest articulation of the covering-law model of explanation, focusing on cases where what is to be explained is some particular fact or event, Carl Hempel and Paul Oppenheim required that the statement of fact to be explained be deducible from a statement of initial conditions plus a law statement.[14] Subsequent refinements extended the model to explanations of lower-level by higher-level laws, and to statistical explanations where the relation of explanans and explanandum is probabilification rather than deducibility. Though the covering-law model has been criticized both as too strong and as too weak,[15] by my lights it has two significant virtues: it recognizes the role of generals (I would add that kind predicates may serve as proxies for explicit statements of law); and it acknowledges that explanatoriness is not a merely pragmatic concept, but an epistemologically substantial one.

Hempel and Oppenheim conceived of explanations as arguments conveying rational credibility on their conclusions. Now, however, it is sometimes said that this idea has been "effectively discredited."[16] Focusing on the statistical version of the covering-law model, Wesley Salmon argued that explanatory power cannot be identified with evidential relevance, because a statistical explanation may not raise the prior probability of its explanandum. Suppose, for example, that we explain that the Geiger counter in the laboratory emitted a single click because it picked up the radioactive dots on the technician's watch-face; but that because of a recurrent malfunction there is a probability of 0.005 that in the course of one second the Geiger counter will pick up a spurious emission, while the probability that it would pick up an emission from the watch is much lower, say 0.001.[17] Taken in their own Bayesian terms, such examples may seem persuasive; but this only reveals the inadequacy of those terms. Our example shows that explanatory power cannot be identified with statistical relevance; but since statistical relevance cannot be identified with supportiveness, it doesn't follow that there is no such connection between supportiveness and explanation as I have maintained.

Of late it is sometimes said that explanation is just a pragmatic or a rhetorical notion. Bas Van Fraassen, an influential exponent of explanation-as-context-dependent, reminds us of Russell Hanson's argument that a physician might explain a death as caused by multiple hemorrhage, while a barrister might explain it as caused by negligence on the part of the driver, a civic planner by the presence of tall shrubbery at the turning, and so forth.[18] Again, Van Fraassen points out, we identify those cumbersome antlers, and not the other contributing factors, as the cause of the extinction of the Irish elk, because the antlers are most salient to us. True enough. But it is salience, not explanation, that is pragmatic: the physician, the attorney, the civic planner, etc. all focus on the particular links in the long causal chain that especially interest them; we focus on that part of the whole causal story about the extinction of the elk that especially interests us.[19]

Nancy Cartwright asks about the camellias she planted in composted manure, some of which flourished, but some of which died.[20] We can explain that some plants died because the soil was hot, while others flourished because the soil was rich, she continues, without knowing what the differentiating factors are, and without being able to specify any exceptionless covering law. But this is only a partial explanation, a kind of promissory note with an unspecified ceteris non paribus clause (perhaps very young, undeveloped plants are damaged by the heat, while older, sturdier plants are able to benefit from the richness). Cartwright's example, like Salmon's, suggests that there are advantages to thinking, as I have done, in terms of degrees of explanatoriness, of better and worse explanatory stories, rather than of explanation, simpliciter.

I chose the phrase "explanatory integration" in preference to the more familiar "explanatory coherence" because, since it requires experiential input, my account of evidential quality is not coherentist. Nevertheless, I can draw on Paul Thagard's work (which, since it gives a distinguished role to "data sentences," is not really purely coherentist anyway),[21] focusing in particular on one feature Thagard stresses as raising degree of explanatory coherence—breadth, consilience,[22] or, as I shall say, scope—and how it intersects with another, specificity. Both scope, breadth of explanatory power, the capacity to unify apparently diverse phenomena, and specificity, identification of specific causal mechanisms, improve explanatory integration. As the connection of explanation and generals reveals, specificity alone won't do. In the margins of his copy of Lamarck's *Zoological Philosophy*—"very poor and useless book"—Darwin objects, "not same theory to plants & animals";[23] his point being that Lamarck's account requires a different explanation of variation in plants, which have no habits to attune to changing environmental conditions, than in animals. But scope alone won't do either; hence Lavoisier's complaint about the phlogiston hypothesis, that it is so vague that "it adapts itself to any and every explanation you like."[24] (We shall see later that it is scope without specificity that gives explanations of the "God did it" type their spurious appeal.)[25]

By 1964, when the Surgeon General's Report asserted a causal link between smoking and lung cancer, 29 controlled studies around the world had established a strong correlation; but, though more than 200 compounds known to be carcinogens had been found in tobacco smoke, the causal mechanism was not known.[26] Women turned out to be more susceptible than men; but, again, it was not known why. Not long ago, however, the specific genetic damage caused by cigarettes was identified, and more recently the role of female hormones in speeding things up has been identified, too. Now there is a unifying explanation, combining specificity and scope in exactly the right way: an account of the damage caused by cigarette smoke, firmly embedded in our knowledge of the physiology of the cell and molecular biology, and tightly interwoven with what we know about heredity, evolution, etc. The present evidence is much more supportive of the claim than the earlier evidence was.

But what are kinds, and what does it mean to say that they are real? Ian Hacking suggests that we think of the real in terms of what we can interact with or manipulate;[27] Steven Weinberg writes that the laws of physics are real in just the way rocks are.[28] But I have borrowed Peirce's term, "generals," in part because its very unfamiliarity reminds us that kinds and laws are—well, *general*; they aren't ordinary physical particulars like rocks, and we can't manipulate them as we can fibers of DNA. I would say, as a rough first stab, that kinds are not simply properties or similarities, but more like congeries of properties held together by laws, i.e., clusters of properties co-occurring because they are lawfully connected; and that a kind is real just in case it is independent of how we believe it to be, i.e., the cluster of properties is lawfully connected independently of our classifications. This suggests why scientists are sometimes able successfully to predict that things of this or that kind will have such-and-such hitherto unsuspected properties;[29] and explains why medical scientists regard homeopathy as hopeless: the "disorders" it purports to cure are bizarre congeries of symptoms—as it were, a tingling in the left little toe accompanied by dreams of robbers—unconnected in any way comprehensible to medicine; i.e., not real generals.

It also suggests why *what* generals are real isn't something we can just read off our language, but something that takes work to find out. A major step in understanding Alzheimer's disease, for example, came when scientists identified a hitherto-unknown kind of molecule, amyloid precursor protein ("APP"), which traverses the membranes of brain cells; and then three kinds of enzyme—alpha, beta, and gamma secretase—which cut APP into shorter forms. Together, the beta and gamma enzymes produce a shorter, stickier protein, beta amyloid, which builds up in the fluid surrounding neurons to form the plaques characteristic of Alzheimer's.[30]

But my first rough stab will certainly need refinement, because the different kinds of kind include not only the physical and the chemical, where it is the least unsatisfactory, but also the biological and the social, where it seems too narrow, too restrictive. I don't mean to return to an older kind of biological essentialism, nor to deny that, since Darwin, we have come to see biological kinds as organized, not simply around clusters of properties, but in part by common descent.[31] As the example of cosmologists' investigations of the first minutes of our universe was chosen in part to illustrate, there are elements of historical contingency even in physics; but this element is much larger in biology. And the larger this element, the more loosely properties are clustered—into meshes of family resemblances rather than tight knots; which is why biological classifications involve a larger element of decision. (Things are even more complicated with the social sciences; but I shall postpone those complications.)

Thus far, I have deliberately steered clear of that old chestnut in philosophy of science: Do kind-expressions have descriptive meaning, or are they purely referential expressions having as their extension all things of the same kind as some unproblematic exemplar: this lemon, this tiger, this piece of gold? Even now, rather than tackling this question head-on, I want to start by stressing that the language of science, though more regimented and better defined than many other stretches of the natural languages of which it is part, is nevertheless quite unlike the artificially precise and abstract languages of formal logic. There are, besides, many words the sciences have borrowed from the surrounding language, and not a few that the surrounding language has borrowed from the sciences. Categories and distinctions designed for more regimented logical languages may not work so well here.

Looking at an ordinary dictionary, you get some sense both of the complexities of scientific language, and of its intricately untidy relations with the larger language in which it is embedded. *Webster's* defines "gold" as "a malleable ductile yellow metallic element that occurs chiefly free or in a few minerals and is used esp. in coins, jewelry, and dentures," and gives as a secondary sense, "something valued as the finest of its kind," as in "heart of gold"; it also refers the reader to the ELEMENT table, which gives the symbol (Au), the atomic number (79), and the atomic weight (196.9665). "Electron," however, has a purely technical definition: "an elementary particle consisting of a charge of negative electricity equal to about 1.602 times 10^{-19} coulomb and having a mass when at rest of about 9.109534 times 10^{-28} gram or about $1/1836$ that of a proton." Then there's the entry for "quark," which goes [with my italics]: "a *hypothetical* particle that carries a fractional electric charge, *is thought to* come in several types (as up, down, strange, charmed, and bottom), and *is held to be* a constituent of hadrons."[32] And here's "phlogiston": "the hypothetical principle of fire *regarded formerly as* a material substance."

Scientific textbooks, also, often include glossaries in which the authors try to encapsulate essential information about the things or stuff discussed. Kind-expressions have descriptive meanings—descriptive meanings into which presumed scientific knowledge is built, whether confidently (as with "electron") or tentatively (as with "quark"), and which shift and change as our knowledge grows; but there is no clear-cut distinction between what is built into the meaning of "X," and what is only presumed knowledge about X's. What used to be thought of as knowledge about X's can gradually become part of the meaning of "X," if it gets firmly entrenched, or may quietly drop out of the meaning altogether, if it turns out to have been wrong. The word "protein" derives from the Greek *protos*, meaning "first, primary"; but scientists (and the rest of us) continue to use the word, with no sense of paradox, even though we now know it is not protein but DNA that is of primary biological importance. Similarly, scientists continue to use "RNA"—ribo*nucleic* acid—though they now know that the stuff is found mostly not in the nucleus, but in the cytoplasm.

But doesn't the view that kind-expressions are purely referential have the advantage of keeping reference stable through changes in our knowledge of the things or stuff in question? It stabilizes reference, all right; in fact, you might say it freezes reference. But it is not clear that this is altogether desirable. For sometimes, when theoretical developments lead to distinctions among kinds previously run together—several kinds of RNA, for example—we want to say, in hindsight, that scientists had formerly referred to RNA indifferently, not realizing the difference between transfer RNA, messenger RNA, and ribosomal RNA. Sometimes there *is no* simple continuity of reference.

Nevertheless, there is an important element of truth in the purely referential view: reference is crucial, for scientists engage in linguistic innovations and shifts of meaning because they seek categories that match real kinds. But we can acknowledge that reference is crucial without denying that kind-expressions have descriptive meaning. The point is reminiscent of Locke, and implicit in Peirce; and it resonates with a nice phrase from an early time-slice of Hilary Putnam:[33] in their constant shifting and readjustments of terminology, the sciences aspire to kind-expressions representing "law-cluster concepts," bunches of properties the co-occurrence of which is mandated by laws of nature.

What about the phenomenon Putnam calls "division of linguistic labor"? A conception of kind-expressions as descriptive as well as referential can comfortably accommodate this phenomenon, though not in quite the way the purely referential view proposes. As everyone realizes where their second language is concerned, but tends to forget when focused on their first, linguistic mastery comes in degrees; and not only in size of vocabulary, mastery of syntax, etc., but, most to the present purpose, in grasp of meanings. Laypersons often have at best a partial grasp of the meaning of scientific terms, while experts have a fuller grasp. Relaxing the connection of meaning and reference, we are willing to describe a seventeenth-century scientist, or a modern layman, as talking about gold, even though he is mistaken about or ignorant of much of what scientists now know to be its characteristics, provided he understands the term well enough to identify gold as, say, the yellow metal coins and jewelry are made of. It is in part because laypersons' mastery of scientific language is usually quite imperfect that scientific evidence is often opaque or downright incomprehensible to those outside the relevant scientific sub-specialty.

Just about every field of human endeavor has its specialized jargon. I think, in this context, of the colorful special vocabulary from the days of sailing ships, many traces of which—"taken aback," "not enough room to swing a cat," "scuttlebutt," "three sheets to the wind," "pipe down"—now enrich ordinary English. There's absolutely nothing wrong with that. There is something wrong, however, with jargon the main purpose of which is simply to impress others with your supposedly arcane knowledge. William Gilbert made the essential distinction long ago, when he announced that he would "sometimes employ words new and unheard of, not (as alchemists are wont to do) in order to veil things with a pedantic terminology and make them dark and obscure, but in order that hidden things with no name . . . may be plainly and fully published."[34] And John Locke had a fine phrase—"affected obscurity"[35]—for that "pedantic terminology to make things dark." Locke thought philosophy especially susceptible to this fault; but by now it seems to afflict the social sciences at least equally, and of late it has made impressive inroads into literary theory too. Like the honorific use of "scientific," this is a kind of backhanded compliment to the natural sciences, where technical terminology and conceptual innovation are essential tools—essential tools, that is, to craft a language fit for their explanatory purposes.

TRUTH AND PROGRESS

The word "science" refers to a loose federation of kinds of empirical inquiry; the goal of science is substantial, significant, explanatory truth. Obvious as this seems to me and, I hope, to you, it is neither so simple nor so uncontroversial as you might think.

Here is Bridgman again: "the objectives of all scientists have this in common—that they are all trying to get the correct answer to the particular problem in hand. This may be expressed in more pretentious language as the pursuit of truth."[36] I agree. If you are trying to find out the answer to some question, your goal is to arrive at a true answer: e.g., that p, that not-p, or perhaps that it's a lot more complicated than that. And scientists, I take it, are trying to discover true answers to the questions they investigate. They sometimes tell us as much: William Harvey, avowing himself "a partisan of truth alone," writes that his investigation of the functions of the heart and the arteries was undertaken "in order that what is true may be confirmed, and what is false set right by dissection, multiplied experience, and accurate observation";[37] Charles Darwin, speculating about the sources of the mammalian type of organization, noticing that "some birds may approach animals, and some of the vertebrata invertebrates," breaks off to warn himself—"Heaven knows whether this agrees with Nature: *Cuidado*";[38] James Watson dreamt of "solving the structure of DNA."

This is not to deny that a scientist may also have other objectives (e.g., to get a grant to do the work, to win the Nobel Prize); it is only to affirm that his proximal goal, insofar as he is genuinely inquiring, is to find true answers to the questions that concern him. Nor is it to deny that, though many scientists think of themselves as engaged in the pursuit of truth, others shy away from the idea, affirming that they seek only to arrive at descriptions consistent with appearances; Van Fraassen reminds us of "the phenomenalism of Ernst Mach, the conventionalism of Henri Poincaré, and the fictionalism of Pierre Duhem."[39] It isn't even to say that scientists who subscribe to instrumentalism or constructive empiricism will necessarily go about their work any differently, or any less well, than scientists of a more realist persuasion. It is only to say that their conception of what they are doing is too modest.

The goal of science can't be to discover truths, Hempel argues in his later, quasi-Kuhnian mood, because scientific claims are rarely decisively verified or falsified, and in those cases where conclusive evidence may be available—existential claims verifiable by positive instances, or generalizations falsifiable by negative instances—this supposedly "conclusive" evidence constitutes proof only on condition that the observation statements concerned are true.[40] But the first argument is a simple non sequitur; and the second relies, additionally, on a confusion of supportiveness with warrant, conclusiveness with decisiveness.

My thesis isn't that scientists seek THE TRUTH, if this means any more than that they seek true answers to their questions; nor that, when they propose answers to their questions, they always or necessarily take those answers to be *proven* true. Very often, a scientist will claim no more for a conjecture than that it is probably, or possibly, or approximately true. Sometimes scientists claim only to offer an idealization which, though true of no actual objects, represents how things would be in certain unrealizable but idealized circumstances. And doubtless scientists are frequently preoccupied with such subordinate goals as getting recalcitrant equipment to work, thinking of a good way to classify data, or getting the wherewithal to take the next investigative step.

The thesis that the goal of science is substantial, significant, explanatory truth has two elements: (1) scientists seek true answers to the questions that concern them; (2) they concern themselves with substantial, significant questions. By now it should be apparent that the second component, like the first, is covertly normative: a lazy scientist anxious to produce something publishable who restricts himself to easily answered but insubstantial questions, though he isn't quite like a dishonest scientist who isn't really trying to find the true answer but only something good enough to get away with, *is* falling down on the job.

I say "substantial, significant, explanatory truth" because the object of the enterprise is to understand things and events in the world, not simply to pile up truths. I don't intend to rule out the possibility that, for example, there might be ultimate laws not themselves susceptible of further explanation; but I shall leave questions about the limits of science for later.[41] I do intend the inclusion of "explanatory" to indicate that "substantial, significant," should be understood in an epistemological sense, for the concern here is with the goal of science *qua* form of inquiry. Whether science ought also to serve other interests—for example, whether it should give priority to questions answers to which are politically desirable, or should shelve questions the answers to which might be politically unpalatable—is an important but separate issue, which, again, I must postpone.[42]

Naturally, I take scientific theories to be, as they appear to be, complex congeries of statements or propositions, genuine candidates for truth or falsity. To be sure, not *every* scientific proposition is either true or else false. In scientific inquiry, as in inquiry of every kind, what we take to be legitimate questions sometimes turn out to be flawed; questions about the properties of phlogiston, for example, turned out to rest on a false presupposition—there is no such stuff. According to old-style instrumentalism, however, theoretical "statements" in science are *never* true or false. In fact, they aren't really statements at all, but only instruments or tools or inference-tickets for making observational predictions. But this position cannot even be stated without a clean distinction of observational versus theoretical statements (or rather, observational statements versus theoretical "statements"); and it seems unmotivated—except, that is, for those wedded to an old-style verificationist theory of meaning.

The main themes of Van Fraassen's "constructive empiricism" are that the goal of science is not to arrive at explanatory theories, not even to arrive at true theories, but only to arrive at empirically (i.e., observationally) adequate theories; and that to accept a theory is not to believe it to be true, but only to believe it to be empirically adequate.[43] Empirical adequacy, Van Fraassen grants, "goes far beyond what we can know at any given time"; nevertheless, the claim that scientists only accept and do not believe their theories "delivers us from metaphysics."[44] This pejorative use of "metaphysics," of course, echoes the older logical empiricists.

Van Fraassen acknowledges that the distinction of observational versus theoretical statements isn't robust, so that we can't isolate the empirical content of a theory by dividing its language into two parts.[45] He proposes to rely instead on the distinction between observable and unobservable things and events, identifying the empirical content of a theory by reference to the empirical substructures of its models. When he goes on to specify the empirical substructures of a theory as those which can be "described in experimental and measurement reports,"[46] it sounds as if he may be covertly relying on the linguistic distinction he overtly rejects; but perhaps this is just a slip. Still, I'm not convinced that the distinction between observable and unobservable things and events is robust enough to bear the weight that constructive empiricism puts on it.[47]

A constructive empiricist who construed "observable" in the narrowest way, as applying only to things detectable by the unaided human senses, would be unable to explain scientists' motivation for devising instruments of observation. A constructive empiricist who (like myself) construed "observable" more broadly, as also applying to things detectable by any available instruments, while avoiding that problem, would still be unable to explain the motivation for devising *new* instruments. A constructive empiricist who construed "observable" more broadly yet, as also applying to things detectable by not-yet-invented but physically possible instruments, would avoid these problems; but would leave realists like myself wondering why scientists should not inquire into why no physically possible instrument could make these or those phenomena detectable by creatures with senses like ours.

Ronald Giere proposes "science without laws" and "realism without truth," urging us to think of scientists not as trying to discover laws—an idea he describes as an artifact of the melange of theology and mathematics peculiar to seventeenth-century science—but as making models; and so as concerned with the fit of models rather than the truth of propositions. Giere seems to use "law" ambiguously: sometimes to refer (as in my usage) to something about the world, but often to refer instead to statements purportedly representing laws in this worldly sense. This ambiguity makes it easier than it ought to be to suggest that, if the law-statements scientists come up with are not exactly or not unconditionally true, there are no laws in the other—the worldly—sense. Giere also uses "model" to include everything from physical models like Watson and Crick's tinker-toy representations of the DNA molecule, through maps, diagrams, etc., to models in an abstract logical sense. Even more confusing, though he is fond of italicizing his slogans, *science without laws*, *realism without truth* (and *naturalism without rationality*, but I'll leave that for later),[48] he also tells us that he doesn't, strictly speaking, mean that science could really do without laws, or without truth. I agree with the part where he takes it back.

Along with metaphors and analogies, models are important helps to the imagination; sometimes, as with Watson and Crick's scale-models of DNA, they also serve as evidential checks; and sometimes, as with computer models of complex processes, they serve as tools for prediction. (In this context I think of those television meteorologists apologetically explaining in the aftermath of Hurricane Irene that all their models predicted that the storm would bypass Miami; only, unfortunately, all the models were wrong.) When scientists construct models of what would happen were such-and-such to occur, they must sometimes go to considerable trouble artificially to set up that hypothetical such-and-such in the laboratory or the particle accelerator. But even when the conditions specified in the model are not actualized until the scientist's activity actualizes them, the goal is truly to characterize a would-be which, if the model is correct, was real all along.

Giere suggests that his model-oriented approach is hospitable to a desirable kind of pluralism; for there may be different useful scientific models of the same phenomenon, as there may be different useful maps of the same territory.[49] So there may; as there may be different useful metaphors or analogies for the same phenomenon. This, however, obliges us to ask when models are different but compatible, and when incompatible. Though it sounds a little odd to describe models as true or false, we can and do describe them as compatible or incompatible with other models, or with the evidence; and as right or wrong. Watson and Crick's three-chain model was wrong, incompatible with Franklin's evidence of the water-content of DNA; their backbone-in models were incompatible with their backbone-out models; and their final, two-chain, backbone-out, like-with-unlike-base-pair model, which Watson described as "too pretty not to be true,"[50] was the *right* model: it represented DNA as a double-helical, backbone-out, etc., molecule, and DNA *is* a double-helical, backbone-out, etc., molecule. We can make sense of the idea of the "fit" of a model, or of its being "right" or "wrong," only in terms of kinds, laws—and truth.

When Arthur Fine says that "the natural ontological attitude"[51] for anyone engaged in any kind of inquiry is to presuppose the reality of the objects of his inquiry, an Innocent Realist like myself will concur—unless, she adds, we are talking about inquiry into whether there really is a Loch Ness monster, an Abominable Snowman, phlogiston, or whatever. However, as the "unless" clause makes clear, this neither answers such further questions as, "yes, but is there *really* a God/phlogiston/a world independent of how we believe it to be?" nor shows them to be illegitimate. The fact that the "natural ontological attitude" is no less natural for the theologian than for the historian or the scientist reveals that there are legitimate questions Fine's view leaves unanswered.

It is pointless to investigate a question if you believe it has false presuppositions. This holds both in the small—it is incomprehensible why anyone would investigate the properties of phlogiston if they didn't think there was any such stuff; and in the large—it is incomprehensible why anyone would try to find out how things are in the world if they didn't think there is a real world which is a certain way and not other ways (which is not to deny that when you aren't sure whether the presuppositions of a question are satisfied, it makes sense to look into them to determine whether the question is legitimate or not).

It is also incomprehensible why anyone would seriously engage in scientific inquiry, or in everyday empirical inquiry, if he didn't think the world is knowable to some extent by creatures with powers such as ours; as it is incomprehensible why anyone would seriously engage in theological inquiry if he didn't think the deity is knowable to some extent by creatures with powers such as ours. ("Seriously," because someone *might* undertake theological inquiry in something like the spirit in which he could, say, set out to map all the geographical details of his 54-volume set of Tolkien novels.) In either case, should the presupposition be false, serious inquiry of that kind—scientific or theological—would be an enterprise as hopeless as searching for a non-existent or unattainable Holy Grail.

Some radical philosophers, sociologists, and rhetoricians of science profess reservations about the legitimacy of the concept of truth. But it is hard to take such professions altogether seriously; for, as Peirce observes, anyone who believes anything, or who asks any question, thereby implicitly acknowledges, even if he explicitly denies, that there is such a thing as truth.[52] Here, for example, is Michel Foucault, grasping for the concept of truth even as he tries to jettison it: "Each society has its regime of truth . . . that is, the types of discourse which it accepts . . . as true; the mechanisms . . . which enable one to distinguish true and false statements . . . the techniques and procedures accorded value in the acquisition of truth."[53]

Whether a (synthetic) description is true or is false depends in part on what it says, which is a matter of human linguistic convention; but, given what it says, whether it is true or it is false depends on whether the things it describes are as it describes them. To say that a statement is true is to say that things are as it represents them to be. It is true that there is a cardinal bird in the feeder outside my study window as I write this just in case there is a cardinal bird in the feeder outside my study window as I write this; it is true that DNA is a double-helical, backbone-out macromolecule with like-with-unlike base pairs just in case DNA is a double-helical, backbone-out macromolecule with like-with-unlike base pairs; it is true that $7 + 5 = 13$ just in case $7 + 5 = 13$; and so on.

To get from here to a fully general account of truth—to generalize Aristotle's "to say of what is, that it is, or of what is not, that it is not, is true; to say of what is that it is not, or of what is not that it is, is false," so that it applies to all statements of whatever form—is a hard theoretical task. Probably the simplest and most direct approach is due to Frank Ramsey: to say that a proposition is true is to say that it is the proposition that p, and p;[54] but this leaves us still in need of an explanation of the sentential quantifiers on which it relies, and so is not quite so simple or direct as it at first appears. But it would be foolish to eschew the concept of truth on the grounds that this work has not yet been completed.

Some statements describe us, and some describe things in the world that depend on us. Whether a statement describing us is true or is false depends on how we are; and whether a statement describing things that depend on us is true or is false depends on how those things are—for such descriptions, those are the relevant aspects of "how the world is." But whether even such descriptions are true or are false does not depend on how you or I or anybody *thinks* the world is. Sometimes, speaking carelessly, we say that something is true for you, but not for me. But what we mean is only that the something—liking Wagner, say, or being over six feet tall—is true of you, but not of me; or else that you believe whatever-it-is, but I don't.

We can describe how the world would be, or would have been, if there were, or had been, no human beings. Before there were human beings and human languages, the English sentence "There are rocks" did not exist; so if sentences are bearers of truth and falsity, it is not the case that "There are rocks" was true before there were people, or that "There are rocks" would have been true even if there had never been people. Nevertheless, there were rocks before there were people, and there would have been rocks even if there had never been people. That is a partial description of how the world would be, or would have been, if there were, or had been, no people.

We learn language by learning to assert or to assent to sentences and sentence fragments in the circumstances in which they are assertible; but human languages are productive, permitting the production of linguistically meaningful sentences of which we may be unable, and perhaps would be unable however long inquiry continued, to determine the truth-value. Some questions about the distant past, for example (which perhaps could have been settled had inquiry been undertaken then) may now be forever impossible for us to answer. Human beings are susceptible not only to error, but to ignorance; and not only to temporary ignorance, rectifiable by another week or year or century or millennium of inquiry, but perhaps, in some instances, to invincible ignorance. But we are also capable of discovering something of how the world is.

When I write of "the world," I mean the one real world—a world at once heterogeneous and integrated, in somewhat the sense of William James's marvelously Janus-faced title, *A Pluralistic Universe*;[55] and I signal my intention to rule out both irrealism, the thesis that there is no real world, and pluralism, the thesis that there is more than one. But much turns on exactly what is meant by "world" or "worlds," so disambiguation is needed.

Appropriately enough perhaps, "pluralism" has several interpretations, in some of which, despite appearances, it is quite compatible with the "one real world" of Innocent Realism. Sometimes "world" is used more or less equivalently to "aspect of the world," and "there are many worlds" means only "the world has many aspects." Of course the world has many aspects; but the pluralist way of putting it may distract attention from the fact that it is precisely because scientists, historians, sociologists, detectives, etc., all investigate various parts or aspects of the same world that their investigations sometimes overlap, and they can borrow from each other.

What about Popper's thesis that there are three worlds—world 1, the world of physical things; world 2, the world of mental states; and world 3, the world of abstract entities, numbers, problems, propositions, and theories?[56] If this were only an awkward expression of the idea that *the* world has three aspects, his might be construed as another of those true-but-misleadingly-expressed quasi-pluralisms; but if it is more than a *façon de parler*, an expression of Popper's commitment to a Cartesian dualism of mind and body,[57] it cannot be so accommodated.

Sometimes, again, "world" means "possible world." I grant that there are real but unactualized possibilities—to acknowledge real generals, after all, is to acknowledge that, as Peirce puts it, there are can-be's and would-be's as well as actually-is's and will-be's that are real;[58] and, of course, I grant that the world might have been different. Some modal realists, however, move from this innocent premise to the far-from-innocent conclusion that the actual world is just one possible world among many. This I can't accept; though possible worlds may be construed as ways the actual world might have been, the actual world cannot.

So I will allow myself an appreciative chuckle when Nelson Goodman distances himself from certain contemporaries of his, "especially those near Disneyland," who have been "busy making and manipulating" possible worlds[59]—and when he objects to the "monopolistic materialist" who insists that everything not false or meaningless must be reducible to physics.[60] But when Goodman claims that there is no one real world, only the many "versions" made by artists, writers, scientists, etc., I am obliged to point out that "version" is a relative term; it requires an "of . . . ." Goodman sometimes writes of "world-versions," or of descriptions under different frames of reference, as being "both true of the same world"; but he also, sometimes, hedges his talk of "the world" with scare quotes. This only disguises the difficulty: the question, what the versions made by artists, writers, scientists, etc., are versions of, is unavoidable; but the right answer—the world, i.e., *the* world—is officially closed to him.

Given Goodman's thesis that scientists, novelists, artists, etc., *make* worlds, his ideas have proven attractive to some, including those inclined to blur the differences between science and literature, who are tempted by a kind of linguistic idealism. We call this collection of stars "The Big Dipper," this body of water "The Pacific Ocean." There is an arbitrariness in the name (there could be a culture in which what we call "The Big Dipper" is called "The Little 'Roo-tail," and what we call "The Pacific Ocean" is called "The Great Grey Water"); and in what we count as the one particular (there could be a culture in which one more star, or one fewer, is collected with those, or in which the part of that body of water in which the tribe fishes has one name, and the part outside its territorial waters has another).

Some are tempted to conclude that we brought the constellation, or the ocean, into existence by naming it. We didn't. Before anyone called the Big Dipper "The Big Dipper," there was nothing called "The Big Dipper"; and before anyone called the Pacific Ocean "The Pacific Ocean," there was nothing called "The Pacific Ocean." But those stars, and that body of water, were there anyway, and would have been there even if there had never been human beings or human languages. Perhaps, looking at the night sky, you are awed by the thought that the Big Dipper existed long before there were people, and will continue to exist, probably, long after human beings are extinct; your thought is entirely coherent.

In any case, it is a mistake—a characteristically nominalist mistake—to concentrate on expressions referring to particulars, such as "The Big Dipper," or "The Pacific Ocean," rather than on predicate-expressions, common nouns, and adjectives representing kinds of thing or stuff, such as "elephant," or "platinum." If we focus too much on singular terms and too little on general predicates, we are liable to exaggerate the arbitrariness of predicates and to forget that one task of science is to discover categories and classifications matching law-clusters in the world.

Like Fine, I resist the "progressivism" built into some recent styles of scientific realism—up to a point. That, or how, science progresses shouldn't be built into our metaphysics, but worked out from our epistemology; it depends on facts about us as well as on facts about the-world-minus-us. Scientific progress has been hard-won; it is far from steady or uniform, and far from automatic. Scientific language is constantly refined, adapted, modified, reworked; what evidence is seen as relevant to this or that question, what kind of account is seen as potentially explaining this or that phenomenon, is constantly readjusted; and there is no guarantee of an inevitable accumulation of truths, or an inevitable replacement of false theories by more nearly true ones, etc. Nevertheless, science progresses.

Like Peirce, you might ask how, when there are "trillions and trillions" of hypotheses that would potentially explain a problematic phenomenon, scientists ever manage to come up with the one that happens to be true. Like Peirce, I would reply that evolution has given human beings an instinct for guessing which, "though it goes wrong far oftener than right, yet the relative frequency with which it is right is . . . the most wonderful thing in our constitution";[61] an instinct, I might add, continuous with other animals' unself-conscious ability to recognize predators, or food.

Like Duhem, you might ask how, given that "logic does not determine with strict precision the time when an inadequate hypothesis should give way," or as I should prefer to put it, given that scientific evidence is never absolutely decisive, scientists ever manage to settle on the right hypothesis. Duhem appeals to the operation of *bon sens*; Polanyi writes of skill, connoisseurship, a scientist's personal judgment.[62] Such answers aren't wrong; but, like talk of a scientist's having a good "nose" for a plausible theory, or of his recognizing an idea as "too pretty not to be true,"[63] they are more allusively metaphorical than one would ideally like.

Ramon y Cajal writes with distaste of the "sybarite researcher" who loves the study of nature primarily for its aesthetic qualities.[64] But Steven Weinberg throws more light on the matter when he suggests that the beauty properly valued in a scientific theory is of a quite particular kind, the sparse simplicity or logical rigidity which, in the most fundamental realms of physics (though not everywhere in science), is a potential truth-indicator; and when he likens a scientist's ability to appreciate such beauty to a trainer's ability to recognize a fine horse. A good nose for a scientific theory, like a good eye for a race-horse, involves an ability to call unreflectingly on tacit knowledge—knowledge which one might, but might not, be able to articulate if called upon to do so.

Traditionally, philosophers of science have asked: is scientific progress gradual or revolutionary? Is it a matter of belief revision or of conceptual change? Is it a matter of accumulating truths, or of repudiating falsehoods and getting nearer the truth? My crossword analogy, like Popper's cathedral analogy, suggests that the least misleading simple answer is: "all of the above." Sometimes there is accumulation of truths (filling in new entries); sometimes there is repudiation of falsehoods (rubbing out old entries); sometimes there is closer approximation to the truth (replacing an entry wrong in more letters by one wrong in fewer). Some steps are larger (replacing a long entry and many intersecting ones) and some smaller (replacing a short entry and a couple of intersecting ones). There are revisions of the truth-values assigned to propositions in a familiar vocabulary, and revisions that involve change in conceptual structures as well as in beliefs (reworking the entries in this part of the crossword in American English instead of British, in Gallego instead of Castellan, or in Malay instead of English).

It is when shifts are large, and especially when they involve major changes in conceptual structures as well as in belief, that we are inclined to speak of a scientific revolution. Since shifts both in belief and in conceptual apparatus may be smaller or larger, the difference between revolutionary changes and others is a matter of degree.[65] This consequence would be disturbing if "revolutionary" implied incommensurability; but since it doesn't, it isn't. There is no guarantee that conceptual revision improves the language of science; but at least sometimes, as scientists introduce new terminology or shift the meaning of old vocabulary, on a small scale or a large, they refine their categories and concepts so as better to match the world. And when they do, they often find ways to build on these successes too; e.g., to move beyond that undifferentiated "nuclein" to DNA, then to A-DNA, B-DNA, and Z-DNA; to RNA, then to messenger RNA, transfer RNA, and ribosomal RNA.

There is no guarantee that at each step new truths are added, or the truth of some question more nearly approximated. Nevertheless, as scientific inquiry proceeds, it often succeeds in adding new truths; and when it does, it generally finds ways to build on those new truths—to find further truths, to devise new instruments or techniques, to revise old categories and classifications, and so forth. Sometimes it finds that previously accepted ideas and theories need to be revised, or have to be abandoned, in light of those new truths. And often it succeeds in arriving at more accurate numbers, in refining and disambiguating its vocabulary, in replacing an approximately true theory by a true one or a closer approximation, or in modifying a theory so that more of it is true; and when it does, it often finds ways to build on these more refined results. Construed in the robustly fallibilist way this suggests, "cumulative," and even "self-correcting," aren't far off the mark.

There is no guarantee that currently accepted theories, even in "mature" sciences, are true. Nevertheless, predictive and technological successes constitute evidence for the theories on which the predictions or the technologies depend. That aircraft seldom burst at the seams is a reason, albeit not an absolutely decisive reason, to think that the theories about metal fatigue on which airplane designers rely are at least approximately true; and the more such successes, and the more interconnected the bits of theory involved, the better reason to think so. Putnam argues that, unless theories in mature science were at least approximately true, their predictive successes would be miraculous.[66] I argue only, more modestly, that true consequences of a claim or theory constitute evidence that is to some degree supportive, and hence to some degree warrants it.

The success of the natural sciences in devising more and more efficient technical helps has been truly extraordinary. Think of the advertisements in scientific magazines for ever more sophisticated apparatus and scientific services (in the back pages of a recent issue of *Science* are ads for custom DNA and RNA labeling, protein, DNA, and RNA sequencing, gene cloning, plasmid purification, "highly purified salt-free oligonucleotides," etc., etc.); or of the ever faster and more efficient ways to share information (the Physics Department library at

Harvard, Holton reports, receives over 5,000 preprints a year, now made more accessible by a custom-designed electronic filing service);[67] or of the ever more sophisticated techniques of computer modelling. But, for all that, there are no grounds for complacency. As science gets more expensive, so that only governments and large industrial concerns can afford to support it, as career pressures mount, as opportunities multiply for a scientist to get rich from his discoveries, as the expert-witness business booms, there is no guarantee that the mechanisms which have thus far nourished scientific imagination and sustained respect for evidence will continue to do so.

AND, IN CONCLUSION

Richard Burian explains in just two words what enables the sciences to filter out the good stuff from a vast effusion of ideas, good, bad, and in between, sophisticated, naive, and half-baked: "reality therapy."[68] That's right—if all the complexities I have spent the last three chapters exploring are understood. Scientific helps to inquiry, fallible and imperfect as they are, enable the imagination, extend evidential reach, and—up to a point—stiffen respect for evidence. And this keeps the sciences—up to a point—in touch with the world.

The natural sciences, that is; but aren't the social sciences different? If the social sciences seek explanations at all, mustn't they be very different from natural-scientific explanations? If there are social kinds and laws, mustn't they be very different from natural kinds and laws? And how could it be maintained that, thus far at least, the social sciences have achieved anything like the impressive successes of the natural sciences? Indeed, a man wouldn't have to be "downright crazy," but at worst a little jaded, to deny that the social sciences have made many true discoveries. It is time to explore the differences—and the similarities.

NOTES

1. Peirce, *Collected Papers*, 5:265.

2. For articulation of my "Innocent Realism," the reader is referred to Haack, "Reflections on Relativism" and "Realisms and Their Rivals: Recovering Our Innocence." Since writing these papers, I discovered that Richard Boyd had already used the word "innocence," in something like the same way, in "Constructivism, Realism, and Philosophical Method."

3. Peirce, *Collected Papers*, 5:172.

4. Kuhn, *The Structure of Scientific Revolutions*, p. 126.

5. Cunningham, "This Story Has Legs."

6. Eddington, *New Pathways in Science*, p. 211.

7. Root-Bernstein, "Darwin's Rib"; my source is Pennock, *Tower of Babel*, pp. 372, 408.

8. Hanson, *Patterns of Discovery*, p. 15. (Why an Eskimo baby, I wonder?—It's not as if an Italian baby, or an Australian, would see the X-ray machine as an X-ray machine!)

9. Holton, *Einstein, History and Other Passions*, pp. 72–73 (citing Galison, *How Experiments End*, chapter 4), and pp. 82–83 (the photograph in question is reproduced on p. 83).

10. Kuhn, *The Structure of Scientific Revolutions*, pp. 63 ff.; Bruner and Postman, "On the Perception of Incongruity: A Paradigm" (the conjecture seems plausible that Kuhn may have picked up the term "paradigm" from this article).

11. In an experiment I saw long ago on television, subjects given orange juice dyed dark purple reported what they were drinking tasted like black currant; was it just their judgment that was skewed, or did the stuff actually taste different? Either seems possible.

12. Kuhn, *The Structure of Scientific Revolutions*, p. 114.

13. On these matters, the locus classicus is of course Grover Maxwell's, "The Ontological Status of Theoretical Entities."

14. Hempel and Oppenheim, "Studies in the Logic of Explanation"; for a summary account see Hempel, *Philosophy of Natural Science*, pp. 48 ff.

15. See, e.g., Kyburg, "Reply" (to Salmon); Salmon, "Statistical Explanation," *Scientific Explanation and the Causal Structure of the World*, and "Four Decades of Scientific Explanation"; Scriven, "Definitions, Explanations, and Theories," and "Explanations, Predictions and Laws." The criticisms are summarized by Kitcher in "Explanatory Unification and the Causal Structure of the World."

16. Collin, *Theory and Understanding*, p. 87, referring to Salmon, "Statistical Explanation"; see also Grünbaum, "A New Critique of Theological Interpretations of Physical Cosmology," p. 35.

17. The example, which combines elements of several of Salmon's own, is taken from Collin, *Theory and Understanding*, p. 87.

18. Van Fraassen, *The Scientific Image*, p. 125, citing Hanson, *Patterns of Discovery*, p. 54.

19. See also Friedman, "Explanation and Scientific Understanding"; Greeno, "Explanation and Information."

20. Cartwright, *How the Laws of Physics Lie*, pp. 51–53.

21. Thagard, "The Best Explanation"; "Explanatory Coherence"; "The Dinosaur Debate: Explanatory Coherence and the Problem of Competing Hypotheses"; *Conceptual Revolutions*, chapter 4; and *Coherence in Thought and Action* (see especially pp. 42 ff., where Thagard's discussion is put in the context of the foundherentist epistemology of my *Evidence and Inquiry*).

22. Whewell's word, from *The Philosophy of the Inductive Sciences*, aphorism 14, *Selected Writings*, p. 257.

23. Darwin, *On Evolution*, pp. 82–86.

24. Lavoisier, *Oeuvres*, p. 640: the phlogiston hypothesis "*s'adapte à toutes les explications dans lesquelles on veut le faire entrer*" (adapts itself to any and every explanation into which one introduces it).

25. A point elaborated in chapter 10 below.

26. See Thagard, *How Scientists Explain Disease*, pp. 32–33, 102.

27. Hacking, *Representing and Intervening*.

28. Weinberg, "Sokal's Hoax," p. 155.

29. A point I owe to Thomas Baldwin.

30. Cowley, "Alzheimer's: Unlocking the Mystery," p. 49. Shortly afterwards, it began to be conjectured that in Alzheimer's patients beta-amyloid traps abnormal amounts of copper; see Hensley, "Alzheimer's Cause May Be Metals Buildup."

31. See Thagard, *Conceptual Revolutions*, chapter 6.

32. The edition of *Webster's* I am using (the ninth) is dated 1991, which may explain its caution about quarks!

33. Putnam, "Is Logic Empirical?"

34. Gilbert, *On the Loadstone and Magnetic Bodies and on the Great Magnet the Earth* (1600), in Hutchins, ed., *Gilbert, Galileo, Harvey*, p. 2.

35. Locke, *Essay* (1690), III:xi:6.

36. Bridgman, *Reflections of a Physicist*, p. 82.

37. Harvey, *Motion of the Heart*, in Hutchins, ed., *Gilbert, Galileo, Harvey*, p. 269.

38. Darwin, *On Evolution*, p. 57.

39. Van Fraassen, *The Scientific Image*, p. 2.

40. Hempel, "The Irrelevance of the Concept of Truth for the Critical Appraisal of Scientific Theories," pp. 77–78. See also chapter 2, p. 40.

41. See chapter 12 below.

42. See chapter 11 below.

43. Van Fraassen, *The Scientific Image*, p. 12.

44. Ibid., p. 69.

45. Ibid., p. 56.

46. Ibid., p. 64.

47. Rather than getting involved in detailed exegesis and discussion of shifts and ambiguities in Van Fraassen's account of observability, I refer readers to Suppe, *The Semantic Conception of Theories and Scientific Realism*, pp. 23–25 and chapter 11.

48. See chapter 11 below.

49. Giere, *Science Without Laws*, especially pp. 5 ff., 84–96.

50. Watson, *The Double Helix*, p. 124.

51. Fine, "The Natural Ontological Attitude."

52. Peirce, *Collected Papers*, 5:211.

53. Foucault, *Power/Knowledge*, p. 131; my source is Windschuttle, *The Killing of History*, p. 131.

54. Ramsey, *On Truth*.

55. Goodman also alludes to this title of James's on p. 2 of *Ways of Worldmaking*.

56. Popper, "Epistemology without a Knowing Subject," pp. 106–12.

57. Popper and Eccles, *The Self and Its Brain*.

58. Peirce, *Collected Papers*, 8:216.

59. Goodman, *Ways of Worldmaking*, p. 2.

60. Ibid., pp. 4–5; cf. the discussion of reductionism in chapter 6 below.

61. Peirce, *Collected Papers*, 5:172–73.

62. Duhem, *The Aim and Structure of Physical Theory*, pp. 217–18; Polanyi, *Personal Knowledge*, chapter 4.

63. Watson, *The Double Helix*, p. 63.

64. Ramon y Cajal, *Advice for a Young Investigator*, pp. 76–77.

65. Thagard, *Conceptual Revolutions*, is illuminating on many issues in this area; but I disagree with him insofar as he treats conceptual change and belief revision as mutually exclusive, and thinks of conceptual revolutions as categorically distinct from other changes.

66. Putnam, *Mathematics, Matter and Method*, vol. 1, p. 73.

67. Holton, *Einstein, History, and Other Passions*, p. 73.

68. Quoted in Bauer, *Scientific Literacy and the Myth of Scientific Method*, p. 89.

6

THE SAME, ONLY DIFFERENT

Integrating the Intentional

> Only if a region of inquiry can be opened up in which both the scientific and the humanist approach play their characteristic roles may we ever hope to gain knowledge of man—knowledge rather than figment, and of man rather than of social atoms.
>
> —Adolf Lowe, "Comment" (on Hans Jonas)[1]

Unifiers think of the natural and the social sciences as essentially alike. Bifurcators think of them as essentially different. Optimistic unifiers—including many Old Deferentialists—think that the social sciences, if not yet as good as the natural sciences, can eventually become so; usually maintaining that the social sciences use the same method as the natural sciences, and that they are in principle reducible to physics. Optimistic bifurcators, eschewing the dream of reduction, think the method of social science is sui generis, and see the social sciences as separate-but-equal, different from but not inferior to the natural sciences. But there are also pessimists of both persuasions. Pessimistic unifiers—including many New Cynics—see the natural and the social sciences as alike permeated by interests, politics, and rhetoric. Pessimistic bifurcators see the social "sciences" as so inferior that they don't really deserve to be classified as sciences at all.

Once again, it's like the blind men and their elephant: *"Though all were partly in the right, yet all were in the wrong."* As my grandmother used to say when she explained a new idea to me: "You know such-and-such? Well, this is the same, only different." The social sciences are like the natural sciences, only not in quite the ways unifiers have traditionally supposed; but also unlike them, only not in quite the ways bifurcators have traditionally supposed. And their prospects are neither as rosy as the optimists hope, nor as gloomy as the pessimists fear.

Are the social sciences like the natural sciences, or are they different? Well, both. All are forms of systematic empirical inquiry; but the social sciences have different subject-matter and, since they appeal to peoples' beliefs, intentions, hopes, fears, etc., to explain their behavior, they are, as I shall say, intentional. Is intentional social science reducible to physics? No—and yes. Social-scientific explanations in terms of beliefs, goals, etc., are not reducible to explanations in physics in anything like the simple way some optimistic unifiers hoped; nevertheless, the intentional social sciences aren't wholly disjoint from the natural sciences, but integrated with them. Does intentional social science investigate the same world as natural science? Of course; but intentional rather than brutally physical aspects of the world. Social institutions are constituted in part by people's beliefs, etc.; they are real, but also socially constructed.

Does intentional social-scientific inquiry use the same method as natural-scientific inquiry? Yes—and no. Like empirical inquirers of every kind, social scientists make conjectures about the explanation of some puzzling phenomenon, check how well those conjectures stand up to the evidence, and use their judgment in proceeding from there. But the explanations sought are of a different kind from natural-scientific explanations; interpreting the evidence requires a different kind of background information; and social-scientific inquiry requires different kinds of "help."

Is intentional social science value-free? Of course not; but, on the other hand, of course. Social scientists often investigate issues that engage our moral or political sensibilities; moreover, what purports to be social-scientific "investigation" quite often shades into something more closely resembling advocacy. Nevertheless, like inquiry of every kind, social-scientific inquiry is subject to certain epistemological values, among them disinterestedness—i.e., in another sense of that multiply ambiguous phrase, "value-freedom." Why don't the intentional social sciences seem to have made anything like the impressive progress of the natural sciences? For a host of reasons, among them that the ideal of respect for evidence is even harder to achieve in social-scientific than in natural-scientific inquiry, and that borrowing mathematical and methodological helps from physics in hopes of looking "scientific" has sometimes proven counter-productive.

It's really quite simple—except for the many complications. But the complications are formidable; so formidable that I can aspire only to sketch the underlying continuities between the social and the natural sciences, and the most important differences.

INTENTIONAL SOCIAL SCIENCE

Like "natural science," "social science" picks out a loose federation of kinds of inquiry; but a federation of kinds of inquiry with different subject-matter than the natural sciences. It is a commonplace that the objects of the social sciences are far more complex than the objects of the natural sciences; and it is true. The hard question is what the special kind of complexity is, exactly, in the subject-matter of the social sciences but not of human biology.

We humans are sign-using creatures capable of forming complex beliefs, intentions, and goals, and of representing the world to ourselves by means of sentences, maps, pictures, and diagrams—creatures whose behavior depends in part on how we represent the world, ourselves, and our place in the world. Moreover, it is in part shared beliefs and intentions that make a collection of people a group, tribe, community, or society, and that make social institutions such as money or marriage possible. Human biology, complex as it is, doesn't have to concern itself with people's beliefs, intentions, hopes, or fears; but psychology, sociology, economics, etc., must do so.

However, medical scientists investigating psychosomatic disorders or the placebo effect *will* be concerned with people's beliefs; so there is an overlap. Natural scientists sometimes need to take account of people's beliefs; social scientists sometimes don't. Within the disciplines classified as social sciences, some parts are nearly indistinguishable from the natural sciences, while others are quite akin to history and even, though more distantly, to literary or legal scholarship.[2] Anthropology and geography, to take two of the most obvious examples, have both physical and sociological sides—sometimes recognized institutionally: Duke has two departments of anthropology; and at Stanford what was formerly one department recently split into Anthropological Sciences, accommodating those whose work is closely akin to earth sciences, the biology of evolution, etc., and Cultural and Social Anthropology, accommodating those with closer affinities to hermeneutics and sociology.[3] Most notably, some parts of psychology investigate creatures not capable of representation, or non- or pre-representational aspects of the human mind, while others study human behavior as mediated by beliefs and goals. Some research—for example, into possible brain-physiological correlates of (supposed) cognitive differences between men and women—straddles the line between the intentional and the non-intentional. It can be enormously difficult to figure out which questions can be answered without reference to people's representations of the world and themselves, and which can't; that is one reason for some of those internecine battles in psychology.

The social sciences are often described as "interpretive." Sometimes this signals a commitment to the idea that they use a distinctive method in which understanding replaces explanation; sometimes it signals an assimilation of social-scientific to literary interpretation. In any case, the multiple ambiguity of "interpretive" creates significant problems. In one sense, all empirical inquiry, natural-scientific inquiry included, is interpretive: it involves the interpretation of evidence, which, though sometimes strong and sometimes weak, is always to some degree incomplete or otherwise imperfect. In another sense, all jointly undertaken inquiry is interpretive: because it involves the sharing of evidence, it requires inquirers to interpret each others' reports of their observations, experiments, and theorizing. So, to flag the characteristic focus on people's beliefs, desires, intentions, etc., while avoiding the pitfalls of "interpretive," I have chosen to write of "intentional" social science.

Intentional social-scientific inquiry always includes people's beliefs, goals, etc., in its purview; but each branch does so in its own distinctive way. Psychologists investigate the role of expectation in perceptual error; economists calculate the mutual interactions of consumer confidence and interest rates; sociologists estimate what increment of cognitive performance can be attributed to charter schools; anthropologists try to understand the significance of a ritual dance in the life of the tribe. (An anthropologist may have to solve simultaneous equations, to investigate the beliefs and motives of the people he studies at the same time as he figures out how to translate their language—interpretation in yet another sense). But it is not these differences, important as they are, but the intentional character shared by all these kinds of social-scientific inquiry, that makes the question of reduction so controversial and so difficult.

THE QUESTION OF REDUCTION

Reductionism, in the ontologico-epistemological sense at issue here, is the thesis that the vocabularies of the other sciences, the social sciences included, can in principle be expressed in terms of the vocabulary of, and their laws derived from the laws of, physics. The Unity of Science program, showpiece of Logical Positivism, was in part an expression of faith in this idea: thus, in *The Logical Struc-*

ture of the World Carnap undertook to show how to derive the statement "it is customary to raise one's hat on meeting a lady of one's acquaintance in the street" from the laws of physics (and those laws, ultimately, from statements about a subject's "elementary experiences"). He didn't succeed. More recently, as sociobiology and brain neurophysiology have boomed, reductionist aspirations have tended to take a more empirical, and less linguistically analytic, turn.

The constants of human nature are brought home to us when we study other humans long ago, or cultures very unlike our own. I recall, a few years back, stumbling across an encyclopedia article on Burkina Faso (the former Upper Volta), illustrated by a photograph of women in a village outside Ouagadougou, almost naked and apparently muddy, talking as they work: the younger women are grinding millet, the caption explained, while the older women sort flower buds for gravy. How absolutely unmistakable, despite vast cultural differences, human commonalities are!

And when we study other creatures, we realize that our own social interactions are in some ways like those of wolves, of lions, even of ants. In a broad sense, other animals besides ourselves have cultures—not cultures in the sense of Gothic cathedrals, Shakespeare, and Verdi, or of fish-and-chips, warm beer, and soccer hooliganism; but behavior or skills shared with and acquired from others of the same species, rather than genetically programmed or compelled by the environment—such as the ritual adopted by pods of killer whales living off Vancouver Island, who line up in formation when they encounter another pod. Primatologists have identified no fewer than thirty-nine traditional forms of chimpanzee behavior, including digging for termites, gathering ants, scooping out marrow, using leaves for seats, cracking nuts against rocks or trees, and using long branches to reach fruit, that qualify, in the broad sense at issue here, as cultural.[4] The continuities with our closest biological relatives can be startling. Kanzi, the smart young bonobo who picked up the sign-language psychologists were trying unsuccessfully to teach his adoptive mother, has been described as uncannily like a human two-year-old. Pointing to Austin (a companion chimpanzee), his trainer told him, "if you give Austin your mask, I'll let you have some of Austin's cereal"; Kanzi promptly handed Austin the mask, and pointed at the cereal box.[5]

Still, it is no disrespect to the dolphins, the chimpanzees, the bonobos, et al., to acknowledge that human beings' capacity for language, representation, and learning goes far beyond that of even the smartest of our closest relatives. Only we humans go in for art, architecture, advertising, bureaucracy, crime, calendars, clothes, cookery, confidence trickery, computing, dancing, drugs, dog-breeding, engineering, fire, farming, furniture, gambling, gossip, hospitals, horse-racing, insanity, jokes, kings, law, literature, logic, mathematics, money, moral codes, music, myth, nations, newspapers, opera, pigeon-fancying, philosophy, puzzles, pottery, qualifications, quantification, religion, sports, shops, schools, science, sorcery, the stock market, transportation, technology, theology, theater, undertaking, verse, visions, war, writing, weather-forecasting, xenophobia, yoga, zoos, etc., etc.

Still, we humans are animals; there are analogues in the behavior of other animals of many of the human behaviors and enterprises on my list; and biological facts certainly constrain what social arrangements are humanly possible. We have such-and-such nutritional needs; our typical life-span is about so many years; we reproduce sexually, and our infants are helpless for a long time; we are social, territorial, language-using animals—and hierarchical, too, as sociologist Vance Packard noticed well before the recent boom in sociobiology. ("I first became interested in social stratification as a farm boy in northern Pennsylvania," he wrote in 1959, "when my father pointed out to me that one of our cows, I believe her name was Gertrude, always came through the gate first at feeding time.")[6] Exactly what the biological constants are, and exactly how tightly they constrain human societies, is a controversial question; but that there are some such constraints seems beyond dispute. Also beyond dispute is the extraordinary variety of human societies, the astonishingly various ways humans have found to live together in groups; and the role of local contingencies of geography or climate,[7] and accidents of history.

Trying to bring as much human behavior as possible within the scope of biological determinants, sociobiologists sometimes describe non-human animals in an anthropomorphic way. E. O. Wilson sounds almost like a Richard Adams[8] when he muses about "what an electric fish thinks" as it orients itself by electric field, or describes ants emitting the pheromones that warn other ants of danger as "say[ing] to other ants, in effect: *danger, come quickly*; or *danger, disperse.*"[9] We, like ants, are social creatures; and there is illumination in seeing that both the pheromone signals of the red harvester ants and, say, the American legal system, can be described, in Wilson's nice phrase, as "semiotic webs." But even the most ambitious sociobiologists allow that there are large and consequential differences between an ant emitting her pheromones and a bank clerk pressing the alarm button and calling out, "Danger! Come quickly!"; even Wilson feels the need to say, "in effect."[10]

Unlike ant behavior, much human behavior is mediated by our beliefs, hopes, fears, etc. The suckling reaction of a human infant can be explained without attributing beliefs or desires to the baby; it has an instinctive tendency to turn its head when something touches its cheek, and an instinctive tendency to suck when something touches its mouth. The explanation of my going to the fridge to get a glass of milk, however, requires reference to my wanting a drink of milk and believing there is milk in the fridge.[11] Some of what we do is purely instinctive; and some is due to panic, anger, confusion, or sheer habit. But how each person is and behaves, constrained by biological universals, mediated by cultural specifics, depends in part on his beliefs and motives. We don't need to be able to specify all the respects in which human social behavior is biologically determined, only to know that it isn't biologically determined in *every* respect, to see that the truth of reductionism will turn on whether beliefs, etc., are reducible to neurophysiological states (or whatever other physical basis a future science might discover). Serious philosophy of social science will require serious philosophy of mind.

For all the remarkable recent successes of the neurosciences, the dream that beliefs and goals will be reduced to neurophysiology and eventually to physics, remains—well, a dream. Indeed, sometimes nowadays it seems the dream has turned into a nightmare: when, for example, despairing of a smooth reduction of beliefs, hopes, and fears to neurophysiological states, and noting that the ceaseless cognitive activity of the ganglia of the sea-slug doesn't involve representations, Paul Churchland concludes that "folk psychology is false, and . . . its ontology is chimerical. Beliefs and desires are of a piece with phlogiston, caloric, and the alchemical elements"[12]—there are none. (How does he ever dare to drive a car, you wonder, or to ride in one?)

And even the strongest advocates of reductionism seem to waver when it comes to the crunch. Discussing the cross-cultural ubiquity of snake-symbolism, Wilson—who is committed to reductionism not only as "the method of science" but also as an ontological truth[13]—promises an analysis of a magician's snake-dream "down to an atom." But no such analysis is forthcoming; in fact, just a few pages later Wilson acknowledges that thus far "the neural pathways of snake aversion have not been explored."[14] Still, assuming a quasi-Lockean conception of meaning as mental images, most of the time he seems confident that a reductionist account of these "nodes" is just around the corner; until, shortly after declaring free will an illusion, he writes that "there can be no determinism of human behavior, at least not in obedience to the simple way physical laws describe the motion of bodies,"[15] because the contents of the mind evolve in accordance with the unique history of the individual. This sounds right; but it also sounds, as Wilson realizes, as if it calls for some compromise of the reductionist agenda.

Representation takes on a life of its own: the capacity for language, or, more exactly, for the use of conventional signs, brings with it the capacity to form complex beliefs and intentions, to pass on skills and information, to think thoughts that have never been thought before. A creature without that capacity may manifest snake-aversion; but only a creature with that capacity is capable of believing, in the fullest sense of the word, that snakes are dangerous. This is not to deny that other animals behave in ways that tempt us to describe them as believing this or expecting that (I say myself, of the cardinal birds who chatter on my deck at 8 A.M., that they think the service is really going downhill, their sunflower seeds should be in the feeder already); nor is it to deny that the capacity for language or sign use is a matter of degree, or that a talented primate may achieve something like the linguistic abilities of a human toddler. It is only to insist that the capacities for language and for belief, in the fullest sense of the word, go together; for in that fullest sense belief involves a characteristic amalgam of verbal and non-verbal dispositions.

A person who believes that snakes are dangerous will have a very complex multi-form disposition, or pattern of dispositions:[16] briefly and very roughly, to shriek at the sight of, and run away from, snakes; to shudder at pictures of snakes; and to assert or assent to sentences to the effect that snakes are dangerous. Assertion and assent may be insincere; but there is no need to worry that sincere assertion can be explained only as assertion-accompanied-by-belief and insincere assertion only as assertion-not-accompanied-by-belief. Someone whose assertion is insincere will speak differently depending on who is listening, and won't act as he speaks: he will assure his enemy that the ice is thick enough to bear his weight, but won't venture on the ice himself—as the saying goes, he won't put his money where his mouth is.

"He walked across the ice," we say, "because he believed it was thick enough to bear his weight." Or think of the standard detective-story ploy, where police trap a suspect by leading him to believe that incriminating evidence is to be found in such-and-such a place, and following him as he rushes to hide or destroy it. Isn't there a danger that my account has turned a real explanation (he did x because he believed that p) into something like a tautology (he did x because he was disposed to do x)? No. It tells us that leading the suspect to believe that p induces a standing disposition which will likely result in a manifestation of the hoped-for incriminating behavior in the presence of other beliefs (that they'll catch me if I don't destroy the evidence) and desires (that I not be caught); but this is no tautology.

With ordinary, garden-variety beliefs, verbal and non-verbal dispositions interlock both causally and referentially: the subject's representing the world to himself this way causally sustains his disposition to act thus and so, and the sentences to which he is disposed to assent are about the things in the world with respect to which he is disposed to act thus and so, in the characteristic semiotic triad—person, words, world. (With mathematical and highly theoretical beliefs, however, any non-verbal dispositions may be very oblique; and then there are those beliefs, as we say, "about" the Holy Grail, the Abominable Snowman, and so forth.)

Far from proposing a Skinnerian don't-look-in-the-black-box behaviorism, I assume that the pattern of dispositions involved in believing . . . , is grounded in an enormously complex neurophysiological configuration; "configuration," rather than "state," because what is involved is an enormously complex web of interconnections among receptors, whatever registers input, and activators, whatever initiates behavior, verbal and non-verbal.[17] However, the pattern of dispositions involved in believing, e.g., that snakes are dangerous, must be neurophysiologically realizable in more than one way; for while my believing that snakes are dangerous involves among other things a disposition to assert or assent to various sentences of English, Ivan's believing that snakes are dangerous involves among other things a disposition to assert or assent to various sentences of Russian. Now, perhaps, you are wondering how Ivan and I can have the same belief—after all, haven't I just said that we have different multi-form dispositions?[18] No. In the relevant sense, we have the *same* disposition: to behave thus and so, and to assert or assent to sentences which, though different, bear the same relation to things and events in the world.

But now (as William James would have said) we come to the meat in the coconut. Thinking about beliefs and their role in human behavior, we are pulled in two directions. From one perspective, the essential thing seems to be the content of a belief; from another, its physical realization. We are pulled towards reductionism when we reflect that, as one habit may sustain or inhibit another, so one belief may be a person's reason for, or against, another; which is to say that the former pattern of dispositions sustains, or inhibits, the latter (to simplify considerably; ordinarily a whole nexus of reasons sustains or inhibits a belief). Both reasoning and the explanation of behavior require that beliefs play a causal role; and the causal interactions of beliefs with other beliefs and goals which produce deliberate behavior are physical goings-on in a person's head, and must be in accordance with physical laws. This makes it seem obvious that believing that *p* just *is* being in this neurophysiological kind of configuration.

Or is it? If we want to know why Jack took a jar from the fridge, sniffed it, threw it away, got into his car, drove to the supermarket, and bought mayonnaise, could an explanation along the lines of "cog A in Jack's brain engaged with wheel B which moved lever C" really do the job? To be sure, a real neurophysiological explanation would be vastly more complex than my caricature; but how would it capture the bearing of Jack's representation *of* the world on his action *in* the world?

The reductionist dream is that a neurophysiological configuration corresponds to the proposition that Carnap had an aunt who lived in Vienna somewhat as a DNA configuration corresponds to cinnabar eyes in fruit-flies. But this seems like the wrong analogy: rather, an explanation of someone's flushing because of the heat in the room, or jumping because of a loud noise, would need to acknowledge not only his internal states but also the connection of those states with the environment; and an explanation of his blushing because of the embarrassing remark he just over-heard would need to acknowledge the connections of his neurophysiological states with these words, and the use of those words in his linguistic community.

Think of an alarm clock, which is certainly a physical thing. The clock's going off is a physical happening brought about when these and those cogs and wheels, or these and those electrical contacts, engage inside the clock. But the explanation of the clock's going off at 7:30 A.M. isn't exhausted by the account of the cogs or the contacts, without reference to human conventions about time; nor is there anything inherent in the cogs or contacts *qua* cogs or contacts that makes them *about* time. Human beings, also, are physical things, and their making these or those noises or marks or movements is brought about by neuro-physiological goings-on. But the explanation of my going to the fridge to get a glass of milk isn't exhausted by a neurophysiological account of the firings in my brain, without reference to the content of my beliefs, etc.; i.e., again, without ref-erence to human conventions, a socio-cultural loop. Nor is there anything inherent in the bits of gunk in my brain, *qua* bits of gunk, that makes them *about* milk; that depends on their connections with milk, and with "milk."

This isn't to say that the physical properties of the material concerned don't matter; of course they do—our brains could no more be made of butter or of brass than a clock could be made of feathers or of fudge. But the belief that *p* is realizable by whatever configurations could be appropriately causally linked to the world, to words in the person's language, and to his motor apparatus; and what is required to identify the relevant families of neurophysiological configu-rations is reference to patterns of linguistic behavior in a person's linguistic com-munity, to reference and meaning, and to the things in the world his beliefs are about: that is, to what makes this the belief that snakes are dangerous, what makes Ivan's and mine the same belief.

It's all physical, all right; but it isn't all physics. So if reductionism were redeemable, it could be only in a quite non-standard form in which the whole socio-cultural-historical story of language, meaning, and reference was told in the vocabu-lary of some hypothetical future physics. And even if this were possible—a very big "if"—a significant difference would remain between the parts of this hypothetical future "physics" which go through that socio-cultural loop, and those which don't.

Rather than reduction, in the strong sense, there is integration of the social with the natural sciences. The Logical Positivists' strong, simple, conception of the way all the truths about the world must fit together has obscured an otherwise obvious fact: heterogeneous truths are no less true, nor necessarily disjointed, for their heterogeneity. A better model might be a map in which a depiction of the roads, towns, etc., is superimposed on a delineation of the contours of the same territory, and integrated in virtue of the fact that the roads go around the lake and through the pass in the mountains, that the town is on, not in, the river, and so on. The natural sciences draw a contour map of the biological determinants of human nature and the biological roots of human culture, on which the social sciences superimpose a road map of marriage customs in New Guinea, the failures of the Soviet economy, the rise of modern science in seventeenth-century Europe, and so on.

THE QUESTION OF REALITY

Those aspects of the world studied by the natural sciences are independent of us; they would have existed, and would have been of these kinds and subject to these laws, even if there had never been any human beings or human languages or natural-scientific theories about them—mostly, anyway: qualifications would be needed with respect to the aspects of the world studied by human biology, and to acknowledge the possibility that this biological species would not be extinct, or this genetically altered variety would not exist, were it not for things human beings have done.

It is probably because as a young woman she worked with heavy plough-horses on a farm in Ireland, Marjorie Grene reminisces, that she was never able to take idealism seriously. Brute physical objects surely are our paradigm of the really real; and this may tempt us to misconstrue reality as causal independence of us, or as mind-independence. But reality can't be defined as causal independence of human beings. There would, after all, have been no human artifacts if there had never been human beings; and yet chairs, books, steam-engines, machine-guns, etc., are certainly real. True, even the synthetic materials on which we now rely so extensively are made from natural stuff. True, also, the possibility of making something that can serve a given function depends on the physical properties of the stuff from which it is made—you can't make a pillow out of granite or a typewriter out of grease.[19] Still, though materials can be recalcitrant, and we can fail in the execution of a design, artifacts are causally dependent on us, and have the characteristics we give them; but they are real nonetheless. Nor can reality be defined as mind-independence; for mental states and processes are mind-dependent, but real nonetheless.

"Real" contrasts, not with "artifactual" or "mental," but with "fictional, a figment." Fictions are however some person or persons represents them to be: Hamlet isn't a real person, but a fictional character, and has the characteristics Shakespeare represents him as having. This is what motivated the first-stab characterization of "real"—"independent of how we believe it to be"—suggested in the previous chapter.[20] But that characterization will need further refinement if it is to capture the difference between natural and social reality.

Social institutions, roles, and rules aren't like rocks, or even roads, nor are they like mental images, or dreams; but they aren't fictions or figments, either. Marriage, banking, the fashion industry, the legal system, science—social roles and rules and kinds—wouldn't exist unless there were human beings and human societies. Human beings don't *physically construct* social institutions as they did the Ziggurat at Ur, the Great Wall of China, highways, skyscrapers, and all the countless artifacts human ingenuity has devised; but such social institutions *are constituted in part by* people's behavior, beliefs, and intentions. They are how they are independent of how anyone in particular believes them to be; but they are not independent of the beliefs, intentions, etc., of members of the society in question generally. That's why a couple "married" by a hedge-priest aren't really married, even if they believe they are; think of poor Audrey in *As You Like It*.

Or take money. There would have been no money if there had never been human beings. Physical tokens of monetary value, such as coins, bills, or credit cards, are artifacts (and not just anything can perform this function: cowrie shells, yes, or bits of metal or paper or plastic, or magnetic traces in a computer, but not mist or asteroids). *Qua* social phenomenon, however, money involves much more than the physical tokens; in fact, as John Searle says, it is an institution of quite staggering complexity.[21] It is beyond my powers, and perhaps your patience, to explore all the complexities here; suffice it to note that among them is that for money to *be* money depends in part on people's believing, intending, and behaving thus and so. That's why ancient Greek coins or cowrie shells aren't money any more, why currencies in which people lose confidence cease to be viable, and how the Euro became money upon the appropriate agreement among members of the European Community. To be sure, the chairman of the Federal Reserve can raise or lower the federal-funds interest rate by announcing that it is now *n* percent; but his being in a position to do so involves a vast, complex structure of institutional facts—which is why you and I can't make the bills we print in the basement money by announcing, or believing, that they are.

It is often assumed that "real" is incompatible with "socially constructed." But social institutions, roles, and rules are *both* real (how they are doesn't depend on how you, or I, or any individual believes them to be), *and*, in a weak sense, socially constructed (they are constituted in part by people's beliefs and intentions). It is because social institutions, etc., are in this sense socially constructed that what social scientists say about the workings of marriage or the banking system or the fashion industry or globalization can have an oblique effect on how marriage, the banking system, etc., actually do work—because, if people know what is said about them, this may change their beliefs, intentions, and behavior. The most obvious kind of example is the self-fulfilling prophecy in economics: unemployment is predicted to fall, or to rise, boosting confidence in, or anxiety about, the economy, and with it employers' willingness, or unwillingness, to take on new workers—and unemployment falls, or rises, as predicted.

This has tempted some social scientists to indulge in a little self-aggrandizement: Anthony Giddens, for example, writing of a "double hermeneutic," and the "reflexivity of modern social life," avers that "social practices are constantly examined and reformed in the light of incoming information about those very practices. . . . [The] characteristic of modernity is . . . the presumption of wholesale reflexivity."[22] And sometimes, when the point about the possibility of social-scientific theories affecting their subjects joins hands with a fashionable linguistic idealism, the self-aggrandizement takes a metaphysical turn, and it is suggested that social institutions are brought into being by social scientists' theorizing. This, the scary form of social constructivism, really is incompatible with the reality of social institutions. Fortunately, however, it isn't true; and sociologists aren't really quite so powerful as some of them like to think they are. Yes, social institutions are partially constituted by people's beliefs and intentions; and yes, social-scientific theorizing can affect its objects. But social scientists no more brought child abuse or schizophrenia or homosexuality into existence by their intellectual activities than biologists brought anthrax into existence by theirs (though by now, to be sure, the idea that medical scientists' activities might bring a new disease into existence is more than science fiction).

In the social as in the natural sciences, explanation and prediction require generality. Optimistic unifiers look for Laws of Society on a par with the laws of physics; pessimistic bifurcators scoff at the idea. True to my neither-of-the-above perspective, I see the truth as lying somewhere in between. David Hume once observed that history would not be possible if human nature weren't essentially the same long ago as now; neither would intentional social science. Fortunately, as Fritz Machlup more recently observed, "[i]n the social world phenomena are not quite so heterogeneous as many have been afraid they are."[23]

There are human commonalities behind cultural differences, a human nature that grounds lawlike generalities true of all human societies. Helmut Schoeck maintains that "[t]hroughout history, in all stages of cultural development, in most languages, and as members of widely differing societies, men have recognized a fundamental problem of their existence . . . the feeling of envy and of being envied." He devotes an entire book to investigating this motive, which he believes "lies at the core of man's life as a social being,"[24] offering a startling range of evidence from anthropology to political theory to mythology and literature, even primatology and ornithology. The fragment that lingers most vividly in my mind is this, from a nineteenth-century book about the Kazak-Kirghiz, nomads who rob caravans travelling across the steppes: rather than let one member of the gang have more than another, "they cut up the objects they have stolen into the most absurd and useless little bits."[25] Schoeck's conjecture exemplifies one kind of potentially illuminating social-scientific explanation, the most generalizing kind: tying historically and geographically distant phenomena together by reference to some underlying aspect of human nature.

Recent attempts to explain the spread of AIDS in sub-Saharan Africa, though less panoramic, illustrate my theme in a different way. The year before the epidemic was first reported in the *Boston Globe* in 1999, two million people in this region had died of AIDS, about 85 percent of the total number for the whole world, and more than 22.5 million people in the region carried the HIV virus. In this part of the world, 55 percent of those with AIDS or HIV are women, and only 45 percent men, whereas in North America 80 percent are men and only 20 percent women. This region, we read in the press, "faces a crisis of shattered mores, where sexuality is no longer governed by traditional norms"; the disease is transmitted primarily heterosexually, with rape and prostitution playing major roles. In 2002, according to statistics compiled by the United Nations AIDS agency and the World Health Organization, there were approximately 3.5 million new infections in the area.[26]

Researchers have found that the rate of HIV infection among migrant workers in South Africa is nearly $2\frac{1}{2}$ times higher than among other workers; according to a report in the *Wall Street Journal*, the HIV rate among AngloGold's miners is 30 percent. Dr. Mark Lurie comments: "If you wanted to spread a sexually transmitted disease, you'd take thousands of young men away from their families, isolate them in single-sex hostels, and give them easy access to alcohol and commercial sex. Then to spread the disease around the country, you'd send them home every once in a while to their wives and girlfriends. . . . That is basically the system we have."[27] A couple of days later, a correspondent observed in the letter column that part of the problem is that there's no television in the miners' hostels. All of this makes perfect sense; but only because of assumptions about human nature unstated because they are too obvious to state: especially in the absence of other forms of relaxation, young men doing physically exhausting and dangerous work who are isolated from their wives and girlfriends for months or years at a time, even if they are aware of the risks, will tend to get together with any women who are available.

The social-scientific components of the explanation of the crisis, no less than the social-scientific components, require a kind of generality. The behavior-patterns, like the infection-patterns, are local, obtaining here and not there; but we can no more explain the behavior-patterns than the infection-patterns without fitting them into some generalizable categories; we need to identify kinds of behavior and mechanisms of motivation as well as a kind of virus and mechanisms of infection.

While some social institutions are universal, many are culturally specific. Everywhere there are differences of status, but only in some cultures are there differences of caste, or Sirs and Lords; everywhere people obtain and distribute food, but only in some cultures are there prices or markets. Real but restricted generalities, rooted in human nature but holding in the context of specific social institutions, allow for the possibility of explanation and even—given appropriate limitations of scope and generous ceteris paribus clauses, and ordinarily only to a probability—prediction.

Unlike natural kinds, social kinds aren't congeries of properties held together by laws of nature but congeries of behaviors held together by people's beliefs and intentions; often very loose congeries, as with the wide range of arrangements anthropologists call "marriage"—a concept the boundaries of which are being extended, in our own society, by advocates of same-sex unions. The looseness of social kinds, and the local, contingent character of social institutions, is the source of some notorious pitfalls of social-scientific inquiry: taking the local and culturally specific (e.g., our society's division of labor between the sexes) for something universal and inevitable; assuming that whatever is true of one variant of a social kind (e.g., the family as constituted in our society here and now) is bound to hold for other forms as well.

QUESTIONS OF METHOD

In the agenda inherited from Logical Positivism it was often assumed that we must choose *either* methodological monism, according to which the social sciences seek explanatory, causal theories and use the scientific method just as physics and chemistry do, *or* methodological dualism, according to which the social sciences seek understanding rather than causal explanation, and use a method of empathy rather than objective observation. This makes me feel like that legendary Irishman who asked for directions to a distant village: "Sure and begorrah, I wouldn't start from here." For one thing, as I have argued, in the sense intended there is no "scientific method." For another, the positivists contrasted their own methodological monism with a methodological dualism according to which, whereas the natural sciences seek causal explanations and depend on objective, publicly checkable observations, the social sciences must rely on a mysterious faculty of understanding, "a policy of deep breathing followed by free association,"[28] in Braithwaite's wonderfully scathing phrase. Whether or not this is adequate to the idea of *Verstehen* in its most exaggerated forms, it hardly seems fair to Max Weber or Alfred Schutz, who seem primarily concerned to insist on the unavoidable necessity of taking actors' conceptions of the world and their actions into account,[29] and who might have agreed with my Irish friend and myself.

As is probably already clear, I think contrasting understanding with explanation is misleading;[30] intentional social science tries *to understand people's behavior by coming up with explanatory hypotheses about their beliefs, goals, etc.*, and seeking out and appraising the worth of evidence relevant to those hypotheses. This much is true of the natural and the social sciences alike; the difference lies in the nature of the explanations, and of the evidence.

Social-scientific explanations typically appeal to motives, beliefs, desires. The motives ascribed will only be comprehensible insofar as they manifest common (though not necessarily universal) human characteristics. When an anthropologist explains that these tribesmen travel long distances in order to find the red clay with which the young men decorate themselves for their initiation rites, we recognize the motivation even though this particular manifestation is unfamiliar; we see the similarity with, say, elaborate and expensive arrangements for a debutante's coming-out party. However, to say that the tribesmen travelled miles because they wanted this special clay (or that Jack went to the supermarket because he wanted fresh mayonnaise) *is* to say, in part, that the tribesmen's desire for the clay, in conjunction with their belief that it was to be found only at a distant river-bank, caused them to walk there (that Jack's desire for fresh mayonnaise and his belief that it was available at the supermarket caused him to drive there). Intentional, belief-desire explanations piggy-back on ordinary causal explanations.

Again, observing people and describing what they do, just like observing a cloud chamber or an X-ray photograph or a CAT scan and making sense of what you see, requires background knowledge. But the kind of background knowledge needed is very different—understanding subjects' beliefs, interpreting what they say, placing their actions in the context of sometimes very culturally specific practices: think of an economist's description of a trader as buying pork futures, or an anthropologist's description of a witch-doctor as casting a spell on his neighbor's cattle, or of a ceremony as rain-dancing, or of a sociologist's description of an artisan as mending the strawberry nets.

Sometimes it is suggested that what is peculiar about the method of intentional social science is that belief-desire explanations presuppose rationality. But "rational" is multiply ambiguous. It is true, obviously, that belief-desire explanations apply only to creatures capable of forming beliefs and desires; and true that such explanations can be correct only when the behavior explained is consonant with the agent's beliefs and desires. It is not true, however, that such explanations can be correct only when the goals and beliefs in question are reasonable. (Part of the explanation of the high rate of HIV infection among very young women in some parts of sub-Saharan Africa, for example, is apparently a widespread belief that sex with a virgin can cure AIDS.) It is true that to make sense of a person's behavior, you need to take account of the way he would describe what he does. It doesn't follow, and it isn't true, that you must presume that his description is correct, or that he agrees with you. The goal is to ascribe to him the beliefs he actually has, and, so far as possible, to understand why he has them: that, say, he lights a fire to warm the hut because he thinks it's snowing, when we know it's really the artificial stuff that, in hopes of learning the native word for snow, we arranged to have dropped from a helicopter.[31] This is no simple matter of maximizing the truth of another's beliefs, as Davidson's version of the Principle of Charity requires, or of maximizing agreement, as Quine's version requires.[32]

A better approach would start out from the thought that all empirical investigation demands the same epistemic virtues: respect for evidence, care and persistence in seeking it out, good judgment in assessing its worth; and that, in a sense, all empirical investigation uses the same method—the method of experience and reasoning: making an informed conjecture, seeing how it stands up to the available evidence and any further evidence you can lay hands on, and then using your judgment whether to drop it, modify it, stick with it, or what. What is distinctive about natural-scientific inquiry isn't that it uses a peculiar mode or modes of inference, but the vast range of helps to inquiry scientists have developed, many of them—specific instruments, specific kinds of precaution against experimental error, specific models and metaphors—local to this or that field or sub-discipline.

As this suggests, insofar as intentional social science is methodologically like physics, it is also methodologically like history, detective work, etc., and like everyday empirical inquiry; while insofar as it is methodologically different from physics—e.g., in its reliance on questionnaires or interviews rather than electron microscopes or cloud chambers for making observations, in its preoccupation with statistical significance—there are also differences between physics and biology. The underlying patterns of hypothesizing, reasoning, and testing are the same for all empirical investigation; but the special techniques overlaid on them will differ from field to field. This is not to say that whatever techniques are used in a field are ipso facto good: the question of the relative worth of clinical and other kinds of evidence in psychology, for example, is certainly a substantive one.[33] Nor is it to deny that among the helps scientists devise are helps to the intellect, including statistical techniques of special relevance to certain of the social sciences.

In those parts of the social sciences, such as physical geography and anthropology, most closely akin to the natural sciences, borrowing and adapting natural-scientific helps is quite appropriate, and has proven fruitful. But in intentional social science not all those natural-scientific helps are appropriate, and some can be counter-productive. Many sociologists, as Robert Merton observes, "take the achievements of physics as the standard for self-appraisal. They want to compare biceps with their bigger brothers."[34] "Physics-envy" has sometimes given us cargo-cult social science,[35] the form without the substance of real inquiry: bits of bamboo in the ears, but no actual radio; "methodology" in spades, but no real effort to discover the truth; symbolic formulae, but no real precision. David Abrahamsen's so-called Second Law of Criminal Behavior, "A criminal act is the sum of a person's criminalistic tendencies plus his total situation, divided by the amount of his resistance," or, as he continues, "$C = (T + S)/R$,"[36] is classic; more recent manifestations are usually more sophisticated, and less overt.

Not that quantification or measurement is always inappropriate in intentional social science—on the contrary, it can be enormously useful; but it is worse than pointless when it disguises banality, when what is being measured or quantified is ill-defined or ambiguous, or when quantification diverts attention from what is important to what can be measured. Hence two key ideas motivating Stanislav Andreski's exasperated description of social science as "sorcery": that in the social sciences the quantification essential to the natural sciences too often serves only to camouflage the obscurity or the vacuity of underlying concepts and categories; and that an excessive preoccupation with methodological technicalities too often substitutes for genuine originality or depth.[37] This is no doubt why, re-reading William James musing about what makes people tick, I am struck by an insight, a penetration, that more recent and more self-consciously "scientific" psychology sometimes seems to have sacrificed.[38]

Even in economics, where numbers are undeniably apropos, combining precision and depth is a difficult balancing act. Sophisticated mathematical models often rely on artificial and unrealistic assumptions about people's motivations; mathematically sophisticated theory and statistically sophisticated empirical work often fail to mesh as you would hope. The problem is not idealization as such (physicists, after all, postulate frictionless surfaces and the like); it is tunnel-vision, blindness to other motives besides the economic.

Because of its mathematical character, economics is sometimes called "the physics of social science." But surely, if any discipline is properly so-called, it is psychology, the discipline to which it falls to investigate the basic contours of human motivation. The misconception arises because we look to mathematical trappings rather than conceptual depth and breadth to identify the most basic social-scientific field. Correcting the mistake, we may begin to suspect that the divisions among the various sub-enterprises of the social sciences are somewhat artificial; and that crossing those disciplinary boundaries—political scientists borrowing economists' methods, economists looking to sociology or psychology, etc.—might be more potentially fruitful, and less potentially dangerous, than crude forms of physics-envy.

Physics-envy, however, is not the only pitfall of intentional social science. There is an equal and opposite danger: transmuting what could and should be inquiry into social phenomena into socio-political advocacy.

QUESTIONS OF VALUE

But *can* social-scientific inquiry be value-free? The question fairly bristles with ambiguities. In several senses of the phrase, the social sciences clearly aren't value-free, nor would we want them to be. Social scientists often investigate questions which engage our moral and political values: race relations, educational provisions, tax policies. They sometimes investigate what people value: how party platforms affect voting patterns, the relative importance of profit-maximization versus empire-building in the motivation of managers of large businesses. And they sometimes investigate whether or not desired values have been achieved: whether target schools succeed in their educational objectives, whether tax reductions increase productivity.

However, since social-scientific inquiry is a kind of inquiry, it falls within the scope of the epistemic values—concern for truth, respect for evidence—relevant to all inquiry. Disinterestedness is one of these epistemic values, requiring the inquirer to seek the truth "regardless of what the color of that truth may be";[39]

regardless, in particular, of what he would like to be the case, or of what it would serve his interests to have believed. In this sense, value-freedom is an ideal for social-scientific as for inquiry of every kind. Matters are complicated by the fact that "disinterested" has several senses: (1) uninterested;[40] (2) having no interest (especially, no financial interest) in a question's coming out a certain way; and (3) not motivated by the desire that an inquiry come out this way rather than that. In the third, crucial sense, "disinterested" is equivalent to "unbiased" or "impartial." In this sense, "disinterested inquiry" is a pleonasm, and "interested inquiry" an oxymoron.[41]

Sometimes it is thought that disinterested inquiry is impossible, because you wouldn't inquire at all if you weren't interested in the answer. Actually, it isn't clear that this is so; often enough people inquire because it's their job, or, as Machlup reminds us, because they have to write a dissertation on *something*.[42] But even if inquiry were always interested in the first sense, it wouldn't follow that it is always interested in the third sense as well. Nor, as Ernest Nagel pointed out long ago,[43] does this conclusion follow from the fact that any inquiry involves selective attention, taking an interest in only some aspects of the phenomenon under study. Sometimes, again, it is thought that if an inquirer is interested in the second sense, i.e., stands to gain financially or otherwise if the inquiry comes out this way rather than that, it follows that he can't be disinterested in the third sense, that he is bound to be partial. This too is a non sequitur. But it *is* human nature to hope that a question will turn out as it would be in your interest for it to turn out; it is harder to be disinterested in sense (3) if you are interested in sense (2).

Impartiality doesn't require that you start with a blank slate, a mind empty of beliefs—in fact, you couldn't inquire at all if you did—but it does require that you have no unbudgeable preconceptions, that you be willing to check out all the evidence, and will change your initial judgment if the evidence turns out against it. As Andreski says, "the ideal of objectivity is much more complex and elusive than the pedlars of methodological gimmicks would have us believe."[44] It is less a matter of technique than of character: of respect for evidence, of a contrite fallibilism, of good faith in inquiry. Remember Crick: "If [Watson and I] deserve any credit at all, it is for persistence and the willingness to discard ideas when they became untenable."[45] Some years before World War I, a political journal asked several prominent French social scientists what they regarded as the most important method of their field. Others returned detailed methodological recommendations; Georges Sorel replied in one word: "honesty." As Andreski observes, "this lapidary answer has lost none of its relevance."[46]

*   *   *

I was careful to describe disinterestedness as an ideal, for real human beings are rarely if ever purely disinterested, completely impartial inquirers; it is rather a case of more or less genuinely seeking the truth of some question, of greater or lesser willingness to seek out and pay attention to evidence that disfavors the conclusion it would be in your interest to reach. But other things being equal, the closer you come to the ideal of impartiality, the greater your evidential reach, the fairer your judgment of the worth of evidence, and the better-conducted your inquiry. People are sometimes tempted to think that bias doesn't matter, so long as competing biases can fight it out. Epistemologically successful "fighting it out," however, requires a community of people who, fallible and susceptible to bias as they may be, are disinterested enough that they can be budged by evidence from their preconceptions, or from what they would like to believe.

The fact that scientific inquiry is a community affair compensates, to a degree, for individuals' falling short of the ideal; and sometimes an obstinate adherent of some approach or line can advance inquiry not despite but because of his obstinacy. It still doesn't follow that impartiality isn't the ideal, let alone that it is just as good, or even better, if each person just pushes his, or his party's, line. That would be as disastrous as having opposing attorneys slugging it out with no judge to referee, and no jury trying to determine the truth of the matter.

There's no denying, however, that disinterestedness can be especially hard to achieve in intentional social-scientific inquiry, especially where the inquirer's moral or political values are deeply engaged; nor that a covert evaluation can easily get built into social-scientific theories. Theorizing about the function of sex roles within the family, for example, or about the causes of inflation, can covertly seem to legitimize, and perhaps thereby help perpetuate, existing social arrangements; definitions of unemployment can disguise how many people are really out of work. More generally, terms referring to this or that social, psychological, political, economic, etc., phenomenon often carry evaluative connotations, positive or negative—and sometimes the terminology shifts with changing political sensibilities, as the "undeveloped" countries to which economists referred in the 1950s and 60s became the "underdeveloped" countries of the 70s and the "developing" countries of the 80s.[47]

Robert Heilbroner discusses a simple but revealing example of built-in evaluation: describing government borrowing as "crowding out" private investment covertly gives priority to private enterprise over government spending, evading the necessity to say why a private casino should be thought more worthy than a public child-care center.[48] And this is no superficial difficulty; it arises from the culturally specific character of many social arrangements, such as polygamy or— the example on Heilbroner's mind—capitalism. Presented as the results of scientific investigation, truths about culturally specific social institutions or roles are easily mistaken for truths about human societies generally.

We expect psychology to help the mentally disturbed, and hope that social-scientific, especially economic, investigation will show us how to solve social problems (a hope, as Merton wryly observes, about as realistic as expecting William Harvey, having just discovered the circulation of the blood, to come up with a cure for heart disease).[49] In principle, questions of ends and means are distinct; and in principle, also, it is clear that, while the task of determining what means would lead to desirable ends is reasonably assigned to the social sciences, deciding what social ends are desirable is a political task.[50] In practice, however, these questions are often almost inextricably intertwined.

In principle, factual questions about the workings of the market, or of polygamy, are distinct from evaluative questions about the relative desirability of market versus feudal, socialized, or mixed economies, or of polygamy versus polyandry or monogamy. In practice, however, these questions too are often almost inextricably intertwined.

In principle, again, descriptive and evaluative aspects of the meanings of key terms are clearly distinguishable. "Crowding out" implies that funds which might have been used for purpose A are no longer available if used for purpose B—the descriptive element; and that purpose A is more worthy—the evaluative element. In practice, however, the descriptive and evaluative aspects of the meanings of key terms are often almost inextricably run together. And when what is at stake engages our feelings very strongly, even if we invent new, neutral terminology, it won't *stay* neutral very long. As Andreski puts it, "no matter how aseptic and odourless when first coined, psychological and sociological terms very quickly acquire overtones of praise or blame in accordance with whether the reality to which they refer is liked or not."[51] If "crowding out" were replaced by "displacing," probably "displacing" would soon enough take on the old, unfavorable connotation.

Sometimes it is suggested that disinterestedness is not desirable in social science. According to some, sociological research cannot just describe the social world, but will inevitably change it as it reveals the ideology by means of which the dominant class keeps underdogs down; indeed, they say, this is precisely the point of social research—to improve the world, not just to describe it. I don't deny that social research can be disturbing, and quite properly so; the example that comes to my mind is Caroline Bird's *The Case against College*, a most unsettling book for one committed to the academic life. And I shall resist the temptation to ask, "How many sociologists does it take to change a lightbulb?" (None; it's not the lightbulb, it's the system that has to change!) It's more important to point out that the thesis that sociological research is inevitably destabilizing presupposes that sociologists uncover truths which the ruling class would prefer not to have generally known; and that the idea that the goal of social research is to improve the world presupposes that we can know not only what would constitute a genuine improvement, but also what steps would likely bring that improvement about. So these radical views presuppose the possibility of the very kind of investigation of social institutions and interactions which I take for granted, but which they officially repudiate.

Sometimes, finally, loose talk about "political discourse" is simply allowed to blur the line between inquiry and advocacy. Political theory, focusing on sovereignty, authority, liberty, justice, fairness, welfare, and the like, takes questions about value-concepts as its subject-matter. But that political theorists are engaged in inquiry into normative concepts doesn't mean that they aren't really engaged in inquiry at all.

Only too often, supposedly disinterested social inquiry *has* been little more than political advocacy in disguise. But to draw the conclusion that the ideal of disinterested social inquiry is humbug is to succumb to the Passes-for Fallacy: that ubiquitous but crashingly invalid argument from the true premise that what passes for truth, evidence, known fact, honest inquiry, etc., is often no such thing, to the false conclusion that the ideas of truth, evidence, knowledge, honest inquiry, etc., are ideological humbug.[52] Honest, well-conducted intentional social-scientific inquiry isn't impossible; just very difficult.

THE QUESTION OF PROGRESS

So why don't the intentional social sciences seem to have made the same impressive progress as the natural sciences? In part, probably, we take a dim view of the achievements of the social sciences because historians, playwrights, etc., have already taught us so much about the complexities of human nature and human society—which makes it harder for a psychologist or a sociologist than for a physicist or a biologist to discover something that seems genuinely new, not already part of our common-sense knowledge, and explains why social scientists sometimes seem to be presenting the familiar, even the banal, in a pointlessly forbidding jargon. Nevertheless, and despite the difficulty of making clear sense of comparative judgments of progress, despite those who reassure us that, in a Kuhnian sense, the social sciences have "progressed" just as the natural sciences have, the feeling lingers that the social sciences, so far anyway, have lagged behind.

Why should that be? In part because the social sciences have sometimes hamstrung themselves by trying to be like physics *in the wrong way*, focusing too much on mathematical trappings, too little on the underlying demands of well-conducted inquiry. But in part, also, as Merton, Wilson, Andreski, and many others have observed, simply because the task of the social sciences is in some ways more difficult and more demanding even than the task of the natural sciences. The social sciences investigate questions about which people have strong personal and political feelings, and, besides, are often under pressure to come up with solutions for social problems for which the public or the government demands speedy remedies, making it harder for social than for natural scientists to remain free of bias. The objects of the social sciences have a peculiar complexity, and, besides, can react to claims and predictions made about them as the objects of the natural sciences cannot, making the task of the social sciences intellectually more difficult.

In many areas of social science, controlled experiments aren't feasible; of course, they aren't feasible in astronomy, either. Social scientists often investigate questions which arouse strong feelings and political prejudices; but so do medical scientists—think of the outrage provoked by the suggestion that abortion might increase the risk of breast cancer,[53] or by the hypothesis that homosexuality might have a genetic basis. In fact, the same is true of *any* kind of investigation which threatens dearly held beliefs about ourselves and our place in the universe. But intentional social-scientific inquiry more often labors under more of these difficulties than natural-scientific inquiry does.

The social sciences don't have the impressive track record of successful prediction that astronomy does. Of course, meteorology doesn't either, and in part for the same reason: predicting a race riot or a general strike, like predicting where a hurricane will make landfall, requires knowledge of an enormously complex mesh of variables. But only in part for the same reason: while there is no barrier in principle to our knowing the whole mesh of meteorological variables, it is doubtful that, even in principle, we could know everything about all the sociological variables. Though we might predict that medical scientists will know the cure for cancer within the next hundred years, we can't predict that they will know that the cure for cancer will be XYZ; for if we could, we'd know it now.[54] More importantly, because of the always-open possibility of unforeseeable conceptual innovations, we can't predict what people's beliefs will be a year or a decade or a century from now. Its subject-matter, in other words, imposes in-principle limits on the possibility of social-scientific prediction.

AND, IN CONCLUSION

It bears repeating that advocacy is not inquiry, and a fortiori not science. As to whether inquiry into social phenomena is properly called "science," it bears repeating, also, that the honorific usage in which "science" and "scientific" are all-purpose terms of epistemic praise is more trouble than it's worth, and is best eschewed. And as for the debate between the optimists and the pessimists, that fine old Leibnizian riddle has it about right: What's the difference between an optimist and a pessimist? They both think this is the best of all possible worlds!

An optimist would hope, given the epistemological importance of cooperation and competition within the sciences, and of their environment within the larger society, that sociology of science could make a significant contribution to our understanding of the scientific enterprise. A pessimist, looking for the inevitable cloud behind every silver lining, would fear that in the actual world such friendly cooperation between epistemology and sociology is hardly to be expected. As we shall see, once again they're both right.

NOTES

1. Lowe, "Comment" (on Hans Jonas), p. 154.

2. I have been helped to see this by Walker Percy, "The Fateful Rift: The San Andreas Fault in the Modern Mind"; Percy, in turn, refers to Peirce, with his conception of Man as a Sign.

3. My source is Stanford's *Humanities and Sciences Quarterly*, summer 1998, "Defining Disciplines: Anthropology Becomes Two Departments."

4. Wilson, *Consilience*, p. 133; Begley, "Culture Club"; de Waal, *The Ape and the Sushi Master*.

5. Wilson, *Consilience*, pp. 131–32. Wilson refers to Savage-Rumbaugh and Lewin, *Kanzi: The Ape at the Brink of the Human Mind*; Wrangham, McGrew, de Waal, and Heltne, eds., *Chimpanzee Culture*; and Fischman, "New Clues Surface About the Making of the Mind."

6. Packard, *The Status Seekers*, p. 24.

7. See Kitto, *The Greeks*, pp. 36 ff., on the role of its mild climate in enabling the extraordinary cultural achievements of ancient Athens.

8. Author of *Watership Down*, the stirring story of a perilous journey undertaken by a group of brave, and not-so-brave, rabbits.

9. Wilson, *Consilience*, pp. 116, 70.

10. Ibid., p. 131.

11. Despite some philosophers' efforts to assimilate the two cases; see Stich, *From Folk Psychology to Cognitive Science*, and Haack, *Evidence and Inquiry*, pp. 162 ff.

12. Churchland, "The Ontological Status of Observables," "Folk Psychology and the Explanation of Behavior," and "On the Nature of Theories"; discussed at length in Haack, *Evidence and Inquiry*, chapter 8.

13. See chapter 4 above, p. 98.

14. Wilson, *Consilience*, pp. 71, 79.

15. Ibid., p. 120.

16. See Haack, *Evidence and Inquiry*, pp. 173 ff.; Price, *Belief*.

17. Perhaps, given his enthusiasm for connectionism, Churchland's bizarre thesis that no one believes anything could be charitably reconstrued as an exaggerated way of saying that to have a belief is to be, not in some simple neurophysiological state, but in some complicated neurophysiological configuration.

18. A question put to me by Corliss Swain.

19. This sentence was written before I saw, in the Kon-Tiki exhibition in Oslo, a stone pillow from Easter Island. But I will let it stand.

20. See chapter 5 above, pp. 131–32.

21. Searle, *The Construction of Social Reality*; the quotation is from p. 3.

22. Giddens, "Nine Theses on the Future of Sociology," pp. 30–31; my source is Windschuttle, *The Killing of History*, p. 206. As we shall see in chapter 7, "reflexivity" has a rather different use among sociologists of science.

23. Machlup, "Are the Social Sciences Really Inferior?" p. 161.

24. Schoeck, *Envy*, p. 3.

25. Ibid., p. 369, citing Levchine, *Description des hordes et des steppes des Kirghiz-Kazaks ou Kirghiz-kaissaks*, p. 343.

26. My sources are "AIDS and the African," *Boston Globe*; "A Devastated Continent," *Newsweek*; Bartholet, "The Plague Years"; Cowley, "Fighting the Disease: What Can Be Done"; and Zimmerman, "AIDS's Spread Inflames Other Crises" (the quotation is from "AIDS and the African," 10 October 1999, A1). Recent press reports suggest that India faces an epidemic of AIDS transmitted by prostitutes and their trucker clients, China an epidemic of AIDS transmitted by unsafe blood transfusions, and Russia an epidemic of AIDS transmitted by drug users and prison-camp inmates.

27. Mark Schoofs, "Undermined"; the quotation is from page A10.

28. Braithwaite, *Scientific Explanation*, p. 272.

29. Weber, "The Interpretive Understanding of Social Action"; and Schutz, "Concept and Theory Formation in the Social Sciences." I am also relying on Brodbeck, introduction to Part 1 of *Readings in the Philosophy of the Social Sciences*, and Heritage, *Garfinkel and Ethnomethodology*, pp. 38 ff.

30. Were I able to undertake it, this would be the place for an engagement with Habermas's philosophy of social science; which, however, will have to be a task for another time.

31. The example is adapted from Burdick, "On Davidson and Interpretation."

32. On the differences between Davidson's and Quine's versions of the Principle of Charity, see Burdick, "On Davidson and Interpretation"; Haack, *Evidence and Inquiry*, pp. 61 ff.; and Haack, "'La teoría de la coherencia de la verdad y el conocimiento' de Davidson."

33. See Meehl, *Clinical versus Statistical Prediction.*

34. Merton, *Social Theory and Social Structure*, p. 47.

35. I believe this nice phrase is due to Richard Feynman.

36. Abrahamsen, *The Psychology of Crime*, p. 37; I owe the example to Barzun, *Science: The Glorious Entertainment*, p. 222.

37. Andreski, *Social Sciences as Sorcery*, chapter 10.

38. James, "On a Certain Blindness in Human Beings"; "The Methods and Snares of Psychology."

39. Peirce, *Collected Papers*, 7:605.

40. Incorrect in British English, but according to *Webster's* (9th ed., 1991) a legitimate sense in American English.

41. See also Haack, "Confessions of an Old-Fashioned Prig."

42. Machlup, "Are the Social Sciences Really Inferior?" p. 165.

43. Nagel, "The Value-Oriented Bias of Social Inquiry" (of course, he doesn't use my terminology).

44. Andreski, *Social Sciences as Sorcery*, p. 103.

45. Crick, *What Mad Pursuit*, p. 70.

46. Andreski, *Social Sciences as Sorcery*, p. 232.

47. I owe the example to Victor Fuchs.

48. Heilbroner, "Economics by the Book," pp. 18–19.

49. Merton, *Social Theory and Social Structure*, p. 49.

50. As Fuchs writes in the first section of "What Every Philosopher Should Know About Health Economics," entitled, "If You Don't Know Where You Are Going, Any Road Will Get You There," "Part of the problem is that we have not decided what it is we want our health care system to do" (p. 186).

51. Andreski, *Social Sciences as Sorcery*, p. 100.

52. See chapter 1, pp. 27–28.

53. McGinnis, "The Politics of Cancer Research."

54. The general argument about prediction in the social sciences is made by Popper in the preface to *The Poverty of Historicism.*

7

A MODEST PROPOSAL

The Sensible Program in Sociology of Science

The persistent development of science occurs only in societies of a certain order, subject to a peculiar complex of tacit presuppositions and institutional constraints. What is for us a normal phenomenon which demands no explanation and secures many self-evident cultural values, has been in other times and is in other places abnormal and infrequent. . . . It is, then, important to examine the controls that motivate scientific careers, that select and give prestige to certain scientific disciplines. . . . It will become evident that changes in institutional structure may curtail, modify, or possibly prevent the pursuit of science.

—Robert Merton, "Science and the Social Order"[1]

[T]he pursuit of science by independent self-coordinated objectives assures the most efficient possible organization of scientific progress. . . . [A]ny authority which would undertake the work of the scientist centrally would bring the progress of science virtually to a standstill. . . . Politics and business play havoc with appointments and the granting of subsidies for research; journals are made unreadable by including too much trash.

—Michael Polanyi, The Republic of Science[2]

Scientific evidence is usually the shared resource of a whole sub-community; inquiry in the sciences, cooperative or competitive, involves many people within and across generations; and science is not conducted in a vacuum, but in a larger social setting which may exert a significant influence on what questions get investigated, what research gets funded, what results get a wide audience—and sometimes on what conclusions get reached. As my quotations from Merton and Polanyi suggest, studies of the internal organization of science and its external environment can help us discern what encourages, and what discourages, good, honest, thorough inquiry, free and accessible availability of results, etc.—a potentially very fruitful cooperation of sociology of science with epistemology. This, briefly and roughly, is what I have in mind when I refer to "the sensible program" in sociology of science.

At this point in time, however, Merton and Polanyi may sound old-fashioned, almost quaint. For some recent sociology of science has seemed dubious, denigratory, or downright dismissive of the epistemological pretensions of the scientific enterprise; and some mainstream philosophers of science, not to mention scientists themselves, returning the favor, seem no less dubious, denigratory, or downright dismissive of sociology of science.

Understandably, perhaps; for, as they have taken on the characteristically disillusioned tone of the New Cynicism, even the titles of books and articles are enough to make you a little nervous: *Interests and the Growth of Knowledge*; *Science as Power*; *Laboratory Life: The Social Construction of Scientific Facts*; *The Manufacture of Knowledge: An Essay on the Constructivist and Contextual Nature of Knowledge*; *On the Margins of Science: The Social Construction of Rejected Knowledge*; *Frames of Meaning: The Social Construction of Extraordinary Science*; *Opening Pandora's Box*; *Science: The Very Idea*; *Where the Truth Lies*; *A Social History of Truth*; *Dismantling Truth: Reality in the Post-Modern World* (including an essay entitled "After Truth: Post-Modernism and the Rhetoric of Science"); *The Golem*, with its slyly knowing subtitle, *What Everyone Should Know about Science*; and so forth. And it's hardly reassuring to notice, between the covers of such books, how routinely references to truth, rationality, reality, etc., are guarded about by scare quotes; how routinely philosophical (aka "rationalist") approaches to science are denounced as bankrupt; how routinely epistemic relativism and assorted other cynical themes are taken for granted; how routinely Kuhn, Feyerabend, and sometimes even—oh dear—Rorty, are cited as authorities on how science works.

Of course, there is no single, simple orthodoxy to which all sociologists of science with cynical sympathies subscribe. Far from it; sub-sects in bewildering variety—the Marxists, the Edinburgh School, the Bath School, the ethnomethodologists, the post-modernists, etc.—argue earnestly among themselves. An outsider like myself, bemused by a rash of important-looking initials—SSK, Sociology of Scientific Knowledge; STS, Science and Technology Studies; DA, Discourse Analysis; NLF, New Literary Forms; and TRASP, for Truth, Rationality, Success, and Progressiveness, referring to the characteristics sociological explanations of the acceptance of scientific theories should most scrupulously ignore—feels almost grateful when Wallis refers to "the sociological sceptic" whose goal is "debunking" science,[3] and Woolgar describes himself and his allies simply as "science critics."[4]

Barnes criticizes Mannheim and Habermas—neither goes far enough, both fail to extend their deeply social, contextualized conceptions of social science to the natural sciences.[5] Aronowitz complains that Bloor's "perspectivist refutation of the notion of objective knowledge" is insufficiently radical.[6] Collins and Pinch chide Kuhn for backsliding from his earlier radicalism.[7] Woolgar complains that the Marxists, for all their political radicalism, are feeble epistemologically; that the relativists haven't pushed their relativism far enough; in fact that nearly all the would-be cynics are "still wedded to an objectivist ontology," when they should be taking Rorty's lead and getting on with the serious work of "dismantling representation *per se*."[8] Tomlinson, who complains that Rorty is just a closet realist with a good line in pseudo-radical patter,[9] presumably finds even Woolgar tame. I am reminded of Fox's exasperated observation that the refrain seems to be "more radical than thou";[10] and of Collins and Yearley's description (of others than themselves, naturally) as caught up in a game of "epistemological chicken."[11]

But with radical sociologists of science, as with the Old Deferentialists, there's not only the part where they say it, but also the part where they take it back. Even the boldest moves in that game of chicken are often accompanied by assurances that the goal is not to denigrate science, only to look at it realistically instead of through philosophers' rose-colored spectacles. So it won't do simply to document the diverse extravagances of cynical sociology of science ("CSS"); not only because the relevant literature is so vast and tangled, and the internecine disagreements so many and so tortuous—though it is, and they are, and I know only too well, from the umpteen versions of this that I've thrown in the wastepaper basket already, how easy it is to get hopelessly ensnared—but also because we need to understand how two incompatible conceptions of sociology of science got so inextricably intertwined, and how sensible sociology of science got lost in the shuffle.

NEUTRALISM, CYNICISM, AND THE "PROBLEM OF REFLEXIVITY"

In the nature of the case, *purely* sociological investigation of science couldn't tell us whether there is such a thing as objectively better or worse conducted investigation, whether natural science is or isn't a rational enterprise, or whether this or that particular natural-scientific investigation is well or poorly conducted. And while a historian of science enjoys the benefit of scientific hindsight, a sociologist of science investigating an on-going scientific investigation can't know how things will turn out, and so may be obliged to remain agnostic about the worth of the scientific work he investigates. According to strict neutralism, as I shall call it, the correct approach to sociology of science must be epistemologically non-committal.

Barry Barnes' observations that "studies of how people actually reason rather than how ideally they should reason" are concerned with "natural rationality," and should be, not "prescriptive," but "tolerant,"[12] might be read in the spirit of strict neutralism. But when he and Steven Shapin aver that "[i]f one aspires to *do* history in a properly 'disinterested' way, it is difficult simultaneously to act as an apologist for science,"[13] the neutralism begins to elide into cynicism. And the more anxious radical sociologists of science seem to describe their work as "naturalistic," "disinterested," or "impartial"—especially when they hint that this is the only alternative to an uncritical scientism, a morally objectionable ethnocentrism—the more unmistakable the elision; for the *naïf* who regards sociological investigation of science as blasphemous, or who assumes that science is a rational enterprise because it is a product of our culture, is surely a straw man.

Strict neutralism is covertly transmuted into cynicism, ostensible neutrality about the *bona fides* of science into epistemic relativism, irrealism, or outright skepticism, "naturalistic, disinterested, impartial" sociology of science into CSS, and "don't assume that natural-scientific inquiry is legitimate" into "assume that natural-scientific inquiry is not legitimate." The mechanism is a kind of conceptual alchemy in which epistemological concepts like knowledge, evidence, warrant, justification, and the like are first stripped of evaluative content and reduced to descriptive sociological concepts, and then a descriptive, anthropological form of relativism is run together with a conceptual, philosophical form.

Because from a purely sociological perspective there is no way to distinguish knowledge, the genuine article, from what is only taken for knowledge, sociologists often write, not of knowledge, evidence, or rationality, but of "knowledge," "evidence," "rationality"; meaning, whatever is *taken to be* known fact, relevant evidence, rational procedure. Shapin is helpfully explicit: in a footnote to a paper presented as launching a radically new social epistemology of science, he writes that he uses "epistemology" in an entirely non-evaluative, purely sociological, sense—not theory of knowledge, but theory of "knowledge."[14] But what is taken to be known fact, etc., is quite often no such thing; so the quotation marks soon take on a cynical tone, no longer just warning but scoffing, no longer just scare quotes but sneer quotes: "so-called 'knowledge,' so-called 'fact'—yeah, right!" And before long it comes to be taken for granted that these content-stripped, covertly cynical senses of "knowledge," etc., *are the only legitimate senses.* Of course, these aren't really legitimate senses of those terms at all; for the concepts of what is taken to be knowledge, relevant evidence, etc., *presuppose* the concepts of knowledge, relevant evidence, etc. Still, the quotation marks make it all too easy to fall for the Passes-for Fallacy,[15] creating the illusion that science offers only so-called "knowledge," so-called "evidence," never the real thing.

The anthropological form of epistemological relativism is a factual claim: that different communities or cultures have different epistemic standards. The philosophical form is a conceptual claim: that it makes sense to talk of stronger or weaker evidence, better- or worse-warranted claims, and so on, only relative to some community or culture.[16] Philosophical relativism doesn't follow from anthropological relativism; and while anthropological relativism falls within the scope of sociology, broadly construed, philosophical relativism does not. But if you confuse the two forms of relativism, and run together knowledge with "knowledge," evidence with "evidence," and so on, it is easy to create the false impression that it could be shown by sociological investigation that knowledge, evidence, rationality, etc., make sense only relative to social context.

Unambiguously neutral sociology of science would be epistemologically inert; it would have nothing, favorable or unfavorable, to say about the epistemology of science. Unambiguously cynical sociology of science, on the other hand, undermines its own aspirations to make objectively warranted claims about the natural sciences; for the philosophical relativism of epistemic standards into which CSS is drawn denies that it makes sense to speak of objectively better and worse evidence, objectively more and less warranted claims. With a sufficiently generous regime of scare quotes it is possible to sustain a strategic ambiguity; but the bulge under the carpet can't be smoothed away.

In the sociology-of-science business the term "reflexivity" refers, not, as in Giddens,[17] to the effect of social-scientific predictions on the social phenomena they are about, but to the thesis that the tenets of sociology of science must apply to sociology of science itself; and rather than a frank acknowledgment that CSS is self-undermining we encounter a whole vast literature devoted to "The

Problem of Reflexivity"[18]—a vast literature in which each new effort to avoid self-undermining really is more radical than the one before. Sociology of science must be reflexive, insist Barnes and Bloor; sociology of science can't be reflexive, reply Collins and Pinch; sociology of science should get out of the business of making claims and turn itself into a form of literature, or a game of textual disruption, suggest Woolgar and Mulkay. But the thing to remember, as we look more closely, is that *neutralism is epistemologically inert, while cynicism is self-undermining.*

THE STRONG PROGRAMME

The "Strong Programme" in sociology of science, at which my title takes a little unkind dig, is associated with the Edinburgh School of Barry Barnes and David Bloor. However, as the commentator realizes who writes, apparently without irony, of "Barnes' modest version of the strong program,"[19] Barnes and Bloor are not entirely of one mind. In fact, *neither* Barnes *nor* Bloor is exactly "of one mind"; Bloor, in particular, creates an almost inextricable tangle of ambiguities—which I shall do my best to disentangle.

Early in *Knowledge and Social Imagery*,[20] Bloor announces that he will use the term "knowledge" equivalently to "accepted belief." Likening his approach to science to Émile Durkheim's approach to religion, he suggests that sociologists have been too ready to treat science as sacred, instead of studying it fearlessly, as Durkheim studied religion, *qua* social phenomenon. The Strong Programme, he tells us, "rests on a form of relativism," a form of relativism allegedly accepted by (of all people!) Popper.[21] True to CSS form, Bloor runs together "knowledge" with knowledge, anthropological with philosophical relativism, and philosophical relativism with fallibilism. Compounding the confusions, he holds that truth (or perhaps "truth") and hence presumably also falsehood (or perhaps "falsehood") is context-relative. "All knowledge is conjectural and theoretical," he writes, "nothing is absolute and final." "*Therefore*"—his word—"all knowledge is relative to the local situation of the thinkers that produce it."[22] Sociological analysis, he claims, reveals the bankruptcy of philosophical or "rationalist" understandings of science by showing that it is scientists' interests that determine what scientific theories are accepted and what rejected. Bloor accepts reflexivity: what the Strong Programme says about science applies to sociology of science itself. Nevertheless, he is confident that the Strong Programme is not self-undermining; the cause of a belief, he argues, is irrelevant to its rationality or irrationality.

It is tempting to argue right off the bat that Bloor's position must be *either* epistemologically irrelevant, *or* self-undermining. If it is true that the cause of a person's belief has no bearing on the rationality or irrationality of his holding it, the thesis that it is interests that cause people to believe things is not self-undermining, but also is of no epistemological significance; and the Strong Programme, far from "reveal[ing] the threadbare fabric" of philosophical understandings of science, as Bloor claims,[23] is epistemologically inert. If, on the other hand, it is false that the cause of a person's belief has no bearing on the rationality or irrationality of his holding it, then the thesis that it is interests that cause people to believe things is epistemologically significant, but also self-undermining; if it is their class interests that cause sociologists of science to believe that it is class interests that cause people to believe things, they are not rational in believing this.

However, this response is too quick; for Bloor's claim is false if "rational" is interpreted as "justified," but true if "rational" is interpreted as "warranted." If a person or group possesses good evidence with respect to a claim, the claim is warranted for them; but if what causes them to accept it is something other than their evidence, such as their class interests, they are not justified in accepting it. So a bloory-eyed optimist might hope that the Strong Programme can have it both ways: it undermines "rationalist" philosophy of science by showing that scientists aren't justified in believing their theories; but it doesn't undermine itself—for, though its proponents are unjustified in believing it, it may nevertheless be warranted for them. However, unless you are leaning over backwards to save the Strong Programme, the glass will look not half-full but half-empty: the Strong Programme leaves everything "rationalist" philosophy of science might have to say about warrant untouched, and at the same time undermines its own claim to be justified.

Moreover, it is unclear whether the claim of the Strong Programme is that scientists' interests sometimes determine theory-choice, that they often do, or that they always do; whether it is that interests are causally determinative, or only influence theory-choice to some extent; and whether it is that only class interests, or interests more generally construed, play a causal role. There are a dozen possible interpretations: it is always/often/sometimes class interests/interests of one sort or another that determine/influence theory-choice.

The weakest of these, that interests of one sort or another sometimes influence theory-choice, is of course true. Sometimes a scientific idea is so deeply entrenched that to question it is to risk your reputation; and this may induce scientists to perceive evidence against it as less powerful than it really is, and thus to give competing ideas less credence than is warranted (think of Avery and the tetranucleotide hypothesis). Professional interests, in other words, sometimes influence what claims scientists accept; for, like all inquirers, scientists are susceptible to wishful thinking. Indeed, it was the possibility that warrant and acceptance should come apart in this way that motivated my distinction of warrant and justification.[24] Now my bloory-eyed friend may try to argue that, if we stick to this weakest interpretation, Bloor's thesis is true and, moreover, poses no real threat to Bloor's own warrant for believing it. But in this weakest interpretation, so far from being novel or exciting, Bloor's thesis is completely uncontroversial; and, if this were all there were to the "Strong Programme," the professions of relativism and allegations about the failure of objectivity, the bankruptcy of rationalist views of science, etc., would be at best rhetorical window-dressing disguising a familiar and unexciting point as new and radical.

Detailed investigation of, say, the effects of drug-company sponsorship of university-based medical research, or of pressures on social scientists to arrive at politically desired conclusions, is undeniably of considerable epistemological as well as sociological interest. However, detailed empirical evidence of the role of interests—in *any* of the interpretations I have distinguished—is conspicuously lacking from *Knowledge and Social Imagery*. In a 1981 article entitled "The Strengths of the Strong Programme" Bloor refers to the papers in Barnes and Shapin's anthology *Natural Order* as having linked various scientific debates—over statistical measures of nominal association;[25] biometry and Mendelism;[26] the "J phenomenon";[27] charm versus color in particle physics;[28] botanical classification;[29] race and intelligence[30]—to the protagonists' interests; but what you find in these papers is ambiguous references to "social or cognitive" interests, and allusions to "connections," "links," or "homologies" the exact nature of which remains unspecified. You begin to suspect that Paul Roth may be right when he complains that SSK case-studies "offer no more than narratives of the post hoc, ergo propter hoc variety."[31]

In any event—even if they correctly report the scientific episodes concerned, a matter that could only be settled on a case-by-case basis—even in principle such case studies couldn't tell us whether the sciences manage to get acceptance and warrant well enough correlated *over the long haul*. Rationality, as Lakatos reminded us, needn't be a step-by-step business. No SSK case-studies could rule out, for example, the possibility that the internal organization of the sciences harnesses scientists' interests to warrant as, according to Adam Smith, a free market harnesses buyers' and sellers' interests to economic success.

Perhaps this is why, in the same article in which he refers readers to Barnes' and Shapin's anthology, Bloor soon shifts back to the high a priori ground, claiming that Hesse's network model of classification, "properly understood, . . . spells out in detail exactly how and why there is a social component to every single classificatory predicate in our language": the conventional character of language "makes the profound involvement of society a pervasive . . . feature of knowledge," because "social interests are coherence conditions imposed on the classificatory network."[32] But it is a mystery how the fact that language is conventional, or that social analogies sometimes play a role in the development of scientific concepts, could show that class interests determine theory-choice; and it is just false that social interests could play the evidential role that Bloor's describing them as "coherence conditions" obscurely suggests.

Perhaps that's why, by the time of the new, 1991 edition of *Knowledge and Social Imagery*, Bloor appeals instead to Shapin's work on the reception of phrenology in Edinburgh in the 1820s. Collins describes this as "the most complete attempt to link . . . political structures to the technical details of a scientific debate";[33] Pickering praises it as a "classic study . . . exemplary work in SSK" providing "yet more documentation of the SSK thesis of social relativity";[34] and Woolgar cites it as among the best of the *genre*.[35] So you might hope here, finally, to find evidence of at least one case where the protagonists' class interests influenced theory-choice. It's encouraging that Shapin realizes that "the mere assertion that scientific knowledge 'has to do' with the social order or . . . is 'not autonomous' is no longer interesting. We must now specify how, precisely, to treat scientific culture as a social product."[36] But it's correspondingly disappointing when he acknowledges that any link between the content of phrenological theory and the interests of the protagonists is "speculation," and that the supposed connection "is exceedingly difficult to discern and to make historically plausible."[37] It is also, as Shapin realizes, quite puzzling. For phrenology, according to which character is determined by brain/skull structure, is an innatist theory; and yet its Edinburgh proponents were mainly of the mercantile middle class favoring egalitarian social reform—the opposite of what you would expect. (At Ellis Island phrenological testing was used to weed out supposedly undesirable would-be immigrants.)[38]

In fact, Bloor himself concedes that any correlation between class and attitudes to phrenology was "chance," observing that both sides were "putting nature to social use, making it underpin their vision of society."[39] And so he misses the opportunity to distinguish the claim that class interests cause a scientific theory to be accepted from the claim that a scientific theory is used in justification of a political agenda; to be more specific about what "class interests" are (could mine, for instance, differ from Bloor's or Shapin's?); to distinguish class from professional interests,[40] and both from cognitive or evidential concerns; and to ask when the desire to promote a political agenda is an incentive to seek out evidence, and when, on the other hand, to ignore or suppress it.

By 1996, in Bloor's *Scientific Knowledge: A Sociological Analysis* (written jointly with Barnes and Henry), though the index entry for "interests, importance in scientific change," lists ten pages where the topic is mentioned, all reference to the Strong Programme has been dropped. There is even one line about "goals and interests of an illegitimate, credibility-reducing kind."[41] But the crucial distinction is only hinted at, never developed; so any hopes that the old firm might now be under new management are soon dashed.

THE RADICAL PROGRAMME

The rival outfit of Collins & Co., based in Bath, is no less committed to a relativism of truth, evidence, etc., to background theory or social context. But Collins is less interested in interests than in scientific failures, false starts, and fumbling, from spoon-bending through worm-running to cold fusion; instead of insisting on reflexivity, he repudiates it; and instead of taking the reality of the objects of natural-scientific theorizing for granted, he flirts with the idea that the "reality" they "investigate" is really scientists' own construction. Collins describes his "Radical Programme" as developing the defensible elements of the Strong Programme, but eschewing reflexivity; Bloor, however, sees Collins' constructivism as making the Radical Programme not less but more radical than the Strong Programme.

Concluding their book on what they thought would be the Kuhnian revolution set off by Uri Geller's demonstrations of supposedly psycho-kinetic spoon-bending, Harry Collins and Trevor Pinch write—they too have a nasty case of scare-quotes-itis—that "a major problem, perhaps the major problem, for the sociology of knowledge is the question of the nature of 'rationality'. . . . If evidence of radical cultural discontinuities can be found in contemporary science . . . , then the scope of the sociology of knowledge is very wide indeed." They don't know, they say, "whether paranormal metal-bending is 'real' or not—nor, as sociologists, do they care";[42] as they write elsewhere, they endorse "a relativistic thesis within which consideration of the 'actual existence' of a phenomenon is redundant."[43] This sounds like strict neutralism. However, with Collins' insistence that "the natural world has a small or non-existent role in the construction of scientific knowledge,"[44] and his and Steven Yearley's observations that "explanations should be developed [on] the assumption that the real world does not affect what the scientist believes about it," and that "all cultural enterprises . . . [have] roughly the same epistemological warrant,"[45] the cynicism behind the neutralist facade seems unmistakable.

And yet much of what Collins and Pinch have to say—that experimental results are not transparently self-presenting, but need interpretation; that evidence is always incomplete and often ambiguous, and how you judge its relevance depends on your background beliefs; that scientific inquiry is nearly always untidy and often enough unsuccessful; that it needs investigating how, in the face of less-than-decisive evidence, a scientific sub-community comes to take a dispute as settled one way or another, or to lose interest in it; that a sociologist studying an ongoing scientific dispute can't know how, in the longer term, the issues will be settled—sounds harmless, and true. All they are saying, Collins and Pinch reassure us in *The Golem*, is that science isn't sacred.

But this isn't, or at least wasn't, all they are saying. It's baffling—until you realize they take for granted that, unless natural science operates by way of strictly deductive arguments,[46] the conclusion is inescapable that knowledge, evidence, rationality, reality, etc., are relative; and that, like Kuhn, they mistake the perspectival character of judgments of relevance for a relativity of relevance itself to background theory. As a result, they go so deeply into debt to the New Cynicism as to threaten the epistemological solvency of their own enterprise.

However, Collins is confident that he can avoid self-undermining by exempting the social sciences from the relativism that afflicts natural-scientific inquiry. "The prescription 'treat sociological knowledge as being like scientific knowledge'," he writes, is "arbitrary, unnecessary and undesirable."[47] It is not the task of sociology of science to decide "whether patterns of explanation applied by sociologists *to science* are equally applicable *to sociology*."[48] Well, yes, strictly speaking the question is a *meta*-sociological one; but surely Collins doesn't really think the firm can avoid bankruptcy by turning the books over to the auditors?

Evidently not; for he goes on to advise that, while the natural world "should be held in doubt" as "problematic," the social world should be treated as "real and as something about which we can have sound data,"[49] "a relatively well-behaved external entity which yields replicable observations."[50] Can he really mean that social-scientific evidence is less incomplete, less ambiguous, than natural-scientific evidence, social-scientific inquiry less messy, less susceptible to false starts and wrong turns, and social reality more real? To say this isn't true would be putting it mildly; it is the opposite of the truth. Surely Collins & Co. can't really think the auditors won't notice if they switch red ink and black?

Evidently not; for Collins goes on to reassure us that he means only that sociological study of this or that natural-scientific episode should remain neutral about the warrant of the theories and the reality of the phenomena in question, and that sociological explanations should not appeal to the truth, rationality, success, or progressiveness (remember "TRASP"?) of the scientific work studied. But on the very next page after advising that the natural world be "held in doubt as problematic," he counsels treating it as "a social construct rather than something real." Suspending judgment about the *bona fides* of the natural-scientific work under study would mean renouncing the aspiration to say anything about the epistemological *bona fides* of the natural sciences; assuming that natural scientists bring the natural world into being makes natural science a kind of myth-making, and thus subordinate to the social sciences. But hard as he tries to fudge the difference between neutralism and cynicism, Collins can't have it both ways, either.

ETHNOMETHODOLOGY OF SCIENCE

Bruno Latour and Steve Woolgar present themselves as ethnomethodologists of science in the quasi-phenomenological tradition of Harold Garfinkel. Unless the peculiar syntax makes you too woozy, it is easy to discern the radicalism implicit in this line of Garfinkel's: "We're going for a reversal of the conventional distinction between the real pulse and the apparent pulse. . . . Our notion of the pulsar locates it at hand as *that which is real in and as of inquiry's hands-on occasions*."[51] "Correcting" the scientists' belief that they discovered the pulsar, Garfinkel explains that, really, they brought the pulsar into being.

Latour and Woolgar follow suit. "Whereas we now have fairly detailed knowledge of the myths and circumcision rituals of exotic tribes, we remain relatively ignorant of the details of equivalent activity among tribes of scientists." Their "laboratory study" consisted of two years spent in Roger Guillemin's laboratory in La Jolla, California, observing the scientists as they isolated minute quantities of thyrotropin releasing factor (TRF) from hundreds of tons of sheep's brains, identified it chemically, and synthesized it.[52] As anthropologists in the laboratory, Latour and Woolgar tell us, they "maintain[ed] an agnostic position" regarding their subjects' beliefs.[53] But, as that reference to "myths" signals, the ostensible agnosticism serves, as usual, to mask a much more radical position—in this case, a strong social constructivism which amounts to nothing less than an outright linguistic/conceptual idealism.

There are three interpretations in which "the" social constructivist thesis is true:

1. Social institutions, etc., are socially constructed insofar as they are constituted by people's beliefs and intentions.
2. Laboratory artifacts are brought into being by means of scientists' physical manipulations.

3. Scientific concepts are brought into being by the intellectual work of scientists.

But explicit in Latour and Woolgar (and implicit in Collins and Pinch) is a much stronger thesis:

4. Not only scientific concepts, but the objects to which they apply—not only the concept of gene or electron, but actual genes and electrons—are brought into being by scientists' intellectual activities.[54]

The fourth and strongest of these forms of social constructivism obviously doesn't follow from any, or all, of the first three; and it isn't true. I shan't repeat my earlier arguments against the idea of the world as just polenta, sitting there homogeneously waiting for us to cut it into lozenges and rounds or mold it into gnocchi, but will try instead to get to the bottom of Latour and Woolgar's confusions.

Apparently the scientists they studied weren't thrilled to overhear themselves described as "a tribe of writers and readers who spend two-thirds of their time working with large inscription devices," and that neuroendocrinology is a "mythology," even if "not necessarily entirely false"; nor much mollified by the concession that they are so skilled that sometimes they "convince others that . . . what they say is true." They liked to imagine that their papers were not ends in themselves, but means of communicating the facts they discovered. An observer to whom the supposed "true meaning" of those papers was opaque, Latour and Woolgar comment, might be "reminded momentarily of an earlier study of religious rituals when, having penetrated to the core of ceremonial behavior, he had found only twaddling and waffling." The benighted natives in their laboratory, if not exactly waffling and twaddling, *are* systematically mistaken about what it is they do: they think they discover facts about independently existing things, when really they construct them—the facts, and the things.[55]

A characteristic passage reconstrues a famous episode "all too frequently transformed into [a story] about minds having ideas."[56] Commenting on Watson's report of the conversation with Donohue that convinced him that his model of DNA with like-with-like base pairs wouldn't do, Latour and Woolgar stress that Watson and Donohue shared an office, that Watson was prompted to go off and cut out new cardboard models of bases to shuffle around, and that he was impressed by Donohue's credentials. Well, yes; but this neglects the crucial fact: Donohue knew more about tautomeric bonds than Watson did—enough to know, and to convince Watson, that the like-with-like model was hopeless. And what Latour and Woolgar don't say, but are clearly committed to saying, is that had Watson been less impressed by Donohue's credentials, *the structure of DNA would have been different.*

Why on earth would anyone think that? To construe facts otherwise than as socially constructed, Latour and Woolgar observe, would hinder the implementation of the Strong Programme;[57] meaning, apparently, that it would restrict the role of sociology of science more than they would like. But this is no argument at all. They tell something of how TRF was isolated, and its chemical structure identified: at one time thought to be perhaps an artifact of the instrumentation, and then not, at one time thought to be a peptide, then not, then again, yes, a peptide. They aren't bad on the messiness of scientific work,[58] and on the very specific dependence of instruments on earlier theory. But none of this brings the conclusion that facts are constructed by scientists' intellectual activities, let alone that TRF is, any closer.

Essential to that conclusion is a kind of verbal shell game. First move: identify facts with statements categorically accepted as true: "A fact is a statement with no modality . . . and no trace of authorship"; "a fact only becomes such when it loses all temporal qualifications." Second move: identify the stuff, TRF, with its recorded traces: "The object initially comprised the *superimposition* of two peaks. . . . [It] was constructed out of the difference between peaks on two curves . . . and the initial label TRF begins to stick"; "an object can be said to exist solely in terms of the differences between two inscriptions."[59] As John Fox comments in an illuminating article on which I have drawn substantially, since Latour and Woolgar also describe TRF as "an unremarkable white powder,"[60] it follows that the perceived difference between two curves is a white powder: which is, as Fox dryly notes, "observably false."[61] Last move: treat the stuff, TRF, as a fact: "We have chosen to study the historical genesis of what is now a particularly solid fact . . . TRF."[62]

"That's a fact" and "that's true" perform closely similar roles; what a true statement states is a fact; and you can't state a fact except by using some words or other. But it doesn't follow that facts are linguistic entities, let alone, as Latour and Woolgar conclude, that facts are nothing but *statements accepted as true*. But, socializing facts in this variant on the by-now familiar scare-quotes maneuver, Latour and Woolgar compound the confusion by identifying scientific concepts and the objects to which they apply—the concept of TRF, or accounts of TRF, with TRF itself. Needless to say, the concept of TRF (which is even less like a white powder than the superimposition of two peaks on a graph is) is an utterly different kind of thing from TRF. An understandably exasperated Barnes, apparently embarrassed by Latour and Woolgar's associating themselves with the Strong Programme, protests that there is all the difference in the world between a cream cake and an account of a cream cake;[63] but Woolgar keeps his cool: "the view that accounts are *constitutive* of reality . . . provides a powerful analytic handle."[64]

The negative argument on which Latour and Woolgar rely is that it's impossible for a realist to explain the "out-there-ness" of supposedly independently existing objects without using words purportedly referring to those objects. True, you can't describe the world without describing it; but it obviously doesn't follow from this tautology that the world is identical with, or created by, our descriptions of it. The positive argument on which Latour and Woolgar rely is that what is a real object, and what an artifact of the instrumentation, emerges only through the process in which scientists make facts. True, those neuro-endocrinologists had to find out whether they were dealing with the effects of real stuff, or of experimental variation and interference; but it obviously doesn't follow that TRF was brought into being by distinguishing it from an artifact of the instruments. In fact this argument couldn't even be stated without presupposing what its conclusion denies: that there is a substantive difference between real stuff and instrumental artifact beyond the sociological distinction between acceptance-as-real and acceptance-as-artifactual.

Latour and Woolgar present themselves as trying, like the scientists they study, to make facts[65]—although, interestingly enough, they don't say that they brought the laboratory, the large inscription devices, or the tons of sheep's brains, into being, as the natives supposedly brought TRF into being. But it's not so easy to repudiate the distinction between representation and thing represented while yourself purporting to represent goings-on in the laboratory. So by the time of *Science: The Very Idea* Woolgar is writing excitedly of "disrupting the ideology of representation";[66] and by the time the final chapter of this book is reprinted in *Dismantling Truth*, he has replaced its last few opaque but purportedly truth-stating pages with a jokey imaginary dialogue in which he interrupts himself in italics (*"I don't much like your tone"*).[67]

It's tempting to respond in kind:

> If Woolgar were to write in verse
> His arguments would be no worse.
> But neither, surely, sad to say,
> Would they thereby become okay.

Still, despite Woolgar's disinclination for "stories about people having ideas," it's clear that what he has in mind is that the role of the agents whose linguistic activities he believes bring TRF and such into existence is disguised in regular prose, but can be brought to the reader's attention by a dialogue form which disrupts the ideology—the false consciousness—of representation.

At this point the ethnomethodologists of science join forces with the NLF party. Earlier, G. N. Gilbert and Michael Mulkay had experimented with "Discourse Analysis," multi-vocal presentations allowing the various protagonists in a scientific controversy to speak.[68] Then, believing that this failed adequately to acknowledge the socially constructed nature of sociological discourse itself, Mulkay began to explore New Literary Forms as a way of avoiding the "implicit commitment to an orthodox epistemology . . . built into the established textual forms of social science,"[69] and to construe sociology of science as "par-ody."

Mulkay's imagined conversations between the parties to natural-scientific disputes, and his imagined Nobel Prize acceptance speech in which all the usually covert jealousies and rivalries are explicit,[70] can be quite entertaining; but (as Pinch and Pinch convey in his parody of NLF parodies),[71] they don't solve the problem. For Mulkay presumably intends that imaginary Nobel acceptance speech to *convey* some of the truths that a straightforward sociological article about Nobel acceptance speeches might *state*; e.g., that such speeches typically suppress the less savory aspects of the real history of a discovery. Similarly, if he is really repudiating the possibility of reference to things and events in an extra-linguistic world, Woolgar must also surrender the aspiration to say, or convey, anything, true or false, warranted or unwarranted, about science.

"You only allowed me into the text . . . when you started to run out of ideas," that italic voice complains. Quite so. But Woolgar isn't fazed: "getting nowhere should be seen as an accomplishment, not a failure"[72]—honestly, I am not making this up! But it's high time I stopped woolgarthering, and moved on.

THE SENSIBLE PROGRAM

Some Old Deferentialists, holding that social factors can be only obstacles to the smooth operation of the logical method of science, would confine sociology of science to an artificially restricted role. New Cynics who hold that scientific "inquiry" is nothing but a form of social negotiation in which scientists bargain their theoretical loyalties for prestige, or nothing but a form of industrial production in which scientists make articles in journals instead of cars or Tupperware or widgets, would give sociology of science an exaggerated importance. The Critical Common-Sensist believes, as usual, that the truth lies somewhere in between. Though they look diametrically opposed, the Old Deferentialists' and the New Cynics' attitudes to sociology of science have something important in common: both neglect the way evidential and social aspects of scientific inquiry interlock; both, overtly or covertly, *contrast* the rational and the social. But this, the Critical Common-Sensist insists, is a false contrast.

Science is a social institution: specifically, like investigative journalism, detective work, theology, and so forth (and unlike banking, prisons, fashion, the automobile industry, trade unions, schools, kinship, churches, the exchange of gifts, and so on), a social institution engaged in inquiry. On a day-to-day basis, scientists may be preoccupied with getting new, improvised, or recalcitrant equipment to work, or with laying hands on a sample of some essential stuff, or with getting hold of this or that information someone else may have; not to mention applying for grants, dealing with editors, or gossiping about what rivals in the profession are up to. Scientific work is rarely even remotely like doing a logic exercise or composing a symphony. But the messy day-to-day stuff ultimately relates, albeit often very obliquely or inefficiently, to the effort to find out how some part or aspect of the world is.

Scientific inquiry demands imagination, time, energy, and integrity; it is susceptible to false starts and wrong turnings, and frustrating when, as often enough happens, things go wrong. Scientists are fallible: they make mistakes; they are misled by ambiguous evidence; they cling too obstinately to ideas that have worked in the past, or in which they have a large professional investment; they rush to embrace glamorous-sounding new conjectures; they invest great ingenuity in devising concepts that turn out not to apply to anything. Sometimes they succumb to pressure to arrive at politically desired conclusions, or to the temptation to fudge or fake for the sake of a renewed grant or a prestigious job. How well scientific inquiry is conducted can be significantly affected by circumstances: e.g., whether others' work on your problem is freely accessible; whether you can get funds for necessary equipment; whether the outfit that pays for the work puts you under pressure to reach results that would serve its financial or political interests; whether you can be sure that rival investigators trying to find the answer to the same question will scrutinize your work; and so on.

So sociology of science is epistemologically relevant; but its relevance is contributory, not exhaustive. Epistemological conclusions can't be established by purely sociological investigation; but *in conjunction with* an understanding of the epistemology of scientific evidence, warrant, and inquiry, sociological investigation of how science works—of the peer-review system, of the politics of science funding, of science education, of litigation-driven science, of the incidence of fraud or plagiarism, and so on—has the potential to illuminate what social factors help, and what hinder, the scientific enterprise. Perhaps it is not necessary for me to point out how the thesis of contributory relevance interlocks with the aspiration to a fruitful cooperation between epistemology and sociology of science; but perhaps it is necessary for me to say explicitly that sociology of science will be better conducted (other things being equal) if the epistemological assumptions that inform it are—yes, *true*.

The sensible program in sociology of science is neither epistemologically inert nor self-undermining. Epistemologically informed, it appreciates the importance of those aspects of the internal organization and social context of the scientific enterprise most central to its proper business, inquiry. It freely acknowledges that its own investigations are subject to the same core epistemological values and constraints as the science it investigates, and indeed all empirical inquiry. And, far from treating science as sacred, it will be as much concerned with when and how science goes badly as with when and how it goes well; it will not neglect the threats that sponsors' interests, pressure to publish prematurely, fear of political repercussions, etc. might pose to the integrity of research. But neither, of course, will it treat science as a confidence trick.

The essential idea of the sensible program, that social and evidential aspects of science interlock, is not in the least new. We find it in the work of epistemologically sophisticated sociologists and sociologically sophisticated scientists, from Merton's and Polanyi's reflections on the organization and environment of science, to the illuminating diagram of Henry Bauer's reproduced as Figure 2,[73] Morton Hunt's investigations of political pressures on research in the behavioral and social sciences themselves,[74] Marcia Angell's commentary on the role of drug companies in financing medical research,[75] etc. Indeed, for all I know, the majority of hard-working journeymen sociologists of science take the sensible program for granted, and CSS is the style only of a rather noisy minority—perhaps, of those most disillusioned by the failures of the brand of Old Deferentialism most inhospitable to sociology of science, the Popperian. The theme of the interlocking of the social and the evidential suggests, not so incidentally, both what is right and what is wrong about the Lakatosian conception, against which some cynical sociologists of science are reacting, of history of science as rational reconstruction: epistemologically illuminating sociology and history of science *will* focus selectively on social mechanisms that help or hinder creativity and respect for evidence, but *won't* artificially impose conformity to a narrowly logical model of rationality.

Consider, for example, how a sensible sociologist of science would approach the topic of scientific fraud. He would recognize fraud as a kind of scientific misconduct; but also recognize it as quite different from, say, sexually harassing your research assistants, cheating the body that funds your work, or failing to get proper consent from subjects—as specifically intellectual dishonesty. As a rough first stab, he might characterize fraud as making knowingly false claims about what results you have achieved; but then he would have to add that while a plagiarist passes off work actually done by others as his own, a fraud

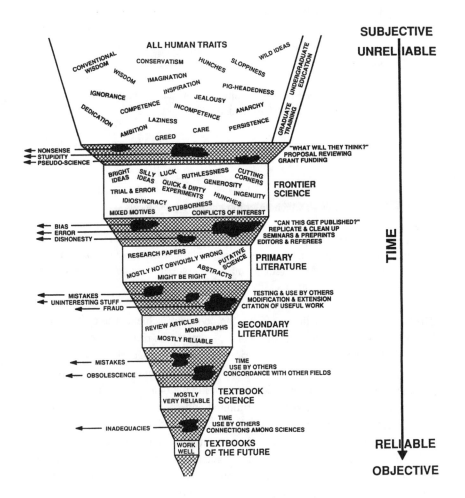

Figure 2. The Knowledge Filter.
(Reprinted, by permission, from Bauer, *Scientific Literacy and the Myths of Scientific Method*, p. 45)

claims to have achieved results neither he nor anyone has obtained.[76] Insofar as it disassociates prestige and achievement, plagiarism weakens the incentive structure of science; but fraud debases the very currency of scientific work, weakening the web of justified mutual confidence on which successful evidence-sharing depends.

In the *Reflections on the Decline of Science in England* that he published in 1830, Charles Babbage distinguished four categories of scientific fraud:

Hoaxing: Fraud practiced with the intention that in due course it be discovered, "to the ridicule of those who have credited it."

Forging: Fraud perpetrated, not to be discovered, but to advance the scientific reputation of the forger.

Trimming: "A form of equitable adjustment," trimming off little bits here and there from observations which differ too much from the mean.

Cooking: There are many recipes, Babbage observes, among them making multitudes of observations and then selecting only those which agree or very nearly agree, and calculating observations which can't otherwise be made to agree by means of two or more different formulae.

(The likely result, he continues, is that the cook will achieve a temporary, undeserved reputation for unrivalled accuracy—"at the expense of his permanent fame.")[77] Of course, Babbage's categories are a little heterogeneous. The difference between forging and hoaxing concerns the perpetrator's intent: the forger hopes to get away with his falsification indefinitely, while the hoaxer intends to reveal it in due course. The difference between trimming and cooking concerns the nature of the falsification: as Babbage remarks, the trimmer's activities, unlike the cook's, at least keep the average of the observations, trimmed or untrimmed, the same.

Fraud involves intent; so you can't get much grip on the concept if, like Woolgar, you are allergic to the idea of minds having ideas. Fraud involves misrepresentation or fabrication of evidence; so you can't get much grip on the concept if you eschew truth, and acknowledge only "so-called 'truth'." In *Where the Truth Lies* (my last example from the looking-glass world of CSS, I promise), Jan Sapp tells the story of Franz Moewus, much of whose work on sexuality in algae, at one time seemingly at the cutting edge of biochemical genetics, is now known to have been fraudulent. This story makes no sense unless there is a difference between truth and fabrication; but in Sapp's book it is sandwiched between assurances that there is no such difference:

> Once we understand that all knowledge about nature is "fabricated" (i.e., constructed out of theory, techniques, and social interaction), and that all scientific reports are based on conscious and unconscious biases, then one cannot judge fraud in terms of timeless truth . . .

and the claim that the history of the Moewus episode

> . . . is about the generation and degeneration of a scientific controversy; about the strengths and weaknesses, strategies, and tactics of the contestants . . . the politics of scientific truth.[78]

In its ordinary use, "fabricated" implies "faked," "deceptive"; but Sapp slides from "scientific knowledge is arrived at by a complex process of devising and constructing models of the world" to "scientific 'knowledge' is all a sham" (and so risks undermining his own pretensions to give a true history).

But even if you aren't handicapped by cynical misconceptions, it can be extraordinarily difficult to distinguish outright fraud from self-deception, over-optimistic interpretation of iffy results, or legitimate interpretation of ambiguous results. In 1936 the statistician Ronald Fisher concluded that the results Mendel reported were just too good to be true—perhaps, he suggested, Mendel had been deceived by an assistant "who knew too well what was expected." In 1978, examining the notebooks in which Robert Andrews Millikan recorded his oil-drop experiments, Gerald Holton noticed that Millikan had selected only his best data for publication ("publish this surely, beautiful!"), and dropped the less favorable ("very low, something wrong").[79] Were Mendel and Millikan frauds, or self-deceived—or just following their keen scientific noses, using their good scientific judgment?

Sometimes it will take intense historical detective-work to determine whether there was fabrication or misrepresentation, and/or whether, if there was, it was intentional. Take the case of Cyril Burt, an educational psychologist and strong hereditarian whose work on intelligence influenced the selective system of secondary education in Britain after World War II (the system under which I grew up). After his retirement, when that education system, and especially the 11-plus examination sorting "academic" children from the rest, came under fire, Burt published a series of articles presenting striking new evidence, from studies of identical twins raised separately, of the heritability of intelligence. In 1972, the year after Burt's death, Princeton psychologist Leon Kamin noticed that, even though new pairs of twins were included in the later studies, the correlations reported in Burt's three surveys, in 1955, 1958, and 1966, were identical to three decimal places—too good to be true. Kamin was a strong environmentalist, profoundly opposed to Burt's political stance; but in 1974 Arthur Jensen, himself a conservative hereditarian like Burt, also concluded that the invariant correlations "unduly strain the laws of chance and can only mean error."[80]

The hints of possible manipulation or fabrication became a full-blown scandal when journalist Oliver Gillie published an article in the *Sunday Times* suggesting that Burt had not only cooked his figures, but even invented fictional research assistants.[81] That Burt was a fraud came to be taken as established fact after his biographer, Leslie Hearnshaw—though suggesting that the dishonesty came only late in Burt's life, and might be explained by his "mental instability"—endorsed it;[82] and by the time Broad and Wade's *Betrayers of the Truth* was published in 1982, Burt could be cited as an Awful Example of scientific dishonesty.[83] Only a few years later, however, Robert Joynson argued that Burt was guilty of nothing worse than the occasional carelessness. Now, after reading Joynson's book,[84] I can honestly say I don't know: not whether Burt cooked or invented figures, or even whether the now-notorious Misses Howard and Conway were real!

Whether or not it is true, Hearnshaw's suggestion that Burt must have suffered some kind of mental breakdown is typical of the attitude of many scientists. Why would a scientist engage in fraud unless he was psychologically disordered in some way? How could he possibly imagine he could get away with it? After all, if your work is of any significance, sooner or later someone will try to build on it, and then misrepresentation or fabrication will surely come to light. Fear of exposure and loss of reputation—the Invisible Boot, in Broad and Wade's nice phrase—is indeed a powerful disincentive. When a scientist ignores this risk, however, it is at least as likely to be out of a sincere conviction that he is right than from some psychological instability.[85] And the risk of exposure is probably closely related to the importance of the claim in question. Fraud in support of a claim that just isn't important enough for anyone to bother trying to build on it or to re-read the evidence for it may never be found out; but fraud in support of a claim significant enough to be worth other scientists' pursuing, or so important that historians of science or scientists themselves will want to revisit the evidence, *will* likely be exposed eventually.

Merton's confidence in "the virtual absence of fraud in the annals of science," thanks to the scrutiny of other scientists,[86] was too optimistic; Broad and Wade's estimate that "for every case of major fraud that comes to light, a hundred or so go undetected," and "for each major fraud, perhaps a thousand minor fakeries are perpetrated,"[87] probably too pessimistic. The sensible sociologist of science I have been imagining, if his taste is for empirical study, might try to test some of Broad and Wade's conjectures: that fraud has become commoner as scientists come under increasing career pressures; that it is commonest in biomedical fields, where precise replicability is not to be expected; and that it has been made easier by an explosion of publications that has put the refereeing system under severe strain. And then he might be drawn into such questions as:[88] Does an author's affiliation, if it is known, affect referees' judgment of his work? Is there a real danger of referees preventing the timely publication of rivals' results? Are referees mostly able and willing to judge papers fairly? Is the situation significantly different now than a few decades ago? What are the consequences of those sometimes very misleading attributions of multiple authorship,[89] or of the incentives to break your work into MPUs (minimal publishable units)? And he would notice a swarm of related questions about the peer-review grant system, and about the influence of industry sponsorship on university research.

If he is of a more theoretical turn of mind, our sensible sociologist might explore what the study of scientific fraud reveals about the role of authority and expertise in science; or he might ask, like Polanyi and Merton, why some kinds of political set-up seem more hospitable than others to the scientific enterprise; specifically why totalitarian regimes seem notably inhospitable.[90] This might lead him to re-examine the Lysenko affair,[91] or the Nazi concepts of Aryan science and Jewish physics.[92]

Even if he focuses exclusively on science in our society, a sensible sociologist would certainly attend to the interactions of science and politics: to political prejudices that bias inquiry; to political sensitivities that discourage honest inquiry, or any inquiry at all; to political disagreements that motivate close scrutiny of others' work. He might find himself thinking about, for example, those recent controversies over a possible genetic predisposition to homosexuality; perhaps wondering whether, by making still-conjectural scientific work widely known, the media may sometimes discourage researchers from considering possibilities too grossly unpalatable to public opinion. And perhaps he would begin to ask himself, as Hunt does, whether here and now, especially where the behavioral and social sciences themselves are concerned, political pressures may pose a real threat to freedom of research; and whether some research might be so potentially damaging that it really is best that it be suppressed.[93]

AND, IN CONCLUSION

So there is plenty of useful work for a sociology of science neither uncritically deferential to the natural sciences nor uncritically cynical about them, and which, rather than competing for control of the territory, cooperates with epistemology to understand the scientific enterprise. And since evidence-sharing, cooperation, and competition all require communication, there is epistemologically useful work, also, for rhetoric of science—for reasonable rhetoric of science, that is; but here too the path will need to be cleared of various metaphysical and epistemological extravagances. This time, however, as we explore the fascinating landscape of the intricate relations between science and literature, the view promises to be terrific.

NOTES

1. Merton, "Science and the Social Order," p. 254.
2. Polanyi, *The Republic of Science*, pp. 8, 16.
3. Wallis, introduction to *On the Margins of Science*, p. 5.
4. Woolgar, *Science: The Very Idea*, p. 99.
5. Barnes, *Interests and the Growth of Knowledge*, pp. 3–4 (Mannheim) 18–19 (Habermas).
6. Aronowitz, *Science as Power*, p. 346.
7. Collins and Pinch, *Frames of Meaning*, pp. 3–4.
8. Woolgar, *Science: The Very Idea*, p. 98.
9. Tomlinson, "After Truth: Post-Modernism and the Rhetoric of Science," p. 53.
10. Fox, "The Ethnomethodology of Science," p. 78.
11. Collins and Yearley, "Epistemological Chicken."
12. Barnes, "Natural Rationality," pp. 115, 124. My source is Roth, "What Does the Sociology of Scientific Knowledge Explain?" p. 98.
13. Barnes and Shapin, introduction to *Natural Order*, p. 11.
14. Shapin, "Homo Phrenologicus: Anthropological Perspectives on an Historical Problem," pp. 42, 65 n. 1.
15. See chapter 1, pp. 27–28, and chapter 6, p. 173.
16. See Haack, "Reflections on Relativism: From Momentous Tautology to Seductive Contradiction."
17. See chapter 6, p. 163.
18. Surveyed in agonizing detail in Ashmore, *The Reflexive Thesis*.
19. Manier, "Levels of Reflexivity," p. 203.
20. Bloor, *Knowledge and Social Imagery*, p. 5.
21. Ibid., p. 158. Barnes also suggests that Popper is a relativist; but unlike Bloor, who seems simply to have confused relativism with fallibilism, his point is that Critical Rationalism is at bottom thoroughly conventionalist. This is true; but it is hardly a reason for accepting relativism.
22. Ibid., pp. 158–59.
23. Bloor, "Sociology of Knowledge," p. 486.
24. See chapter 3, pp. 73–76.

25. MacKenzie, "Statistical Theory and Social Interests: a Case Study"; Barnes and MacKenzie, "On the Role of Interests in Scientific Change"; MacKenzie, *Statistics in Britain 1865–1930*.

26. Barnes, *Interests and the Growth of Knowledge*, pp. 59–63; Barnes and MacKenzie, "Scientific Judgement: the Biometry-Mendelism Controversy."

27. Wynne, "C. G. Barkla and the J Phenomenon: A Case Study of the Treatment of Deviance in Physics."

28. Pickering, "The Role of Interests in High-Energy Physics: The Choice between Charm and Colour."

29. Dean, "Controversy over Classification: A Case Study from the History of Botany."

30. Harwood, "Professional Factors."

31. Roth, "What Does the Sociology of Scientific Knowledge Explain?" p. 96.

32. Bloor, "On the Strengths of the Strong Programme," p. 211. There is no reference to Quine or Duhem in *Knowledge and Social Imagery*; there is one reference to Hesse in the book, and another in the afterword added to the 1993 edition, but neither is apropos.

33. Collins, "Stages in the Empirical Programme of Relativism," p. 10, n. 7.

34. Pickering, *Science As Practice And Culture*, p. 6.

35. Woolgar, *Science: The Very Idea*, p. 114.

36. Shapin, "Homo Phrenologicus," p. 42.

37. The quotation is from Shapin, "The Politics of Observation," p. 166; on p. 167 he entitles the final section of the paper "Social Interests and Esoteric Knowledge: Some Speculations."

38. I owe the point to Cornelis de Waal.

39. Bloor, *Knowledge and Social Imagery*, pp. 170, 171. See Slezak, "The Social Construction of Social Constructionism," for details of changes in the second edition of Bloor's book—described as "stylistic alterations" and correction of spelling errors—weakening his claims so as to avoid Laudan's critique in "The Pseudo-Science of Science."

40. Harwood, to his credit, does.

41. Barnes, Bloor, and Henry, *Scientific Knowledge: A Sociological Analysis*, p. 128.

42. Collins and Pinch, *Frames of Meaning*, pp. 185, 184.

43. Collins and Pinch, "The Construction of the Paranormal," p. 262.

44. Collins, "Stages in the Empirical Programme of Relativism," p. 3.

45. Collins and Yearley, "Journey into Space," pp. 372, 384; my source is Roth, "What Does the Sociology of Scientific Knowledge Explain?" p. 99.

46. See Collins and Pinch, *Frames of Meaning*, p. 5: "we show the inadequacy of any conception of major scientific disagreement in purely cognitive terms such as logical incompatibility."

47. Collins and Pinch, *Frames of Meaning*, p. 190, n. 1.

48. Collins, "What is TRASP?" p. 216.

49. Ibid., p. 217; I notice that here there are no scare quotes around "real."

50. Collins, "Special Relativism: The Natural Attitude," p. 141.

51. Garfinkel, Lynch, and Livingston, "The Work of a Discovering Science Construed with Materials from the Optically Discovered Pulsar," p. 137. Note the (presumably ironical) double reference to "discovery."

52. See Roll-Hansen, "Studying Natural Science without Nature?" pp. 169–70.

53. Latour and Woolgar, *Laboratory Life*, pp. 17, 31.

54. See Sismondi, "Some Social Constructions" (my source is Roll-Hanson, "Studying Natural Science without Nature?" p. 173); cf. Musgrave, "Idealism and Antirealism."

55. Latour and Woolgar, *Laboratory Life*, pp. 69–70, 75, 129.

56. Ibid., p. 171.

57. Ibid., p. 105.

58. As is Knorr-Cetina's *The Manufacture of Knowledge*.

59. Latour and Woolgar, *Laboratory Life*, pp. 85, 106, 125–26.

60. Ibid., p. 108.

61. Fox, "The Ethnomethodology of Science," p. 73.

62. Latour and Woolgar, *Laboratory Life*, p. 177.

63. Barnes, "On the 'Hows' and 'Whys' of Social Change," p. 492. In the introduction to *Science in Context*, however, Barnes—or perhaps one may hope it is his co-editor, Edge—writes that "specific orderings are constructed not revealed, invented not discovered" (p. 5).

64. Woolgar, "Critique and Criticism: Two Readings of Ethnomethodology," pp. 507–508.

65. Latour and Woolgar, *Laboratory Life*, p. 34.

66. Woolgar, *Science: The Very Idea*, p. 106.

67. Woolgar, "The Ideology of Representation and the Role of the Agent"; the quotation is from p. 140.

68. Gilbert and Mulkay, *Opening Pandora's Box: A Sociological Analysis of Scientists' Discourse*.

69. Mulkay, *Sociology of Science: A Sociological Pilgrimage*, p. xvii.

70. Mulkay, "Noblesse Oblige," in *Sociology of Science*.

71. No, this isn't a typo; it is a reference to an article Pinch attributes to "Pinch and Pinch," entitled, "Reservations about Reflexivity and New Literary Forms or Why Let the Devil Have All the Good Tunes?"

72. As reported in Collins and Yearley, "Epistemological Chicken," p. 305. In 2000 I was startled to read, in *Oxford Today*, that Woolgar now holds a chair in business at Oxford.

73. From Bauer, *Scientific Literacy and the Myth of Scientific Method*.

74. Hunt, *The New Know-Nothings*.

75. Angell, "Is Scientific Medicine for Sale?" (discussed in chapter 11, pp. 320, 322).

76. Goodstein is careful to make this distinction ("Conduct and Misconduct in Science," pp. 34–35). Federal agencies including the Public Health Service and the National Science Foundation, however, write of "fabrication, falsification, plagiarism," and

Nylenna et al., in "Handling of Scientific Dishonesty in the Nordic Countries," write of "breaches of research and publication ethics."

77. My source is Hyman, ed., *Science and Reform: Selected Works of Charles Babbage*, pp. 122 ff.

78. Sapp, *Where the Truth Lies*, pp. 20, 297.

79. My source is Broad and Wade, *Betrayers of the Truth*, pp. 34–35.

80. Kamin, *The Science and Politics of I.Q.*; Jensen, "Kinship Correlations Reported by Sir Cyril Burt"; Hearnshaw, *Cyril Burt, Psychologist*, chapter 12. See also Broad and Wade, *Betrayers of the Truth*, pp. 203 ff.; Gould, *The Mismeasure of Man*; and Grove, *In Defence of Science*, pp. 106 ff.

81. Gillie, "Crucial Data Was Faked by Eminent Psychologist."

82. Hearnshaw, *Cyril Burt, Psychologist*.

83. As he is in Sapp's book, published in 1990.

84. Joynson, *The Burt Affair*. My thanks to Edward Jayne for drawing this book to my attention.

85. Medawar, "The Strange Case of the Spotted Mice" (on Hixson's *The Patchwork Mouse*) is especially illuminating on this; see also his "Scientific Fraud" (on Broad and Wade's *Betrayers of the Truth*). Both reviews are reprinted in *The Threat and the Glory*.

86. Merton, "The Normative Structure of Science," p. 276.

87. Broad and Wade, *Betrayers of the Truth*, p. 87.

88. See Hull, *Science as Process*, on this topic.

89. See Anderson, *Impostors in the Temple*, chapter 4. According to Begley, "Science Breaks Down When Cheaters Think They Won't Be Caught," in two recent cases—the work of Jan Hendrick Schon of Lucent Bell Labs, and the supposed discovery of a new atom at Lawrence Berkeley National Laboratory—fraudulent scientists had apparently duped their collaborators. The first of several retractions of papers based on Schon's data was published in *Science* on 1 November 2002.

90. Polanyi, *The Republic of Science*; Merton, "Science and the Social Order."

91. See Soifer, *The Tragedy of Soviet Science*; Counts and Lodge, *The Country of the Blind*, chapter 6 (which includes numerous quotations from contemporary papers and speeches); Broad and Wade, *Betrayers of the Truth*, pp. 186–92, are informative in brief space. Aronowitz, who holds that "ideology is ineluctable in all intellectual labor" (*Science as Power*, p. 204), maintains that the Lysenko affair is misunderstood when taken to show the dangers of political interference in science.

92. See Beyerchen, *Scientists under Hitler*; "What We Now Know about Nazism and Science."

93. See chapter 11, p. 324.

STRONGER THAN FICTION

Science, Literature, and the "Literature of Science"

> The scientist addresses an infinitesimal audience of fellow composers. His message is not devoid of universality but its universality is disembodied and anonymous. While the artist's communication is linked forever with its original form, that of the scientist is modified, amplified, fused with the ideas and results of others, and melts into the stream of knowledge.
>
> —Max Delbrück[1]

As scientists investigate how the world is, they create an enormously complex labyrinth of signs—words, chemical formulae, mathematical symbols, graphs, panels of bar-codes, computer images, and so forth: signs that contribute significantly both to the successes of the sciences, and to their failures. Metaphors and analogies stretch scientists' imaginative powers— or, sometimes, lead them astray. Neologisms and shifts and twists of meaning gradually create a vocabulary that comes closer to real kinds, improving the degree of support of theories—or, sometimes, not. Papers, conference presentations, textbooks, letters, and now e-mail messages, etc., etc., enable the pooling of evidence essential to the scientific enterprise—or, sometimes, impede the communication of well-founded ideas, or make ill-founded ideas seem plausible.

Of late, noticing that writing plays a significant role in the scientific enterprise, literary scholars and rhetoricians have turned their attention to the "literature of science." And, indeed, given the importance of signs, interpretation, communication to the scientific enterprise, a reasonable rhetoric of science surely has a useful contribution to make. Mainstream philosophers of science and scientists themselves, however, sometimes reveal the same uneasy hostility to those "goddamn English professors,"[2] and to the whole enterprise of rhetoric of science, as some do to sociology of science. Rhetoric of science, Max Perutz writes in exasperation, seems to him nothing better than "a piece of humbug masquerading as an academic discipline."[3]

How did we come to such a pass?

The New Cynicism has been influential in rhetoric of science as in sociology of science, its familiar tone readily recognizable though now transposed from sociological into textual-literary-rhetorical jargon. The relativist and irrealist extravagances that Perutz protests are rooted in the characteristic falsehoods that have informed the radical wing of rhetoric of science—remember, as I give my list, that unlike truths, falsehoods don't have to be mutually compatible: What determines whether a scientific claim is accepted is how attractively it is presented. There is no real difference between presenting good evidence and simply being persuasive. What passes for objectively good evidence is, simply, whatever is found persuasive by the scientific community—often the effect of metaphors drawing on familiar social phenomena. There is no real difference between science and myth, fiction, or fable, no distinctive or more intimate relation of scientific language to the world. There is no real difference between scientific and literary texts—all are simply imaginative constructions rather than representations of an extra-textual reality. There is no extra-textual reality. And so forth.

The phrase "there is no real difference," recurring in my list of falsehoods, reveals that radical rhetoric of science prefers assimilation to discrimination, tending, especially, to assimilate science to literature, and scientific to literary texts. So perhaps, if we could get clearer about the differences between science and literature and between scientific and literary texts, we might begin to see what a reasonable rhetoric of science would be, and do.

SCIENCE AND LITERATURE; SCIENTIFIC AND LITERARY TEXTS

Scientists engage in writing, and novelists, playwrights, etc., engage in inquiry. But the word "science" picks out a loose federation of kinds of inquiry, while the word "literature" picks out, not any kind of inquiry, but a loose federation of kinds of writing. In its broadest usage, "literature" refers to writings of just about any kind—as when we speak of "keeping up with the literature" on our subject. And the word is also used honorifically, to refer to aesthetically admirable writing on whatever topic.[4] This aesthetically honorific use of "literature" is no less potentially confusing than the epistemically honorific use of "science." But when, in what follows, I write of "imaginative literature," the concern will be mainly with its fictional character, not its literary merits.

The inquiry in which writers of imaginative literature engage, whether a matter of informal observation and pondering over the quirks of human nature or of systematic research into a place or time, is essential to their enterprise—*but as a means to the end of writing* edifying, entertaining, provocative, expressive, moving, illuminating . . . novels, plays, etc. And the writing in which scientists engage is essential to their enterprise too—*but as a means to the end of finding out* significant, explanatory truths, well warranted by evidence, about the world and how it works.

Science, like literature, requires imagination. A scientist imagines structures and classifications which, if he is successful, are real; and explanations, laws, and theories which, if he is successful, are true. Imagination, and imaginative exploration of imagined potential explanations, is essential. But to go beyond mere speculation, appraisal of the likely reality or truth of his imaginative creation—itself often requiring imagination in the design of experiments, instruments, and so on—is essential too. This is part of the point Einstein was making when he described the liberty of choice of scientific concepts and theories as "not in any way similar to the liberty of a writer of fiction," but more like that of "a man engaged in solving a well-designed word puzzle";[5] and of the point Peter Medawar is making when he writes that "scientific reasoning is a kind of dialogue between the possible and the actual, what might be and what is in fact the case."[6]

Of course, when a writer inquires, he is subject to the demands of evidence, as scientists are. But as he goes about his distinctive business, a writer of imaginative literature imagines people, events, and stories, which, if he is successful, are illuminating about real human beings, real human doings, and real human and moral possibilities. Imagination, and imaginative exploration of imagined scenarios and characters, comes first; and an author may also imaginatively explore linguistic means and modes, and imaginatively "test" his imagined characters and events for believability. But it would be something of an embarrassment to discover that the characters or events he thought he had imagined were, actually, real—not quite as devastating as it would be for a scientist to discover that the stuff or phenomenon he thought he had been investigating *wasn't* real, but disconcerting nonetheless.

To be sure, novels, plays, etc., are very often set in real places; they may include real events as well as imaginary ones, and may involve real people, whether overtly acknowledged or more or less thinly disguised—think of those discussions in the press about whether Saul Bellow's novel, *Ravelstein*, is a betrayal of his friendship with the late Allan Bloom, on whose life it is based.[7] Some novelists like to leave the reader guessing: Edward Rutherfurd's *London* is straightforwardly about London; but is Italo Calvino's *Invisible Cities* about Venice? Still, though nearly all novels combine them, the essential difference between the fictional and the real remains. Irving Wallace writes in the afterword to *The Prize* that his descriptions of Stockholm and the Nobel ceremonies are factual; however, he continues, "the characters who people these pages . . . are make-believe; and the entire plot [is] purest fabrication. If the characters or situations have . . . any counterparts in real life," he continues, "the resemblance must be accepted as surprising coincidence."[8] Except in such an unlikely eventuality as Wallace worries about, the fictional characters a novelist creates are not only imaginative but also imaginary. By contrast, the imaginative constructions of science, when they are successful, are precisely *not* imaginary or fictional, but real.

Scientists, like writers, use metaphors. Sometimes a literary metaphor—the army and the church as the Red and the Black, a great river as a book[9]—is the thread that ties a novel or story together; and sometimes a scientific metaphor—the chaperone molecule, parental investment, the Invisible Hand—is an intellectual tool a scientist uses as he works towards an account of a physical or social structure or kind. Once again, however, it is a case of "both . . . , but." A subtle literary metaphor will be extended, expanded, explored, but remain a metaphor as the writer finds new ways to play on it. Hermia insults Helena: "Thou painted maypole!" Helena appeals to Lysander: "O, when she is angry, she is keen and shrewd; she was a vixen when she went to school; and though she be but little, she is fierce"; and Lysander joins in against Hermia: "Get you gone, you dwarf; you minimus, of hindr'ing knot-grass made; you bead, you acorn."[10] Or one writer may pick up another's metaphor; as in *The Man*, where Wallace plays on Stendhal's metaphor, describing how Douglass Dilman, the first black president of the United States, is slandered in a press report alleging that Wanda Gibson, his "comely Negro paramour," is in collusion with Soviet agent Franz Gar.[11] A sturdy scientific metaphor will likewise be extended, expanded, explored, and passed from hand to hand—but with the ultimate goal of enabling a literally true articulation of a theory about cellular processes, reproductive behavior, markets, or whatever.

Scientific texts differ from literary texts in ways that reflect the differences between science and imaginative literature. "Ways," in the plural, because although a scientific report is a very different kind of thing from a novel, there is no single, simple distinction of two kinds of text, scientific and literary, or even scientific and "other"; rather, there are umpteen distinct but related dimensions on which a text can be located, with scientific texts tending to cluster in one area of the grid, literary texts in another.

When I write of "scientific texts," I refer to scientific articles and reports, as distinct from scientists' memoirs and such. Such texts are intended, not to tell the whole untidy story of how a scientific result was achieved, but to present the result and the evidence for it.[12] A paradigmatic scientific text is putatively truth-stating; putatively referential; about stuff, things, and events in the natural world or, quite often, artifacts created by scientific activity; evidence-presenting; aimed at an audience of other scientists; and written in a direct, explicit, dry, closed style. A paradigmatic literary text is unlike a paradigmatic scientific text in just about all these ways. Other kinds of text—food labels, directions on medicine bottles, appliance instructions, library recall notices, history books, statutes, contracts, zoning regulations, the Constitution, tables of logarithms, customs and immigration forms, recipes, telephone books, historical novels, science fiction, market-research questionnaires, magazine articles, philosophical treatises, philosophical dialogues and plays, letters to the editor, television and film guides, the Book of Common Prayer, etc., etc.—are like the paradigmatic scientific text in some ways, unlike it in others; like the paradigmatic literary text in some ways, unlike it in others; like each other in some ways, unlike each other in other ways.

The thesis that scientific texts are putatively truth-stating reflects, as it is meant to, my modest style of realism. An old-fashioned instrumentalist would deny that scientific theories are true or false; and a new-fangled constructive empiricist would deny that, in presenting their theories, scientists are making truth-claims. But because I am skeptical of any sharp distinction between observational and theoretical statements, or between observable and unobservable things and events, I take scientists' avowed intent to discover how things are seriously.[13]

I say "putatively truth-stating," rather than simply "truth-stating," because I want to include texts which state what the author takes to be truths, but are in fact falsehoods. But I mean to exclude fraudulent scientific texts, where, feigning a truth-stating intention, the author presents as true what he knows or believes to be false. (This means that in the case of a jointly authored text where one author is sincere and the other knowingly deceptive there will be no good answer to the question, whether the text itself is putatively truth-stating but mistaken, or fraudulent;[14] e.g., in the case of that notorious paper by Benveniste and his colleagues describing the effect of high-dilution homeopathic remedies—not only because it is unclear whether Benveniste is self-deceived or knowingly faking, but also because some of the several authors of this paper may be faking it, while others are sincerely self-deceived.)

To be sure, works of imaginative literature teach us truths; novels, plays, etc., convey truths (or sometimes falsehoods) about what makes real human beings tick. But that doesn't mean that they too must be putatively truth-stating; a novel may convey truths (or falsehoods) by means of sentences which, being about fictional characters, are *not* true—though sometimes a novelist will break off the story briefly to state the truths it illustrates: as George Eliot breaks off the story of *Daniel Deronda* to reflect that "[i]t is a common sentence that Knowledge is power; but who hath duly considered . . . the power of Ignorance?"[15]— such as Gwendolen Harleth's fatal shallowness.

It would be a whole other enterprise fully to spell out "convey"; but an example will suggest how imaginative literature can illustrate truths about real human beings. Alison's Lurie's novel, *Imaginary Friends*, conveys without stating some of the same truths[16] that Leon Festinger's textbook on cognitive dissonance and his studies of millennial sects convey by stating.[17] Festinger describes experiments and social-scientific studies of people's reactions to inconsistency or "dissonance" among their beliefs and attitudes, predicting inter alia that, if their prophesies are falsified, members of millennial sects won't give them up, but reinterpret them and/or start proselytizing more energetically. Lurie tells the story of two sociologists who pretend to join the spiritualist sect they are studying, and, with the others, wait in Verena Robert's aunt's parlor for Ro of Varna to come to earth in his spaceship as the Message promised. When, apparently, nothing happens, the real members of the sect don't lose faith; Ro *has* come, his spirit has entered one of them. Which one? the senior sociologist, who soon starts to believe it himself, and by the time the novel ends is off his head altogether—or could he be faking it? Lurie leaves the reader not quite sure.

A novelist, I shall say (using "pretend," as distinct from "feign," to mark the difference between fiction and fraud), without intending to deceive his readers, invites them to imagine, to suspend disbelief, to pretend. That's the point of the "once upon a time" of children's stories, signaling: "let's pretend." As this reveals, the practice of fictional pretended-reference is parasitic on the practice of regular reference to real people, events, etc., and the practice of truth-conveying on the practice of truth-stating.

The oblique way that literature conveys its truths has led some to suppose that the truths that literature conveys must be superfine, peculiarly literary truths. This is a mistake; the truths that literature conveys are regular truths, sometimes as startlingly familiar as the truths about *Homo academicus* that Malcolm Bradbury conveys in his story of James Walker and his colleagues and students at Benedict

Arnold University,[18] or David Lodge in his story of Morris Zapp and his colleagues and students at Euphoric State and the University of Rummidge.[19] Perhaps there are moods that can only be evoked, emotions that can only be expressed, by obliquely literary means; but there are no truths that can be fictionally conveyed but cannot, even in principle, even at clumsy length, be stated. Here I'm with Frank Ramsey: "if you can't say it, you can't say it, and you can't whistle it, either." (This, however, is by no means to say that all truths fall within the scope of the sciences to discover.)

Like scientific texts, food labels, theological treatises, newspaper articles, historical narratives, biographies, police reports, tourist brochures, advertising copy, bus timetables, etc., etc., are putatively truth-stating. What distinguishes scientific texts from others in this much broader category depends on what the "other" is. For example, a scientific text is aimed at an audience of the scientist's peers, a piece of scientific journalism at the educated public; a scientific text is putatively evidence-stating as well as putatively truth-stating, a food label is not.

Historical texts are generally like scientific texts in being putatively evidence-presenting as well as putatively truth-stating. The difference between them is primarily a question of subject-matter—though it won't quite do simply to say that a historical text, unlike a scientific one, is concerned with past events, since works of cosmology or of evolutionary biology are also about past events. Some historical texts have a claim to count as literary, in the aesthetically evaluative sense, as well as historical; most historical texts are narrative in form. Most novels are also narrative in structure; but it doesn't follow that the distinction between history and imaginative literature dissolves. There are borderline cases: Elsa Joubert's *Poppie*, for example, which she describes as a novel "based on the actual life story of a black woman living in South Africa today." But a clear difference remains between clear cases: a history of the Millerites, putatively truth-stating and putatively referential, narrates putatively real events involving real people; *Imaginary Friends*, neither putatively truth-stating nor putatively referential, narrates imaginary events involving imaginary people.

The style most appropriate to a text obviously depends on its purpose and its intended audience. A piece of scientific journalism will be less detailed and more idiomatic than an article in a scientific journal; a medicine-bottle label will give top priority to clear directions for taking the stuff, a pharmacist's list of drugs to indications, side-effects, and contra-indications.

In official scientific writing—as everyone knows whose first chemistry lesson, like mine, was devoted to the use of the passive voice—a studiedly neutral "style of no style" has become the norm. Still, when you compare scientists' official presentations and papers with their conversations and letters, you realize there is also a large element of convention in this official, neutral style. A series of experiments in mutagenesis officially referred to as "tests of the triplet code by means of frame-shift mutants in bacteriophage T4"[20] were known in the trade as "the Uncles and Aunts experiments"—the relation involved was one up and sideways.

That stiff style and toneless tone of much official scientific writing is in part a badge of respectability, and can sometimes be a substitute for genuine rigor rather than an absolute necessity given the author's purpose—a ubiquitous problem where social-scientific writing is concerned, and perhaps not entirely insignificant even in the natural sciences. After all, like other people, scientists have been known to publish to decorate their vitae rather than with serious communicative intent.

But insofar as the goal is to inform other inquirers as efficiently as possible, the highest priority must be the explicit, the literal, the cognitive, the direct. Emotional associations of the words, and so forth, are at best beside the point; what matters most, as Frege observed long ago,[21] is the meaning, the reference, the cognitive content. So in his professional writings a scientist properly aims to be as explicit as possible, to close his text to interpretations other than the one he intends, to express the intended content and no other. The interpretation of a text is always a cooperative effort between author and reader; but in the case of scientific texts, the author will properly try to minimize the constructive role required of his readers.

It is very different where the primary purpose is to be aesthetically pleasing; when not only sound and rhythm, emotional nuance, and the breadth and depth of associated ideas, but also a certain implicitness, an openness to interpretive resonance and to the reader's constructive contribution, can be part of the pleasure. To paraphrase Delbrück: in scientific communication what matters is not the form, the particular words, but the content which, if the scientist is successful, will become established scientific knowledge. When aesthetic pleasure is primary, however, the *how* is no less important than the *what*. (Perhaps that is why, though in scientific writing metaphors and similes are essentially interchangeable, because equivalent with respect to what is expressed, from an aesthetic point of view metaphor, with its far greater syntactic flexibility, often has an edge over the more stilted "like" construction.)

The spareness and tautness of the best scientific writing can be aesthetically pleasing, but fancy literary forms would get in the way; imagine how much harder it would have been if Watson and Crick had tried to explain the structure of DNA in iambic pentameter. No, I haven't forgotten Lucretius' scientific poem, *De rerum natura*, of which a new English translation was recently published, in iambic pentameter, yet! But I stick to my guns: the science Lucretius presents is simply not comparable in complexity or sophistication; and if it had been, the literary form would have been a serious obstacle to communication.

Sometimes there is a kind of mismatch between the purpose of a text and its style. We implicitly criticize a work of imaginative literature when we call it "didactic" because it conveys the truths it wants to get across a bit too obtrusively directly, illustrates them a bit too obtrusively crudely. In *The Ragged-Trousered Philanthropists* Robert Tressell allows his characters to make chapter-long socialist speeches; names his fictional employers "Sweater," "Grinder," and "Didlum," and his rival building firms "Rushton's," "Dauber and Botchit," and "Smeariton and Leavit"; the local Mugsborough newspaper is *The Obscurer*; and, just in case you were in any danger of missing his attitude to religion, one of his ministers is a Mr. Bosher, and the other, Mr. Belcher, dies when his internal gases finally explode! When we describe a scientific article as "rhetorical," it is usually a criticism of a different but related kind of mismatch. As I. A. Richards observes, "we distrust a scientist when we think he is influencing us by his manner";[22] we expect him, in his professional writings, to tell us as straightforwardly as possible what he takes to be the truth of the matter he has been investigating, and to present the evidence that this is indeed how things are.

Still, as Augustine once said—elegantly, and truly, too—a thing is not necessarily true because badly uttered, nor false because spoken magnificently.[23] And, while a dubious or lazy scientist wanting to persuade others of the *bona fides* of his work may occasionally succeed by means of rhetorical flourishes, another and perhaps more effective strategy is to hide behind that blandly neutral official style. If it weren't for one line in Benveniste's paper about how a "solution" so dilute that it contains not a single molecule of the supposed "solute" *works because the water remembers that it once contained bee-venom*, a careless or inexpert reader would hardly notice anything untoward.

There's nothing wrong with such poetic playfulness on scientific ideas as Neil Rollinson's "Quantum," which begins:

> If we could live long enough
> we'd be able to watch the effects
> of cold fusion turn the diamond
> on your wedding-ring to a nub of iron.
> I should have given you something purer,
> more incorruptible than a diamond,
> a single proton perhaps, with its up and down
> quarks and its gluons, a love heart
> inside it, your name and mine.

Nor is there anything wrong with putting texts to different uses than their authors intended—with Ogden Nash making a verse of the mangled-English instruction manual for a Taiwanese toaster-oven, John Betjeman making a poem of the bus timetable, or even the wallpaper company making a design of a text in Japanese. There isn't necessarily anything wrong, even, with giving a different interpretation to someone's words than he intended—with the newspaper borrowing Shakespeare's "Now Is the Winter of Our Discontent" as a headline referring to the miners' strike, say, or my using Francis Bacon's observations about the factitious despair of Descartes' hyperbolic skepticism as a commentary on some philosophical extravagances of our times.[24]

But there is something not quite right when literary critics, accustomed to attuning themselves to the nuances, resonances, and multiple layers of meaning of literary texts, playing in the same way with carefully defined and delimited phrases from mathematics, physics, biology, etc., fudge the difference between understanding a text or phrase, and putting it to a new use. Gross and Levitt complain about N. Katherine Hayles' literary conceits on non-linearity, chaos theory, etc.,[25] and Sokal and Bricmont about Lacan's mongreloid psycho-analytic-topological extravagances and Irigaray's feminist musings on "$E = MC^2$."[26] John Sturrock is indignant: "far better wild and contentious theses . . . than the stultifying rigor so inappropriately demanded by Sokal and Bricmont."[27] If I could get a word in edgewise it would be to observe, simply, that inquiry, whether it be literary, psycho-analytic, feminist, or scientific, requires not only imagination, but also checking how one's imaginative constructions stand up to evidence.

There's nothing wrong with writing imaginary scientific dialogues, speeches, etc., or works of fiction about science and scientists. Far from it—you can learn a lot from William Cooper's *The Struggles of Albert Woods* about the role of personality and ambition in science; or from Wallace's novel about the jealousies, rivalries, politics, and personal dramas that may lurk behind the bland facade of official Nobel acceptance speeches. But there is something wrong when sociologists of science like Mulkay and Woolgar try to avoid the "problem of reflexivity" by resorting to New Literary Forms. Novels like Cooper's or Wallace's convey truths about science, truths which could be stated directly. For Cooper or Wallace, who are not afflicted by cynical misconceptions about truth, evidence, etc., this is no problem. But for Mulkay, Woolgar, & Co., who *are* so afflicted, it is no solution; for if there are no warranted truths about science to be stated in sober sociological articles, there are no warranted truths about science to be conveyed by fanciful sociological dialogues, either. The NLF consortium can't avoid bankruptcy by only *pretending* to balance its books!

There's nothing wrong with looking at scientific and other putatively truth-stating texts from a stylistic point of view. But there is something wrong when the putatively truth-stating, evidence-presenting character of scientific, historical, etc., texts is fudged, blurred, or twisted—which, unhappily, is what has happened when literary scholars and radical rhetoricians of science have tried to make a killing by cooking Watson and Crick's books.

RADICAL RHETORIC OF SCIENCE

Watson's memoir, *The Double Helix*, is not itself (in the main) a scientific book, but an autobiographical-historical one. It is a very personal story; as Watson says, rather than writing with the benefit of hindsight, he tried to describe how things seemed at the time to his very young, very brash earlier self. Sir Lawrence Bragg, who wrote a kindly introduction, nevertheless acknowledged that "those who figure in the book must read it in a very forgiving spirit." Crick found the book vulgar and infuriating: "such a naive and egotistical view . . . as to be scarcely credible"; and toyed with the idea of getting his revenge by writing his own memoir, to be entitled *The Loose Screw*, or maybe *Brighter Than a Thousand Jims*, and to open "Jim was always clumsy with his hands."[28] (Watson had begun: "I have never seen Francis Crick in a modest mood.")

When Rosalind Franklin's biographer, Anne Sayre, criticizes Watson's book, it is on the straightforward grounds that some of the truth-claims Watson makes are in fact false.[29] The prickly, dowdy, bluestocking "Rosy" of Watson's book, she tells us, bears little resemblance to the real person. The real Franklin wasn't, as Watson says, Maurice Wilkins' research assistant, but an independent researcher; she wasn't, as Watson says, adamantly anti-helical, but had already recognized the helical structure of the B form of DNA; and she didn't even wear glasses, so Watson's speculations about whether she would be better-looking if she took off her spectacles and did something with her hair are pure fantasy. When Sayre describes Rosy as a "fiction," she is using the term, as it is sometimes used, pejoratively. She is well aware that Watson's book is *not* fiction, in the more usual sense, but is intended to give a true account of real events involving real people, or at least a true account of how Watson perceived those events and people at the time. Her point is, simply, that Watson's putatively truth-stating autobiographical-historical text is, in various respects, false. So far, fair enough.

However vulgar and infuriating it may be, and however inaccurate about Franklin, Watson's book, like not a few works of history or autobiography—like Pepys's diary, to which Bragg, along with several reviewers, compared it—is written with great verve and dash, and tells a story of unique fascination. After it was published in a critical edition by Norton in 1980, it soon began to attract the attention of literary scholars, not all of whom treated it as soberly as Sayre. Writing of *"The Double Helix* as Literature," John Limon toys with the idea of parsing its title *The Double He-lix*. At one time, the working title of the book was *Base Pairs*;[30] and Limon, relishing the conceit this suggests to him, duly searches the text for evidence of a sexual relationship between Watson and Crick. "In one [photograph], Crick, at the right and somehow elevated, points upward with some sort of instrument towards the DNA model; Watson, lower left, looks as nearly straight up as he can without tilting his head back. Between them, their DNA offspring grows to the ceiling." Evidently disappointed, however, Limon finds that "five pages later is the reverse photograph. Now . . . [Watson] is well above and much bigger than Crick. . . ."[31] This seems—well, an odd preoccupation, to put it mildly.

Limon never goes quite so far as to deny outright that *The Double Helix* is intended to be truth-stating; but his whole discussion seems designed, not to explore how much wiggle-room Watson's avowed intention to describe how things seemed to him at the time allows—a good question, to which I should like to hear Sayre's answer—but to fudge the issue. Alluding to Bronowski's comment that Watson's book would have been called *Lucky Jim* "if [Kingsley] Amis had not been so inconsiderate as to make this title famous in advance," Limon tells us that the two books "have the same basic structure: failure-failure-failure-stunning success," the same fairy-tale "logic of disasters."[32] It never seems to occur to him that Watson's book might have this structure because the scientific work had this structure, let alone that the logic of falsification or confirmation might be more to the point. Observing that Watson could have set his book in Oxford rather than Cambridge, could have changed the names of the protagonists, could have made the topic cancer research, Limon writes of "Watson's non-fiction novel"—or, as by the end of his paper he prefers to say, his "factual novel." But it is misleading, to put it mildly, to assimilate Watson's autobiographical memoir to, say, C. P. Snow's *The Search*, which really is a work of fiction, but loosely based on the early work in X-ray crystallography by W. T. Astbury's group in textile sciences at Leeds.

Still, probably Limon's extravagances strike you, as they do me, as only foolishness, not to be taken too seriously. But when rhetoricians in the grip of bad metaphysics, epistemology, and philosophy of language begin to turn their attention from Watson's memoir to his and Crick's scientific papers, things get hairier, and scarier. S. Michael Halloran is a straightforwardly old-fashioned epistemic relativist offering a vulgar-Kuhnian "rhetoric of scientific revolution." Noting that Watson and Crick's first paper in *Nature* has been described as marking the beginning of a revolution in biology, taking for granted that this was a revolution in the Kuhnian sense—that it involved a change of world-view and a new paradigm incommensurable with the old, a new paradigm promoted by "presumably, rhetorical means"[33]—Halloran asks: why was it this 1953 paper, rather than Oswald Avery's 1944 paper,[34] that is seen as decisively identifying DNA as the genetic material, and as marking the start of the revolution? And answers: because of Watson and Crick's rhetorical strategies, such as their daring to use "we" instead of the passive voice, to describe their model as "of considerable biological interest," and so on.

To be sure, their paper is, as this kind of thing goes, quite lively. They worked hard to make it accessible, and, yes, appealing; and it is pervaded by a kind of barely suppressed excitement (is it any wonder?—they had, after all, solved the structure of DNA, discovered the secret of life, as Crick was telling everyone who would listen down at the pub!). But the idea that it is the liveliness of the writing, rather than the importance of the discovery, that gives this paper its unique place in the history of molecular biology, is bizarre. To be sure, Avery's work was also of "considerable biological interest." But the main reason it didn't set off a revolution wasn't (as some rhetoricians would have us believe) that its style is too dry, nor (as some sociologists would have us believe) that Avery wasn't part of the Phage group;[35] it was rather, as we saw earlier,[36] that in 1944 the tetranucleotide hypothesis was so entrenched that even Avery himself hesitated to draw the obvious conclusion from his work.[37] Nor is it at all plausible to think, as Halloran assumes, that Watson and Crick's paper began a revolution in Kuhn's sense of the term; where is the incommensurability with Avery's and all the others' indispensable preliminary work, or with the Pauling-Corey model that they criticize as incompatible with known chemistry?

Sometimes scientists are a little carried away with enthusiasm for their ideas: Darwin's family teased him about his description of a larval cirripede with "six pairs of beautifully constructed natatory legs, a pair of magnificent compound eyes, and extremely complex antennae"—just like an advertisement![38] And no doubt scientific theories have sometimes come to be accepted because of the rhetorical skill with which they were presented. But Watson and Crick's work is not plausibly regarded as such an instance.

Even Halloran, however, seems mousy compared with Alan Gross, who announces: "I do not intend merely to rehearse the claim . . . that Watson and Crick use persuasive devices to convince scientists of the correctness of their structure; rather, I want to support a more radical claim: that the sense that a molecule of this structure exists at all, the sense of its reality, is an effect only of words, numbers and pictures judiciously used with persuasive effect."[39] At first you might think he is just equivocating between the triviality that scientists were persuaded of the reality of a molecule with such-and-such structure by means of the words, diagrams, and photographs in Watson and Crick's papers, and the bizarre claim that the idea that there even *is* such a molecule is an illusion created by words, photographs, etc. But there is more to it than that.

The first move is to stress "Watson's autobiographical distortions": his exaggeration of his own ill-preparedness for the work, of Franklin's hostility to helical models, of the danger that Pauling would find the solution first.[40] Watson, Gross writes, "consistently casts himself in the role of the youngest son in the fairy tales" (this reading, he concedes, "is not mine *de novo*," but was already suggested in Bronowski's review).[41] To be sure, there are parallels between Watson's narrative and those folk-tales; but then, as we say, "life imitates art," and truth sometimes really *is* stranger than fiction.[42] Now, however, Gross adds that Watson chooses "psychological over literal truth";[43] and, citing Richard Lewontin's observation that Watson's book "speaks to the secret dreams [of scientists] in a familiar vocabulary,"[44] and ignoring the many scientists whose reviews were severely critical, concludes that scientists found *The Double Helix* persuasive because of the "fit between the view Watson presented and their own preferred view of reality."[45] This is the hook on which he will hang his claim that the rhetoric of Watson's memoir, and of his and Crick's paper about DNA, is the same: scientists find the words, photographs, etc., persuasive because they fit with "their preferred view of reality."

The full force of this phrase appears only when, in a epilogue entitled "Reference without Reality," Gross endorses what he calls "motivational realism," the thesis that the possibility of genuinely referential, objectively true theories "is the psychological anchor that makes a life in science meaningful";[46] i.e., that scientists couldn't do their work unless they believed that it is possible to arrive at true theories about things and events in the world. Perhaps needless to say, this is no realism at all.

Acknowledging that this motivational "realism" requires supplementation, Gross adds a rhetorical "realism" supposedly compatible with metaphysical realism and at the same time closely resembling the "radical relativism" of Goodman's *Ways of Worldmaking*.[47] It's hard to see how such a logical feat is possible, since according to Goodman there is no one real world, only the many "versions" made by artists, scientists, etc.[48] So it comes as no surprise to find that the supposed compatibility of Gross's rhetorical "realism" with metaphysical realism is achieved by "redescrib[ing] realist analyses in rhetorical terms"; i.e., content-stripping so as to exclude any reference to experience, to facts, to reality, or even to coherence, except qualified by "scientists agree that . . . ." Thus: "scientific truth *is seen as* a *consensus* concerning the *coherence* of a range of utterances."[49] Perhaps needless to say, this is not what scientific (or any other) truth is; and Gross's rhetorical realism is no more genuine realism than his motivational realism is.

Though his transmutation of "true" obscures matters somewhat, apparently Gross's view is that, though scientists would be unable to do their work unless they believed they were describing an independent reality, this belief, while psychologically necessary, is not, in the usual sense of the word, true. The idea that a molecule of this double-helical structure exists at all has, as Gross might put it, only psychological, not literal, truth. Or, in plain English: scientists only *think* it exists.

Gross doesn't deny that scientific texts are putatively truth-stating and putatively referential; scientists, he agrees, believe they are describing a world independent of their theory-construction. But this, he thinks, is an illusion. Watson and Crick are not feigning, exactly, nor pretending; they are not outright frauds, nor simply writers of science fiction. But they are, in effect, self-deceived fabulists. The difference between them and Benveniste (assuming Benveniste is genuinely self-deceived, and not an outright fraud) is only their greater success in getting the agreement of others in the scientific community.

Gross's view of science is like an atheist's view of theology. A belief that a deity exists is the psychological anchor that makes a life in theology possible; however, an atheist would add, since this belief is false, theologians (though not exactly frauds, and not exactly writers of religious fiction) are in effect self-deceived fabulists. But science is different from theology in a crucial respect: that there is a real world which is a certain way and not other ways is a presupposition not only of scientific inquiry, but of all empirical inquiry, including the most ordinary of everyday investigation into whether that check has cleared, what caused that leak in the roof, etc.—as it is, also, of inquiry into Watson and Crick's texts.[50]

Like the falsehoods that feed the ambitions of cynical sociology of science, the falsehoods that encourage the extravagances of radical rhetoric of science pose problems of reflexivity. If Gross intends his rhetorical (ir)realism as a completely general thesis, it would undermine his own claim to make true statements about scientific texts. But if he intends his rhetorical (ir)realism as a thesis specifically about scientific discourse, some argument would be needed why, while other texts, his own included, succeed in referring to real things and in stating truths about them, scientific texts do not. But he apparently sees no need for any argument that the objects of natural-scientific study have only "psychological reality," but the objects of rhetorical study are really real.

It is customary to distinguish a narrow use of "rhetoric" in which it contrasts with "logic" and refers to emotive language and other non-rational means of persuasion, and a broad use in which it refers to the art of prose discourse, logic included.[51] But this distinction evanesces if, like Halloran, you take for granted that Watson and Crick could not be presenting evidence for their model, only an inevitably circular "argument"—a term he uses in a way guaranteed to raise a logician's hackles, to mean something like "persuasive performance." As it also evanesces if, like Gross, you presuppose that logic is only special-purpose rhetoric, that "the incompleteness of rhetorical deduction [*sic*] is different in degree, not kind, from the incompleteness of scientific deduction," and that "rhetorical analyses show how the sciences construct their specialized rhetorics from a common heritage of persuasion . . . [creating] bodies of knowledge so persuasive as to seem unrhetorical—to seem, simply, how the world is."[52] Blurring the distinction between rhetoric in the broad sense and in the narrow, radical rhetoricians of science, like cynical sociologists of science, are able to hide their deep skepticism about the scientific enterprise behind a mask of neutrality.

The great strategic advantage of blurring the distinction between inquiry and discourse, of focusing exclusively on scientific texts and ignoring their relation to the world, is that rhetoric takes its place as master-discipline. Gross isn't shy about this; apparently stung by Perutz's bluntness, in the preface to his second, 1996, edition, he contrasts his own ground-breaking book with staid monographs like Charles Bazerman's *Shaping Written Knowledge*, and boasts that what was needed when *The Rhetoric of Science* was first published was precisely "a book [like mine] with an authoritative title by a major press which made bold claims for the place of rhetoric in the understanding of science."[53]

Only two years later, however, he writes that its inability to solve our most pressing political problems doesn't alter the fact that "science is the best and only way of finding the truth about the material world."[54] Of course, this isn't much of a concession if he still thinks truth is just "a consensus concerning the coherence of a range of sentences." But it appears, at least, to be a remarkable volte-face. So there, as the Underground Grammarian would say; let nothing you dismay! But it's high time we moved on.

REASONABLE RHETORIC OF SCIENCE

A reasonable rhetorician of science will be sensitive, as Bazerman is, to the differences among different kinds of text, and aware of the relativity of style to purpose and audience, of the evolution of scientific language with the growth of scientific knowledge, of reference and truth as achievements at once empirical (constrained by the world) and social (the work of a vast inter-generational community).

Every scientist, if he is not to start on his part of the crossword alone and from scratch, depends on others—on the results of experiments others have performed, on instruments others have devised, on conjectures others have made, on vocabulary others have introduced: a pooling of evidence made possible in part by means of articles, presentations, textbooks, etc. The ideal is for the degree of credence given a claim within the relevant sub-community to depend on the quality of the evidence it has, for transmission of information within the scientific community to maximize epistemological efficiency.

How efficient the transmission process is will depend in part on who controls the journals, how honest the refereeing process is, who writes and talks to whom, and whether there are effective means of finding relevant material; but also in part on such broadly rhetorical factors as how intelligibly, clearly, and explicitly scientists communicate with each other. "With" is important; efficient transmission of results depends on the audience as well as on the presenter. When Marshall Nirenberg first read his paper on the coding problem in Moscow in 1961, only a few people showed up; a photograph of the tiny audience shows several people apparently asleep. But at least one person present was "bowled over"; and he told Crick, who arranged for Nirenberg to give the paper a second time, at the very end of the program, in the great hall, now to an audience of hundreds. "His delivery this time," Crick later wrote, "electrified the audience."[55]

There is a real distinction between modes of communication that promote the epistemologically desirable correlation, and those that impede it, illustrated by the contrast between these scenarios: (1) a scientific claim comes to be accepted within the relevant sub-community because strong evidence is clearly communicated in a journal article or conference presentation; (2) a scientific claim comes to be accepted in the absence of good evidence because it is promoted by means of emotive language, snazzy metaphors (and/or glossy photographs, melodramatic press conferences, etc.).

But in practice there is rarely such a clean division of cases; and the usual way of drawing the distinction—"mere rhetoric," on the one hand, versus "logic," on the other—leaves a lot to be desired. As Donald McCloskey insists, the distinction isn't a simple one;[56] but the failure of simplified attempts to draw it doesn't show that there is no legitimate distinction to be made. Casting aspersions on one's opponents' competence, for example, which at first sounds definitely in the epistemically inefficient category, can't automatically be classified with mere name-calling; for warrant depends in part on each scientist's grounds for confidence in the competence of the others on whose work he depends—a thought that puts me in mind of the trouble I sometimes get myself into when I take up an airline magazine crossword where the passenger before me left off.

Similarly, liveliness of writing can't automatically be classified with mere emoting; for a paper or presentation too dull to hold the audience's attention will fail to transmit information successfully.

The epistemologically desirable kind of communication cannot be a matter of mere *Sprachethik*, of norms of good conversational conduct—at least, not as such norms are ordinarily understood. McCloskey writes of listening, of paying attention, of not raising one's voice;[57] however, as Crick observes, reflecting on his working relationship with Watson, successful collaboration requires that "you must be very candid, one might almost say rude, to the person you are working with"; when "politeness creeps in . . . this is the end of good collaboration in science."[58]

But neither is clear presentation of evidence very aptly described as "logic." Formal-logical cogency is necessary—if the Pauling-Corey model of DNA were correct, such and such consequences would follow; but these consequences are incompatible with known chemistry; ergo . . . , etc.; but it is not sufficient. Not only does evidence ramify in all directions in a structure more like a crossword puzzle than a logical proof, but—the point that will most interest the serious student of scientific language—because of the connection of supportiveness of evidence with explanatoriness, and of explanatoriness with kinds, the vocabularies of the sciences shift and mutate as classifications are revised and refined.

So rhetoric of science will bear not only on questions about the transmission of information, but also on other questions even closer to the core of scientific inquiry, about the way the vocabulary of science grows and takes on information. Progress isn't always a matter of shifting around the truth-values assigned to propositions in a fixed vocabulary; it often involves changing, refining, modifying, or thickening the vocabulary itself. Scientific inquiry is hampered without good terminology, and good scientific terminology is itself an achievement of inquiry, dense with theory.

A simple example: after the early observations of crystals of the pigment of blood, in 1864 Felix Hoppe-Seyler gave the protein crystallized from blood its name: "hematoglobin" or "hemoglobin." As Judson puts it, "the word comes apart like the molecule":[59] "heme," from the Greek word for blood, adopted to refer exclusively to the iron-bearing, non-proteinaceous component; and "globin," for the colorless proteinaceous part. A less simple example: in 1869 Friedrich Meischer discovered a substance in the nucleus but distinct from the proteins, the function of which he thought was to store phosphorous; he called it "nuclein." By 1889 Richard Altmann had succeeded in obtaining protein-free nuclein, and called this phosphorous-containing acidic component "nucleic acid." Only in 1944 did DNA acquire its remarkably informative name, "deoxyribose nucleic acid,"[60] and not until 1952 was it generally acknowledged to be the genetic material. Later again, "pentose nucleic acid" was found to be, more specifically, ribo nucleic acid, subsequently acknowledged to be acid*s*, in the plural (and found mostly not in the nucleus, but the cytoplasm); and then, almost a century after "nuclein" was coined, we had "transfer RNA," "messenger RNA," and so on (see box on the linguistic archaeology of "messenger RNA," right).[61]

Theory is fallible; and scientific terminology may be bad as well as good, and may sometimes fail to pick out anything real. As theories are modified, meanings will shift, and translation of a later theory into the vocabulary of an earlier may be possible only by way of clumsy circumlocution. This argues in favor of a modest version of what philosophers of scientific language call the "meaning-variance" thesis: scientific terms take on new content as inquiry proceeds; and as new theories replace old the meanings of some terms retained through theory-change will shift, so that the same form of words will not always have the same meaning in the new theory as in the old. It doesn't follow, as some have thought, that supposedly rival theories are always logically incomparable, and so not really rivals. Supposedly rival theories are either intertranslatable (perhaps only by way of long-winded paraphrase), or not. For two theories to be incompatible, some claim made by the one must translate into the negation of some claim made by the other. So if the theories are not intertranslatable, they are compatible with each other; but if they are intertranslatable, they may be logically compatible with each other, or incompatible.

Metaphor is one source of novel scientific terminology; but this is by no means its only role in science. Some scientific metaphors, though cognitively inessential, oil the wheels of communication agreeably. Hershey and Chase's experiments identifying DNA as the genetic material were known as the "Waring blender experiments," because that was what they used to break the virus coat off bacteria infected with viruses. Wollman and Jacob's experiments (also using a Waring blender, which Wollman had bought as a present for his wife who, however, "was completely disgusted with this instrument," which had no place in a decent French kitchen) came to be known as the "coitus interruptus" experiments, because they interrupted the process at regular intervals. As the experiments proceeded, Jacob and Wollman began to conjecture that the "male" transfers a long piece of his chromosome into the "female" on a very specific time schedule, broken when mating is interrupted; hence Jacob's nice phrase, "the Spaghetti Hypothesis."[62]

THE LINGUISTIC ARCHAEOLOGY OF "MESSENGER RNA"

messenger RNA (1961): an RNA that carries the code for a particular protein from the nuclear DNA to a ribosome in the cytoplasm and acts as a template for the formation of that protein.

ribosome (ca. 1958): any of the RNA-rich cytoplasmic granules that are sites of protein synthesis.

RNA (1948): any of various nucleic acids that contain ribose and uracil as structural components and are associated with the control of cellular activities.

DNA (1944): any of various nucleic acids that are localized esp. in cell nuclei, are the molecular basis of heredity in many organisms, and are constructed of a double helix held together by hydrogen bonds between purine and pyrimidine bases which project inward from two chains containing alternate links of deoxyribose and phosphate.

cytoplasm (1874): the protoplasm of a cell external to the nuclear membrane.

protein (ca. 1844): any of numerous naturally occurring extremely complex combinations of amino acids that contain the elements carbon, hydrogen, nitrogen, oxygen, usu. sulfur, and occas. other elements.

nucleus (1704): a cellular organelle that is essential to cell functions . . . , is composed of nuclear sap and a nucleoprotein-rich network from which chromosomes and nucleoli arise, and is enclosed in a definite membrane.[*]

****

Every symbol is a living thing, its meaning inevitably grows, incorporates new elements and throws off old ones. Science is continually gaining new conceptions. . . . How much more the word *electricity* means now than it did in the days of Franklin; how much more the term *planet* means now than it did in the time of Hipparchus. These words have acquired information.

—C. S. Peirce

You must remember, at that time even suitable terms were hard to find. I was trying to say that there was something else besides genes; later I began to call them controlling elements.

—Barbara McClintock,
recalling the work on genetic control systems she began in 1944–45

[*] Definitions and dates are taken from *Webster's Ninth Collegiate Dictionary* (1991).

But some scientific metaphors go beyond merely picturesque speech; they are serious speculative instruments. The cognitive usefulness of such metaphors, in scientific inquiry as elsewhere, is to direct inquiry into new avenues; their worth, therefore, depends on the fruitfulness of the intellectual territory to which those avenues lead.[63] Some scientific metaphors call on familiar social phenomena; but a reasonable rhetorician will resist the temptation to judge their worth by reference to the desirability or otherwise of the social phenomenon in question. If, for example, he takes an interest in William Hamilton's metaphor of parental investment—described by Robert Trivers as "the most important advance in evolutionary theory since the work of Charles Darwin and Gregor Mendel"[64]—he will be less interested in the merits of capitalism than in how the metaphor is cashed out (sorry!) in literal terms, and how well the resulting account stands up. However, as I soon discovered, this is no easy task for an outsider; for the parental investment metaphor seems to have mutated as it passed from hand to hand, spelled out now this way, now that. Hamilton himself, so far as I can tell, doesn't use the phrase; but in the introduction to his 1964 article offering a mathematical model of inclusive fitness that will allow "a limited restraint on selfish competitive behaviour and [the] possibility of limited self-sacrifice," he writes of "sacrifices involved in parental care."[65] Trivers defines parental investment as "any investment by the parent in an individual offspring that increases the offspring's chances of surviving (and hence reproductive success) at the cost of the parent's ability to invest in other offspring," including the cost of producing the reproductive material, but also time or effort expended on behalf of the offspring. "The general rule," he says, "is this: females do all the investing and males none of it." The parental investment idea, he suggests, explains inter alia the sexual insecurity of basically monogamous male pigeons, and their tendency to stray to other females.[66]

Martin Daly and Margo Wilson, who attribute the parental investment metaphor to Ronald Fisher and Trivers, observing that "every animal is an investor," and that the "ultimate payoff is always . . . the inclusion of the investor's genes in future gene pools," also suggest that, because of the vastly greater size of ova relative to sperm, that "the female's parental investment in each offspring surpasses the male's." However, they acknowledge, "the application of the economic model is seldom straightforward in practice," and male and female reproductive strategies aren't always so different.[67] Richard Dawkins is also preoccupied with expensive eggs versus cheap sperm, but acknowledges other costs such as food, risk, and time devoted to nest maintenance. "Individuals of either sex 'want' to maximize their total reproductive output," he writes, and "males are in general likely to be biassed towards promiscuity and lack of parental care." Depending on which female counter-ploy evolves, we can expect to find "he-man" species (fast females, philandering males) and "domestic bliss" species (coy females, faithful males); in practice, Dawkins adds, we find all intermediates between the two, including some in which fathers do more child-care than mothers. He will not, he continues, discuss what might predispose a species toward one form of breeding system rather than another.[68] He and David Barash, pursuing the metaphor in another direction, both stress that past parental investment is, as economists say, "sunk cost"; the crucial question is: "What is the best course of action *now*?" not: "How much have I already invested?"[69]

E. O. Wilson defines parental investment as "behavior towards offspring that increases the fitness of the latter at the expense of the parent's ability to invest in other offspring"; thus apparently including only the costs of child-rearing, not the costs of courtship or the relative size of egg versus sperm. The parental investment metaphor, he says, has given rise to new insights into "sex ratios, marriage contracts, parent-offspring conflict, grief at the loss of a child, child abuse, and infanticide."[70]

All this suggests—an important point, I think—that a serious scientific metaphor may lead different people in different directions, perhaps some better and some worse. Despite the apparent differences about what costs parental investment includes and what investment-maximization explains, the metaphor has evidently played a significant role in enabling understanding of how evolution might explain "altruistic" behavior. Still, noticing how Trivers jokes that the salary Harvard University paid him for lecturing on reproductive success was inadequate to sustain any of his own,[71] and how Dawkins talks constantly of what organisms "should 'want'," I wonder if the metaphor hasn't also helped to disable critical thinking about the interaction of the social and the biological in human sexual behavior: e.g., about why, if female coyness is the biological default-position, many societies find it necessary to impose harsh sanctions on women who aren't coy enough.

To get to the bottom of all that, however, even supposing I could, would be a whole other enterprise. The point I want to stress here is that a reasonable rhetoric of science will acknowledge that a metaphor may be a better guide *or* a worse; that progress may be enabled by epistemically efficient transmission of information, *or* hampered by epistemically inefficient transmission; that scientific terms may take on information, *or* misinformation, and sometimes (most often in the social sciences, but perhaps not only there) evaluative coloration; and so on. After all, science is a thoroughly human enterprise: fallible and imperfect, ragged and uneven, but for all that, remarkably successful, as human enterprises go.

AND, IN CONCLUSION

Science and literature are both human enterprises, of course, though by my lights quite different ones. This is not to deny that they use some of the same human capacities; of course they do—like ballet and baseball, or chemistry and cooking. And neither is it to make any judgment about their relative worth. Is literature or science more valuable? The question seems to me as unanswerable as, "Would you rather be blind or deaf?"

Worries about the "two cultures," or, more recently, the "culture wars," the "science wars," and all that, focus in the main on perturbations in an academic teacup. The most troubling of these, to my mind, is less those English professors' ignorance of specific scientific ideas and theories, than a weakening of their grip on the differences between inquiry and fiction; and the fact that some of the goddamn philosophy professors have been tempted to follow suit, transmuting their enterprise from a kind of inquiry—which is what, if it is to be worth anything, philosophy must be—into just "a kind of writing."[72]

But other questions about the role of science in our culture are not academic in either sense of the word. Admiration for the achievements of the sciences by no means precludes acknowledgment that there are legitimate worries about the place of science in society, about access to scientific information, funding priorities, regulation of potentially dangerous research, application of scientific results, and so forth. These questions cannot responsibly be left to scientists alone; but resolving them wisely requires good information about the relevant scientific work—no easy matter. Even scientists themselves, however well informed about developments in their specialty, have to rely on others in other areas of science; and the rest of us have to rely on scientists and scientific journalists able and willing to communicate scientific developments to the lay public. Similarly, as the courts rely more and more on scientific evidence, the problem of how to discriminate decent work from flimsy speculation becomes more and more acute. But now I am wandering into a whole new bramble-patch.

NOTES

1. From Delbrück's speech accepting the Nobel Prize in 1969; my source is Judson, *The Eighth Day of Creation*, p. 614.
2. This now-famous phrase is Glymour's, from *Theory and Evidence*, p. ix.
3. Max Perutz, "The Pioneer Defended," p. 58.
4. The concept of the aesthetic is more complex, and has a more complex history,

than my rather casual use suggests; Harries, *The Broken Frame*, explores some of the complexities I shall set aside.

5. Einstein, "Physics and Reality," p. 295.

6. Medawar, *Induction and Intuition in Scientific Thought*, p. 48; my source is Judson, *The Eighth Day of Creation*, pp. 226, 640.

7. Jones, "Odd Outing"; Tanenhaus, "Bellow, Bloom and Betrayal."

8. Irving Wallace, *The Prize*, p. 702.

9. Stendhal, *The Red and the Black*; Twain, *Life on the Mississippi* (I owe the latter example to Asma Uddin).

10. Shakespeare, *A Midsummer Night's Dream*, act 3, scene 2, lines 296, 323–26, 328–29.

11. Wallace, *The Man*, p. 453.

12. Which by no means makes it, as Medawar suggests, a "fraud."

13. See chapter 5, pp. 135 ff.

14. A possibility drawn to my attention by Michael Neumann.

15. Eliot, *Daniel Deronda*, p. 268.

16. Or falsehoods; as I recently learned (thanks to Adrian Larner) the Catholic Apostolic Church, founded early in the reign of Queen Victoria with twelve disciples, held that Christ would come in glory before all the original disciples died; but when the last of those disciples died and Christ had not come, they put an advertisement in the *Times* to announce that they were winding up their church because they had been mistaken.

17. Festinger, *A Theory of Cognitive Dissonance*; Festinger, Riecken, and Schachter, *When Prophecy Fails*.

18. Bradbury, *Stepping Westward*.

19. Lodge, *Changing Places*.

20. Crick, "The Genetic Code"; Judson, *The Eighth Day of Creation*, pp. 462, 482, 485.

21. Frege, "On Sense and Reference."

22. Richards, *Science and Poetry*, p. 29.

23. My source is Robert Merton, *Social Theory and Social Structure*, p. 70; however, Merton gives no reference.

24. Haack, *Manifesto of a Passionate Moderate*, p. vi.

25. Gross and Levitt, *Higher Superstition*, pp. 98 ff.; Hayles, *Chaos Bound*.

26. Sokal and Bricmont, *Fashionable Nonsense*; Lacan, "Of Structure as an Inmixing of Otherness Prerequisite to Any Subject Whatever"; Irigaray, "Sujet de science, sujet sexué?"

27. Sturrock, "Le Pauvre Sokal," p. 8.

28. Judson, *The Eighth Day of Creation*, p. 183.

29. Sayre, *Rosalind Franklin and DNA*, especially pp. 17 ff.; see also Judson, *The Eighth Day of Creation*, pp. 147 ff.

30. Bronowski, "Honest Jim and the Tinker-Toy Model," reprinted in Stent's edition of *The Double Helix*.

31. Limon, "*The Double Helix* as Literature," p. 36.

32. Ibid., pp. 30–31.

33. Halloran, "Towards a Rhetoric of Scientific Revolution," p. 229.

34. Avery, MacCleod, and McCarty, "Studies of the Chemical Nature of the Substances Inducing Transformation in Pneumococcal Types."

35. See Judson, *The Eighth Day of Creation*, p. 57.

36. See chapter 3, p. 78, and chaper 4, p. 103.

37. See Judson, *The Eighth Day of Creation*, p. 39.

38. Darwin, *Autobiography and Letters*, p. 105; the quotation is from *Origin of Species*, p. 440.

39. Gross, *The Rhetoric of Science*, p. 55.

40. On the question of how correct Watson's perception was that Franklin was antihelical, and that he and Crick were in a great race with Pauling, see Judson, *The Eighth Day of Creation*, pp. 141 and 151 (Franklin) and 154, 161–62, and 193 (Pauling).

41. Gross, *The Rhetoric of Science*, p. 60. Bronowski's name does not appear, however, in the index of Gross's book.

42. Thanks to Michael McCracken for reminding me of the relevance of these two sayings.

43. Gross, *The Rhetoric of Science*, pp. 59, 61.

44. Lewontin, "'Honest Jim's' Big Thriller About DNA," p. 2.

45. Gross, *The Rhetoric of Science*, p. 62.

46. Ibid., p. 200.

47. Ibid., p. 201–206; the term "rhetorical realism" is mine.

48. See chapter 5, pp. 141–43.

49. Gross, *The Rhetoric of Science*, p. 204.

50. On the presuppositions of scientific and everyday empirical inquiry, see chapter 5, pp. 124–25; on the presuppositions of theological inquiry, see chapter 10, p. 267.

51. See Vickers, *In Defense of Rhetoric*.

52. Gross, *The Rhetoric of Science*, pp. 12, 206–207.

53. Ibid., p. viii (Perutz is quoted on the page before).

54. Gross, "Learned Ignorance," p. 95.

55. Judson, *The Eighth Day of Creation*, p. 501.

56. McCloskey, *Knowledge and Persuasion in Economics*, pp. 273 ff., criticizing Maki, "Two Philosophies of the Rhetoric of Economics," and "Diagnosing McCloskey."

57. McCloskey, *Knowledge and Persuasion in Economics*, pp. 99 ff.

58. Crick, *What Mad Pursuit*, p. 13.

59. Judson, *The Eighth Day of Creation*, p. 409.

60. See Portugal and Cohen, *A Century of DNA*.

61. Olby, *The Path to the Double Helix*; Portugal and Cohen, *A Century of DNA*.

62. Judson, *The Eighth Day of Creation*, pp. 394–95. As the scare quotes indicate, the use of "male" and "female" in this context is an extended one: bacteria with F factor or F plasmid are referred to as "male," those without as "female" (I owe this point to David Wilson).

63. See also Haack, "Dry Truth and Real Knowledge: Epistemologies of Metaphor and Metaphors of Epistemology"; Boyd, "Metaphor and Theory Change." In a general

way, I find Boyd's conception of the role of metaphor in science sympathetic; but I don't subscribe, as he does, either to Black's account of metaphor, or to the Putnam-Kripke model of scientific language.

64. Trivers, *Social Evolution*, p. 47.

65. Hamilton, "Genetical Evolution of Social Behaviour," part 1, p. 1.

66. Trivers, *Social Evolution*, p. 207 (definition of parental investment; females do all the investing), and 203 (pigeon behavior).

67. Daly and Wilson, *Sex, Evolution, and Behavior*, pp. 31–32, 56–57.

68. Dawkins, *The Selfish Gene*, chapter 9.

69. Barash, *Sociobiology and Behavior*, p. 189.

70. Wilson, *Consilience*, p. 169.

71. Trivers, *Social Evolution*, p. 52.

72. Rorty, *Consequences of Pragmatism*, p. 92; see also Haack, "As for That Phrase, 'Studying in a Literary Spirit' . . . ."

9

ENTANGLED IN THE BRAMBLE-BUSH

Science in the Law

[T]here are important differences between the quest for truth in the courtroom and the quest for truth in the laboratory. Scientific conclusions are subject to perpetual revision. Law, on the other hand, must resolve disputes finally and quickly. The scientific project is advanced by broad and wide-ranging consideration of a multitude of hypotheses, for those that are incorrect will eventually be shown to be so, and that in itself is an advance. Conjectures that are probably wrong are of little use, however, in the project of reaching a quick, final and binding legal judgment—often of great consequence—about a particular set of events in the past.

—Justice Harry Blackmun[1]

Science is entangled with society in a host of ways: with industry, education, the press, religion, and the law. And science is entangled with the law in a host of ways: in legal regulation of scientific research; in the law of patents and intellectual property; in constitutional cases over creationism in science classes;[2] in legal wranglings over whether recently discovered ancient human remains must be given to the Native American tribes who claim them, or may be investigated by scientists to determine their origin;[3] and—the topic on which I shall focus here—in rules and decisions about what scientific testimony is admissible in court. My concern, more precisely, will be the evolving mesh of rules, precedents, and practices the law has devised to handle scientific evidence and to reconcile the values, practices, and priorities of science with its own.

Even this relatively limited topic presents formidable complexities; for the law really is, as Carl Llewellyn's marvelous metaphor of the bramble-bush suggests, living, constantly growing, and—though it produces valuable fruit—thorny and tangled, sometimes seemingly impenetrable. Moreover, in criminal cases and civil, the courts need to call on scientific testimony of just about every kind imaginable: forensic identification evidence such as fingerprints, hair or blood or DNA analysis, identification of bullets or tool-marks or bite-marks; psychological evidence about the reliability of eyewitness testimony or of memory, about insanity, mental incompetence, battered wife syndrome, rape trauma syndrome, recovered memories, or future dangerousness; toxicological and epidemiological evidence of the risks of drugs and medical procedures, or of exposure to this or that chemical; diagnoses of the causes of tire blowouts and boiler explosions; etc., etc.

DNA evidence vividly illustrates both the potential and the problems. Who could have imagined, when DNA was first identified as the genetic material a half-century ago, that DNA analysis would by now have come to play such a role in the criminal justice system, and in the public perception of the law? Even twenty years ago, forensic scientists could tell only whether a blood sample was animal or human, male or female, and, if human, of what type—the least common blood type being found in 3 percent, and the commonest in 43 percent, of the U.S. population. DNA analysis has made vastly more accurate identification possible (reportedly, the FBI laboratory estimated the probability of a random match of the stain on Monica Lewinsky's dress to President Clinton's semen as one in 7.87 trillion). First introduced in a criminal case in the United States in 1986, its power to enable justice has been startling; by spring 2002, DNA testing had exonerated more than one hundred prisoners, including a significant number on death row.

Despite many complications, practical and legal—samples may be lost or destroyed; testing is expensive; in some jurisdictions it is necessary to litigate to get test results considered after the appeals process has been exhausted—the legal system has begun slowly to adapt. Indeed, DNA evidence is so powerful that it has already brought about some modification of the law's traditional emphasis on "quick, final and binding judgments": some states have enacted legislation mandating post-conviction DNA testing, and some have extended or eliminated the statute of limitations where DNA evidence may be available; others have legislation pending.[4]

Of course, policemen, and laboratories, are fallible; sometimes, for example,

samples are not presented blind, but in a way that suggests a positive identification is expected. Moreover, DNA analysis can rule out a suspect with greater certainty than it can identify a perpetrator, which requires sometimes tricky assumptions about the reference class. So it is hardly surprising that DNA evidence is not
only a powerful tool for exonerating the innocent and identifying the guilty, but
also (as the O. J. Simpson case brought home to all of us) potentially powerfully
confusing. Juries may be baffled when asked to weigh an extremely high probability, assuming the test was properly conducted, that an innocent defendant
would not have the same DNA profile as the perpetrator, with a much lower, but
non-negligible, probability that the laboratory made a mistake or the sample was
contaminated.[5] A Carnapian judge in Canada would not allow DNA experts to
testify to the random-match probability—fearing that, rather than weighing this
along with other evidence, the jury would simply take it as the probability that the
defendant was guilty.[6] In the U.S., not only have some sophisticated criminals
learned how to avoid leaving any traces that might identify them, but at least one
prisoner has apparently tried to exploit the potential for confusion by petitioning
for, and obtaining, a DNA test—which, however, confirmed his guilt.[7]

Most recently it has been those dramatic DNA exonerations that have made
the headlines: "The Death Penalty on Trial," "DNA: The Great Detective," and
so forth. Only a few years ago, however, it was the epidemiological, toxicological, clinical, etc., evidence in major tort cases that was in the news. And here, in
the wake of an influential book, *Galileo's Revenge*, in which Peter Huber stressed
the greed of plaintiffs' attorneys hoping to earn huge contingency fees by winning cases with "junk science,"[8] it was not the potential of scientific evidence to
assist the legal fact-finding process, but the danger that juries would be bamboo-
zled by flimsy scientific speculation, that was to the fore. Huber's critics, such as
Kenneth Cheseboro, were more inclined to stress the heartlessness of huge cor-
porations hoping to avoid compensating the victims of their profitable but dangerous products.[9] But by 1997, as *General Electric Co.* v. *Joiner* went before the
Supreme Court, headlines in the *Wall Street Journal* announced: "Judges
Become Leery of Expert Witnesses"; they are "Skeptical of Unproven Science"—the "Testimony of Dilettantes."[10]

The history of efforts to ensure that, when the legal system relies on scientific testimony, it is not flimsy speculation but decent work, is long and tortuous.
In fact, as legal rules and standards for the admissibility of such testimony are
formulated and reformulated, interpreted and reinterpreted, repudiated and resur-
rected, as one rule is judged too restrictive and replaced by another that is ostensibly less so but turns out to submit proffered scientific testimony to far more
serious scrutiny, it begins to seem like nothing so much as a formal dance: two

steps forward, three steps back, one step forward—or perhaps, as the responsibility is shifted up and back from juries to judges to the scientific community, like a game of Pass the Parcel.

Most commentators focus primarily on the legal, political, practical, and institutional problems of scientific testimony—on excesses and abuses of the adversarial process, on staggering jury verdicts and enormous contingency fees in the tort system, and so forth. However, to judge by how frequently, in the history of the law on scientific testimony, implicit or explicit assumptions about the nature of scientific evidence and the character of scientific inquiry have been crucial, there is something about scientific testimony that encourages and enables the operation of such unsavory motives as Huber and Cheseboro describe; the problems, that is, are in part epistemological. In fact, I believe, the epistemological problems are fundamental; but they are magnified and amplified by those legal, political, practical, and institutional factors.

Learned Hand glimpsed this when he wrote, a century ago now, that the expert witness is "an anomaly": an anomaly from a legal point of view because, unlike other witnesses, he is allowed to offer his opinion; and an anomaly from an epistemological point of view because, when the parties present opposing expert witnesses, the jury must decide "between two statements each founded upon an experience foreign in kind to their own," when "it is just because they are incompetent for such a task that the expert is necessary at all." With expert witnesses, Hand continued, we set the jury "to decide, *where doctors disagree*."[11]

Hand appreciates one of the epistemological difficulties: though scientific inquiry is continuous with everyday empirical inquiry, there is usually no way of judging the worth of scientific evidence without substantive knowledge of the field in question. When a lay person, or even a scientist from another specialty, tries to judge the quality of evidence for a scientific claim, he is liable to find himself in the position of the average American asked to judge the reasonableness of entries in a crossword puzzle where, though some of the clues are in pidgin English, the solutions are all in Bengali, and require a knowledge of Islam. Hand also appreciates one of the ways in which the adversarial system can amplify this epistemological difficulty: when experts are hired by the parties, there is a built-in incentive to present evidence in a slanted way, and to omit or obfuscate awkward facts. But when Hand urges court-appointed experts as a simple solution, he reveals that he hasn't appreciated all the epistemological difficulties. Perhaps he thinks that experts disagree only because they are hired by the parties to do so; but there is no guarantee that specialists in a scientific field won't sometimes disagree for legitimate reasons. And it doesn't seem to occur to him that sometimes warranted answers may be simply unavailable to anyone.

A few epistemologically ambitious commentators have argued, in effect, that the true account of scientific knowledge is such-and-such, and, in consequence, the right legal formula or procedure for handling scientific evidence in court is thus and so.[12] My purpose here is nothing so ambitious. Legal efforts to winnow decent scientific evidence from the chaff, I agree, have often been based on false assumptions about science and how it works. Unfortunately, however, when we correct those false assumptions, no easy or obvious solutions suddenly appear. But we may begin to appreciate why it has proven impossible to find a legal form of words that will ensure that only decent scientific evidence is admitted, or a simple way to delegate the responsibility to scientists themselves; and, perhaps, accepting that mistakes are inevitable, to wonder what kinds of mistake we should be more willing, and what less willing, to tolerate.

I shall begin with a brief legal history; and then, very cautiously approaching that legal bramble-bush with my philosophical pruning-shears, try to identify some of the misunderstandings about science that have impeded legal efforts to distinguish decent scientific evidence from junk, and some of the ways in which the legal system magnifies and amplifies the inevitable difficulties posed by scientific testimony.

THE LAW ON SCIENTIFIC TESTIMONY: A BRIEF HISTORY

Once upon a time, in cases where expert knowledge was required, jurors with the necessary expertise were specially selected—e.g., a jury of butchers when the accused was charged with selling putrid meat; and sometimes specially qualified persons would be summoned to help determine some matter of fact which the court had to decide—e.g., masters of grammar for help in construing doubtful words in a bond. Hand reports that the first case he can find of expert testimony as an exception to the rule that the conclusions of a witness are inadmissible was in 1620, in *Alsop* v. *Bowtrell* (see timeline, next page).[13] And only considerably later, with the gradual evolution of the adversary system and the rise of the hearsay rule, do we find expert witnesses called by the parties and subject to cross-examination; Stephan Landsman points to *Folkes* v. *Chadd* (1782)—in which Lord Mansfield "placed the court's seal of approval on the whole adversarial apparatus including contending experts, hypothetical questions, and jury evaluation"[14]—as an early example of expert testimony in this modern sense.

TIMELINE OF SELECTED CASES AND RULES

1620. *Alsop* v. *Bowtrell* [England]: first case of expert testimony as exception to the rule that the conclusions of a witness are inadmissible.

1923. *Frye*: novel scientific evidence is admissible only if it is "sufficiently established to have gained general acceptance in the particular field to which it belongs."

1975. Federal Rules of Evidence: 104(a): courts' gatekeeping role; 401: relevant evidence is admissible unless otherwise provided by law; 403: unfair prejudice, confusing or misleading the jury, and waste of time are grounds for exclusion; 702: expert evidence, including scientific evidence, is admissible subject to exclusion under 403; 706: a court may appoint its own expert witnesses.

1983. *Barefoot*: defendant's constitutional rights were not violated by admitting psychiatric testimony of his future dangerousness (though the American Psychiatric Association reports that two out of three such predictions are wrong). The rules of evidence "anticipate that relevant, unprivileged evidence should be admitted and its weight left to the fact-finder."

1985. *Downing*: multi-factor test "in the spirit of the Federal Rules": expert testimony should be relevant, not confuse or mislead the jury, and should be reliable.

1993. *Daubert*: FRE supersede *Frye*. Interprets FRE as requiring court to screen proffered scientific evidence for relevance *and reliability*. In judging reliability, courts should look to the methodology, not the conclusions, of proffered experts. Factors that may be considered: falsifiability (taken to be the mark of real science); known or potential error rate; peer review and publication; acceptance in the scientific community.

1996. Judge Sam Pointer (to whom federal silicone breast-implant cases are consolidated) sets up Rule 706 National Science Panel to look into the scientific evidence.

1997. *Joiner*: the standard of appellate review for decisions on admissibility of evidence is (not a "hard look," but) abuse of discretion. There is no real distinction between methodology and conclusions; courts may look to both in judging reliability.

1999. *Kumho*: the *Daubert* ruling applies to all expert testimony, not only to scientific expert testimony. *Daubert*'s list of factors is "flexible"; it is within courts' discretion to determine which factors are appropriate in the case at hand, and to use all, any, or none of them.

2000. Revised Federal Rules of Evidence: 701: allows limited opinion evidence by lay witnesses; 702: reliability as well as relevance required for admissibility of expert testimony.

Between roughly 1850 and 1920, the only test for the admissibility of expert, including scientific, testimony was whether the witness was qualified, or, in a more sophisticated version, whether he had expertise beyond the range of knowledge of the average juror. But then *Frye* v. *United States* (1923) gradually began to set a new standard for the admissibility of scientific evidence.

Mr. Frye was charged with murder, and had confessed. Later, however, he repudiated the confession; and took, and passed, a polygraph test (or more exactly, a discontinuous test of systolic blood pressure changes under questioning; the technology was in an early and primitive stage).[15] But the trial court judge excluded this evidence, holding that deception tests were inadmissible unless there is "an infallible instrument for ascertaining whether a person is speaking the truth or not."[16] On appeal, the D.C. court confirmed the exclusion of this testimony, ruling that novel scientific evidence "crosses the line between the experimental and the demonstrable," and so is admissible, only if it is "sufficiently established to have gained general acceptance in the particular field to which it belongs."[17]

In its first decade, *Frye* was not cited even once; and in its first quarter-century it was cited in only eight federal and five state opinions, primarily in reference to lie-detector evidence. But by the early 1980s it was being cited over and over, and twenty-nine states had adopted the "*Frye* rule" as their standard for admitting novel scientific testimony.

In *Frye*, the D.C. court had offered almost no argument for its ruling; in 1980, however, in *United States* v. *Addison*, the same court argued that "the requirement of general acceptance in the scientific community assures that those most qualified to assess the validity of a scientific method will have the determinative voice."[18] Rather than requiring only that scientific experts be qualified, the *Frye* test has courts rely obliquely on the verdict of the appropriate scientific sub-community to determine whether scientific evidence proffered is solidly established work or unreliable speculation, making general acceptance a surrogate for a direct legal requirement that admissible scientific evidence be reliable. But the *Frye* rule is indeterminate in many ways: what, exactly, must be accepted by what proportion of what group? how is a court to discriminate the doctor who legitimately disagrees from the crank? and how is it to go about determining whether proffered evidence is generally accepted? No wonder that, as *Frye* was applied and contested in the courts, its effect was sometimes more and sometimes less restrictive, with voice-print evidence, for example, sometimes admitted, sometimes excluded.[19]

Sometimes the relevant community was very narrowly defined. In *People* v. *Williams* (1958), the prosecution's own experts conceded that the medical profession was mostly unfamiliar with the use of Nalline to detect narcotic use, but the court upheld the admissibility of its evidence all the same; the Nalline test was "generally accepted by *those who would be expected to be familiar with its use*," and "in this age of specialization more should not be required."[20] And sometimes *Frye* was set aside to admit testimony about techniques too new to be generally accepted. In *Coppolino* v. *State* (1968), the prosecution was allowed to introduce the results of a test for the presence of succinylcholine chloride or its derivatives in human tissues, devised by the local medical examiner specifically for this trial—and so not known to, let alone generally accepted in, any scientific community. The appellate court cited *Frye* but, ruling that the trial judge did not abuse his discretion, nevertheless upheld the admissibility of this evidence.[21]

Despite the many inconsistencies in its application, however, the commonest criticism of the *Frye* rule was that it was too restrictive, that it kept out testimony juries should be allowed to hear. Dean Charles McCormick's influential treatise on the law of evidence, published in 1954, disapproved of courts' screening proffered testimony for reliability, and urged that "any relevant conclusions supported by a qualified expert witness should be received unless there are distinct reasons for exclusion."[22]

The Federal Rules of Evidence, enacted in 1975, encapsulated this less restrictive approach. Rule 104(a) affirms the gatekeeping role of the court in ruling on admissibility of evidence. But Rule 401 states that relevant evidence—evidence which has any tendency to make the existence of any fact of consequence to the determination of the action either more or less probable than it would otherwise be—is admissible unless otherwise provided by law. Rule 702 states that expert evidence, including but not restricted to scientific evidence, is admissible subject to exclusion under Rule 403. Rule 403, specifying the grounds for exclusion, mentions the danger of unfair prejudice, confusion of the issues, or misleading the jury, but does not mention general acceptance in the appropriate scientific community. Rule 706 allows a court to appoint its own experts.

In *Barefoot* v. *Estelle* (1983), the question before the Supreme Court was whether a Texas defendant's constitutional due-process rights were violated by the admission of psychiatric testimony concerning his potential future dangerousness. As required by Texas statute, one of the questions put to the jury in the sentencing phase was whether there was a probability that Mr. Barefoot would commit further criminal acts of violence. The State called two psychiatrists, neither of whom had examined Mr. Barefoot: Dr. Holbrook testified, "within reasonable psychiatric certainty," that Mr. Barefoot would be violent in future; and Dr. Grigson (notorious throughout Texas as "Dr. Death") testified that there was a *"one hundred percent and absolute"* certainty of this.[23] However, according to an amicus brief filed with the Supreme Court by the American Psychiatric Association, two out of thee predictions of future dangerousness are mistaken, mostly in the direction of over-prediction.

The majority of the Supreme Court affirmed the appeals court's denial of habeas corpus. "Neither Barefoot nor the Association [the APA] suggests that psychiatrists are always wrong with respect to future dangerousness, only most of the time," Justice White observed;[24] and to accept the petitioner's argument would call into question all the other contexts in which the legal system relies on predictions of future behavior. It is not the judge's role to exclude relevant evidence, he continued, but up to the jury to determine which experts are credible; state and federal rules of evidence "anticipate that relevant, unprivileged evidence should be admitted and its weight left to the fact-finder, who would have the benefit of cross examination and contrary evidence by the opposing party."[25]

But Justice Blackmun wrote in an outraged dissent that the majority's holding that the psychiatric testimony in this case was admissible because it was subject to cross-examination and impeachment "is too much for me. . . . In a capital case, the specious testimony of a psychiatrist, colored in the eyes of an impressionable jury by the inevitable untouchability of a medical specialist's words, equates with death itself."[26] He continued: "[I]n the mind of the typical lay juror, a scientific witness has a special aura of credibility. . . . It is extremely unlikely that the adversary process will cut through the facade of superior knowledge."[27]*

For almost twenty years it remained unclear whether the Federal Rules superseded or complemented *Frye*. Scholars debated whether the Federal Rules were compatible with the *Frye* test: some arguing that they weren't, because they didn't mention consensus in the relevant community, and some arguing that they were, because they didn't mention consensus in the relevant community (!).[28] Some courts suggested that the *Frye* general acceptance test be replaced by a less restrictive "substantial acceptance" standard, and the 1987 edition of a textbook on the Federal Rules suggested irenically that the *Frye* test be reconstrued under Rule 403 as "an attempt to prevent jurors from being unduly swayed by unreliable scientific evidence."[29]

By the mid-1980s stress on reliability was growing. In *United States* v. *Downing* (1985), vacating a trial court decision to exclude expert psychological evidence about the weaknesses of eyewitness testimony, Judge Becker proposed,

---

* Mr. Barefoot was executed on October 24, 1984. After repeated reprimands, Dr. Grigson was expelled from the American Psychiatric Association in 1995 for his irresponsible testimony in Texas capital cases. Though in accordance with an amendment of September 2001 the state may not offer evidence to establish that the race or ethnicity of the defendant makes it likely that he or she will engage in future criminal conduct, Texas law continues to require testimony as to future dangerousness in the sentencing phase of a capital case (Texas Criminal Code, art. 37.071).

in "the spirit of the Federal Rules," a three-factor test for admissibility which added a requirement of reliability to the explicit requirements in the Rules that the evidence be relevant and that it not confuse or mislead the jury. And by 1992, when Huber had popularized the perception that the courts were being flooded with "junk science," initiatives were under way to include a reliability requirement in the FRE. These were still hanging fire, however, when the Supreme Court pre-empted them by reinterpreting the FRE in the landmark scientific-evidence case, *Daubert* v. *Merrell Dow Pharmaceuticals* (1993).[30]

The *Frye* rule arose in a criminal case, and for most of its history was cited only in criminal cases. *Daubert*, however, was a civil case in which the trial court relied almost exclusively on *Frye* in ruling the plaintiff's expert evidence inadmissible. The plaintiffs were two minor children and their parents, and the claim was that the children's birth defects were caused by their mothers' having taken the morning-sickness drug Bendectin during pregnancy. But the plaintiffs' expert evidence (based on animal studies, pharmacological studies of the chemical structure of Bendectin, and an unpublished "re-analysis" of previously published human statistical studies) was disqualified under the *Frye* test. The Ninth Circuit Court of Appeals confirmed the trial court's decision to exclude; the proffered evidence, which was not peer-reviewed, was not generally accepted in the relevant scientific community.

But in 1993, reversing the exclusion of Mr. Daubert's proffered expert testimony, the majority of the Supreme Court repudiated the *Frye* test as an "austere standard, absent from, and incompatible with, the [Federal Rules]." However, while holding that the Federal Rules superseded *Frye*, and even as it stressed the Rules' preference for admissibility, the Supreme Court interpreted the Rules as requiring not only relevance but also reliability: "[U]nder the Rules the trial judge must ensure that any and all scientific testimony or evidence admitted is not only relevant, but reliable."[31]

Jurors, whose job it is to determine sufficiency, are to concern themselves with expert witnesses' conclusions; but judges, whose job it is to determine admissibility, must focus "solely on principles and methodology" to make "a preliminary assessment of whether the reasoning or methodology underlying the testimony is scientifically valid and . . . properly can be applied to the facts in issue."[32] In determining whether what is offered is really scientific knowledge— *knowledge*, not mere opinion, and genuinely *scientific* knowledge, "with a grounding in the methods and procedures of science"—a key question will be "whether it can be (and has been) tested."[33] Justice Blackmun's opinion for the majority quotes law professor Michael Green: "Scientific methodology today is based on generating hypotheses and testing them to see if they can be falsified; indeed, this methodology is what distinguishes science from other fields of human inquiry,"[34] and refers to Popper and Hempel. Retaining something of the *Frye* test in the liberalized form of indications, rather than necessary conditions, of admissibility, the ruling also mentions peer-review, a "known or potential error rate," and "widespread acceptance"—the so-called *"Daubert* factors."

However, pointing out that there is no reference in Rule 702 to reliability, urging that the question of expert testimony generally not be confused with the question of scientific testimony specifically, and criticizing his colleagues' excursus into philosophy of science as unnecessary and unwise, Justice Rehnquist observed:

> I defer to no one in my confidence in federal judges; but I am at a loss to know what is meant when it is said that the scientific status of a theory depends on its 'falsifiability,' and I suspect some of them will be, too. . . . I do not think [Rule 702] imposes on them either the obligation or the authority to become amateur scientists.[35]

As Hand might have put it, *Daubert* set federal judges to decide "where doctors disagree." Judge Kozinski, to whom *Daubert* was remanded, grumbled that, though he and his colleagues on the federal bench are "largely untrained in science and certainly no match for any of the witnesses whose testimony we are reviewing," the Supreme Court's decision obliged them to determine whether those witnesses' testimony is "good science" derived by the "scientific method."[36] In ruling, as he had before, that Mr. Daubert's proffered expert testimony was inadmissible, Judge Kozinski argued first that most of the proffered testimony wasn't relevant, since the experts didn't even claim that it was more likely than not that Bendectin caused the birth defects; and then that Dr. Palmer, the only witness who did claim this, said nothing about his methodology, and so failed the reliability requirement. Introducing a new *"Daubert* factor" of his own, Judge Kozinski suggested that litigation-driven science is less reliable than other science, because it is potentially tainted by the parties' interests and less subject to the scrutiny of others in the field. Most to the present point, his ruling clearly relied less on scientific judgment than on legal legerdemain.*

---

* Merrell Dow had already taken Bendectin off the market a decade before *Daubert*; appeals courts had ruled in their favor, but litigation costs were prohibitive. In 2000, however, after seventeen years of research—much of it litigation-driven—the FDA again declared the drug safe. Duschesnay, Inc., which since 1975 has marketed a generic version of Bendectin, Diclectin, in Canada, has asked the FDA for permission to sell Diclectin in the United States, and has been required only to prove that the drugs are chemically identical.[37]

It is difficult, to put it mildly, to square Justice White's construal of the Federal Rules in *Barefoot* (relevance is sufficient) with the much more restrictive interpretation the Supreme Court gives them in *Daubert* (reliability is also necessary).[38] And in fact, despite the stress in the *Daubert* ruling on the Federal Rules' preference for admissibility, since *Daubert* federal judges have tended, on the whole, to be more active and more restrictive gatekeepers than before. So it is ironic that the Florida Supreme Court, which in *Flanagan* v. *State* (1993) had chosen to stick with *Frye* rather than adopt *Daubert*,[39] argued in *Hadden* v. *State* (1997) that *Frye* is preferable because it is more stringent.

United States v. *Bonds* (1993) was briefed in accordance with *Frye* but decided in accordance with *Daubert*. DNA analysis from the FBI laboratory was ruled admissible on the grounds that it was testable, and had been tested; the court acknowledged that the lab's error-rate was disturbingly high, but argued that it is "implicit" that the FBI rate of error is "acceptable to the scientific community."[40] In other cases DNA evidence was sometimes admitted, sometimes not, error rates sometimes taken seriously, sometimes not.[41] Again, polygraph evidence has been both admitted and excluded under *Daubert*; and in *United States* v. *Scheffer* (1998) the Supreme Court left the admissibility of such evidence to courts in each jurisdiction to decide.[42] Nevertheless, though there have been many inconsistencies in its application, and despite the rhetoric about *Frye*'s imposing too "austere" a standard, the overall effect of *Daubert*, especially in civil cases, was in the direction of more rigorous judicial screening of expert testimony.

One of the many subsequent cases applying the Federal Rules as interpreted in *Daubert*[43] was the case of Robert Joiner. Mr. Joiner had worked for the Water and Light Department of the City of Thomasville, Georgia, since 1973. Among his tasks was the disassembly and repair of electrical transformers in which a mineral-based dielectric fluid was used as a coolant—dielectric fluid into which he had to stick his hands and arms, and which sometimes splashed onto him, occasionally getting into his eyes and mouth. In 1983 the City discovered that the fluid in some of the transformers was contaminated with PCBs—considered so hazardous that their production and sale has been banned by Congress since 1978—and began requiring employees to wear protective gear.

In 1991 Mr. Joiner was diagnosed with small-cell lung cancer; he was 37. He had been a smoker for about eight years, and there was a history of lung cancer in his family. He claimed, however, that had it not been for his exposure to PCBs and their derivatives, furans and dioxins, his cancer would not have developed for many years, if at all. On this basis he sued Monsanto, which had manufactured PCBs from 1935 to 1977, and General Electric and Westinghouse, which manufactured transformers and dielectric fluid. His case relied essentially on expert witnesses who testified that PCBs alone can cause cancer, as can furans and dioxins, and that since he had been exposed to all three, this had likely contributed to his cancer.

Removing the case to federal court, GE et al. contended that there was no evidence that Mr. Joiner suffered significant exposure to PCBs, furans, or dioxins, and that in any case there was no admissible scientific evidence that PCBs promoted Joiner's cancer. The district court granted summary judgment, holding that the testimony of Joiner's experts was no more than "subjective belief or unsupported speculation."[44] The court of appeals reversed, arguing that Rule 702 displays a "preference for admissibility," and in the present instance, the question of admissibility was outcome-determinative: if the scientific evidence offered were excluded, Mr. Joiner would simply have no case. So, as Judge Becker had held in *Paoli* (1994), a "particularly stringent standard of review" should apply to the trial judge's exclusion of expert testimony.[45] But in 1997, reversing the admissibility of Mr. Joiner's expert evidence, the Supreme Court held that the appropriate standard was abuse of discretion; and it was *not* an abuse of discretion for the district court to have excluded Mr. Joiner's experts' testimony.[46]

The legitimacy of the distinction between methodology and conclusions now became a hotly contested issue. The *Daubert* Court, taking the distinction for granted, had interpreted trial judges' gatekeeping role as requiring them to focus solely on methodology, not conclusions. But, Mr. Joiner's lawyers argued, the district court had no objection to the methodology of the studies cited, only to the conclusions that their experts drew; and this was a reversible error. GE's brief argued that the court of appeals treated *Daubert*'s requirement of scientific methodology "at such a superficial level as to leave it meaningless—calling for no more than the invocation of scientific materials";[47] and that Mr. Joiner's experts made the mistake of supposing that "multiple pieces of evidence, each independently being suspect or weak, provide strong evidence when bundled together"—the "faggot fallacy."[48] Mr. Joiner's lawyers replied that his experts "were applying a methodology which is well established in the scientific method. It is known as the weight of evidence methodology. . . . There are well-established protocols for this . . . published as the EPA's guidelines. There are similar guidelines for the World Health Organization."[49] GE's lawyers, they continued, never challenged Mr. Joiner's experts' methodology before; indeed, they used the "weight of evidence" methodology themselves.

Rather than challenging Mr. Joiner's claim that the district court failed to restrict its attention to methodology as *Daubert* requires, the majority of the *Joiner* Court sustained its ruling that there was no abuse of discretion by holding that *"conclusions and methodology are not entirely distinct from each other."*[50] Justice Stevens, however, protested that this was neither true nor helpful. "The difference between methodology and conclusions is just as categorical as the distinction between means and ends." The district court ruling on reliability was "arguably not faithful" to the statement in *Daubert* that the focus must be on methodology rather than conclusions. The majority "has not adequately explained why its holding is consistent with Federal Rule of Evidence 702 as interpreted in *Daubert* v. *Merrell Dow Pharmaceuticals*."[51]

The evidence on which *Daubert* was focused was scientific evidence; and as Justice Rehnquist had anticipated in his dissent, whether and if so how *Daubert* was to be applied to other, non-scientific expert testimony soon became a contested issue.

For example, on appeal in *Berry* v. *City of Detroit* (1994), Judge Guy took the *Daubert* factors to apply to non-scientific as well as scientific experts, ruling the testimony of a retired sheriff as to the inadequacy of the City's police training and disciplinary procedures inadmissible: his theories had not been tested; he had no peer-reviewed publications; there was no reason to think his ideas were generally accepted; and his methodology was as suspect as his conclusions. But in *United States* v. *Starzecpyzel* (1995) Judge McKenna ruled that handwriting analysis is a practical skill rather than a science, falling under Rule 702, not *Daubert*; so the *Daubert* factors are largely irrelevant, as is Popper's philosophy of science. The proffered testimony was therefore admissible, although—to combat the danger that they would otherwise give it too much weight—the jury should be instructed that it was not *scientific* testimony. And in *Moore* v. *Ashland Chemical Inc.* (1998), in an analysis relying on *Daubert* and *Joiner*, the majority of the appeals court held that the district court did not abuse its discretion in excluding a physician's testimony as to the cause of the plaintiff's respiratory disorder; it was "not grounded in science as required by Daubert and its progeny, and, therefore [*sic*], was not sufficiently reliable for the jury to consider." But dissenting Judges Dennis, Parker, and Stewart, concerned that since *Daubert* the federal courts had been "balkanized" on how to handle expert testimony outside "hard science," protested that the majority ruling "subverts the liberal thrust of the Federal Rules of Evidence and the principles enunciated in Daubert by locking the gate on causation evidence derived through the . . . methodology of clinical medicine."

The Supreme Court tackled this balkanization in *Kumho Tire Co., Ltd.* v. *Carmichael* (1999), where the question concerned the admissibility of the proffered testimony of an expert on tire failure (based, I am amused to read, on his "visual-inspection methodology") that an automobile accident was caused by a manufacturing defect. *Daubert*'s reliability requirement, the *Kumho* court ruled, applies to the testimony of experts of every kind, including accountants, engineers, etc., as well as scientists; there is no clear dividing line between scientific and "other specialized" knowledge, and the key word that establishes the level of evidentiary reliability is not "scientific," but "knowledge."[52] *Daubert* is "flexible," and its list of factors neither necessarily nor exclusively applies to all experts or in every case; e.g., general acceptance is no indication of reliability if the discipline involved is unreliable.[53] Where it is appropriate to use one or more of the *Daubert* factors, the court should do so; but it is within the court's discretion to determine this.[54]

Joiner established that abuse of discretion was the appropriate standard of review for questions of admissibility of expert evidence, and that judges might look to conclusions as well as methodology; *Kumho* put it within the courts' discretion to apply any, all, or none of the *Daubert* factors. Nevertheless, *Daubert* has not been entirely dismantled, for its interpretation of the FRE as requiring reliability as well as relevance as necessary conditions of admissibility remains, and *Kumho* stresses the need to look specifically at the expert's reliability regarding the task at hand.

In his concurring opinion in *Joiner*, reflecting that *Daubert* requires judges "to make subtle and sophisticated determinations about methodology," when they "are not scientists, and do not have the scientific training that can facilitate the making of such decisions,"[55] Justice Breyer had urged that judges make more use of their power under Federal Rule 706 to appoint scientists to advise them. In fact, already at the time of *Joiner*, *Daubert* had prompted wider use of this power.

In 1992, the FDA had banned silicone breast implants, formerly "grandfathered in." They were not known to be unsafe; but manufacturers had not, as required under FDA regulations, supplied evidence of their safety. Understandably, the ban caused a good deal of anxiety, and provoked a wave of fear, greed, and litigation. In 1996, Judge Sam Pointer of the U.S. District Court in Birmingham, Alabama, who had been in charge of all several thousand federal implant cases for more than six years, convened a National Science Panel—an immunologist, an epidemiologist, a toxicologist, and a rheumatologist—to review evidence of the alleged connections between silicone implants and various systemic and connective tissue diseases.[56]

Judge Pointer's carefully phrased remit asks

to what extent, if any and with what limitations and caveats do existing studies, research, and reported observations provide a reliable and reasonable scientific basis for one to conclude that silicone-gel breast implants cause or exacerbate any . . . "classic" connective tissue diseases . . . [or] "atypical" presentations of connective tissue diseases. . . . To what extent, if any, should any of your opinions . . . be considered as subject to sufficient dispute as would permit other persons, generally qualified in your field of expertise, to express opinions that, though contrary to yours, would likely be viewed by others in the field as representing legitimate disagreement within your profession?[57]

Two years and (only) $800,000 later,[58] after selecting from more than two thousand published and unpublished studies those they thought most "rigorous and relevant," in December 1998 the panel submitted a long report, concluding that the evidence studied and re-analyzed (apparently the forty or so studies submitted by each side plus about one hundred others, including unpublished studies, Ph.D. dissertations, and letters) did not warrant the claim that silicone breast implants cause these diseases. However, in some respects "the number and size of studies is inadequate to produce definite results"; animal testing "may not fully predict the human effects"; some evidence suggests that silicone implants are not entirely benign (they can cause inflammation, and droplets can turn up in distant tissues); and, while most people in the field would agree with their conclusions, a few might not.[59]

Despite Judge Pointer's efforts to ensure that his experts were unimpeachably neutral, the plaintiffs' lawyers objected that his rheumatologist had undisclosed connections with one of the defendants, Bristol-Meyers Squibb, while a member of the panel: in August 1997 he signed a letter soliciting up to $10,000 in support of a rheumatology meeting he co-chaired, stating that "the impact of sponsorship will be high, as the individuals invited for this workshop, being opinion leaders in their field, are influential with the regulatory agencies"; in October 1998 he signed a $1,500-a-day fee arrangement with BMS, and in November 1998 he received $750 for participating in a company seminar.[60] In April 1999, averring that there was no actual bias—though acknowledging that there might be a regrettable appearance of bias—Judge Pointer ruled against the plaintiffs' motion that the panel's report be excluded.

The wave of silicone breast implant litigation seems to have subsided—though only gradually and raggedly, of course. While the Pointer panel was at work, Judge Jones of the district court in Oregon had appointed his own technical advisors; and in *Hall* v. *Baxter Healthcare Corp.* (1996) granted the defendants' motion to exclude the plaintiffs' proffered causation testimony, but deferred the effective date of his opinion until the release of the National Science Panel report.

State court judges in *Dow Chemical Co.* v. *Mahlum* (Nevada, 1998) and *Minnesota Mining and Manufacturing Co.* v. *Atterbury* (Texas, 1999) didn't pay attention to the report. And in *Meister* v. *Medical Engineering Corp.* (2001), the U.S. District Court in Washington, D.C., denied the defendants' motion to reconsider its prior *Daubert* ruling in light of the National Science Panel report; in due course the jury found for the plaintiff and awarded $10 million in damages. But after reading the report the court granted a judgment in favor of the defendant notwithstanding the verdict, a judgment upheld by the court of appeals. Panel members had presented eight days of videotaped testimony, which was to be shown in federal trials; most of the cases were settled while the tapes were still being edited.[61]

In April 1999 about two dozen Massachusetts superior court judges attended a two-day seminar on DNA at the Whitehead Institute for Biomedical Research. A report in the *New York Times* quotes the director of the Institute: in the O. J. Simpson trial lawyers "befuddle[d] everyone" over the DNA evidence; but after this program, "I don't think a judge will be intimidated by the science." Judges will "understand what is black and white . . . what to allow in the courtroom."[62] And in May 1999 the American Association for the Advancement of Science inaugurated the CASE (Court Appointed Scientific Experts) Project to make available "independent scientists who would educate the court, testify at trial, assess the litigants' cases, and otherwise aid in the process of determining the truth," and who would be "as candid, objective, and disinterested as possible."[63] Duke University's Registry of Independent Scientific and Technical Advisors also aims to provide independent scientific experts.

The bramble-bush is alive and well, growing new fruit, and new thorns, almost every day. Well, perhaps not *new* fruit, exactly. In December 2000, the Federal Rules of Evidence were revised: Rule 701 now allows limited opinion evidence by lay witnesses; and Rule 702 now provides that qualified experts may testify if their testimony is relevant, provided (1) it is "based on *sufficient* facts or data," (2) it is "the product of *reliable* principles and methods," and (3) the witness "has applied the principles and methods *reliably* to the facts of the case" (my italics). Though the new Rule 702 merely makes explicit what the Supreme Court held in *Daubert* was implicit in it all along, it has already, in *Rudd* v. *General Motors Corp.*, been interpreted as tightening standards of admissibility.

So there; let nothing you dismay! As the Underground Grammarian's marvelously world-weary comment comes once again to mind, I realize that the epistemological points that most need emphasizing in what follows are negative, identifying misunderstandings which have hampered legal efforts to discriminate decent science from junk, and overly simple responses to the formidably difficult problems of handling scientific evidence.

THE LAW ON SCIENTIFIC TESTIMONY:
A BRIEF EPISTEMOLOGICAL COMMENTARY

The strength of the *Frye* rule, from an epistemological point of view, is its recognition that non-scientists are often not in a good position to judge for themselves whether a scientific claim or theory is warranted, but will have to rely on what scientists tell them; and, if they are prudent, should ask whether scientists in the field agree, and how sure they are. Among the weaknesses of the *Frye* rule, from an epistemological point of view, are its presuppositions that there is a definite point at which scientific claims or techniques cease to be "experimental" and become "demonstrable"; that a claim or technique achieves this "demonstrable" status only when it is generally accepted in the community; and that only generally accepted, and hence "demonstrable," claims and techniques should be admitted.

On its face, the third assumption seems extremely restrictive. In fact, it's not clear that it would *ever* permit novel scientific evidence to be admitted; for how could a theory or technique that has already made it to "general acceptance" be genuinely novel? In principle, it seems, *Frye* would confine the courts to textbook science. But it takes only a moment's reflection to realize that, in practice, how restrictive *Frye* would be depends on exactly what is required to be accepted by what proportion of what community. The narrower and more homogeneous the relevant community is taken to be, the likelier it is that there will be agreement; the broader and more heterogeneous the community, the likelier that there will be disagreement. Moreover, how good an indication of reliability general acceptance in a community is depends on the character of the community concerned. In fields where work is rigorously conducted and evidence efficiently shared, general acceptance is likely to be reasonably well correlated with warrant. But some of the communities to which *Frye* may direct a court's attention consist exclusively or almost exclusively of people who make their living applying the very techniques in question—more like a technicians' trade union than the RNA Tie Club or the Phage Group; and general acceptance in such communities is a poor proxy for reliability. Even in rigorous and epistemologically efficient communities, when it is an individual rather than the community whose evidence is, in the relevant sense, the best, general acceptance may be too conservative a test—as, evidently, the court in *Coppolino* believed.

And the first two presuppositions are over-simplified at best. Rather than a sharp line, there is a continuum from unwarranted through poorly warranted to well-warranted claims. How warranted a claim is at a time depends on the quality of the best evidence at that time—usually, but not always, the shared evidence of a scientific sub-community. But the degree of credence given a claim in the scientific community is only an imperfect indicator of its degree of warrant; which is only an imperfect indicator—albeit the best we can have—of its truth.

Interpreted as requiring relevance but not reliability, the Federal Rules may seem epistemologically relatively unproblematic. But it is worth noting that, while the question of relevance in *Barefoot* was a legal one, judgments of *scientific* relevance are no easier for a lay judge or jury to make than judgments of reliability are; whether this experimental result is relevant to such-and-such a theoretical claim, for example, depends on the truth of other scientific questions.

Still, the epistemological problems became acute when, with *Daubert*, the Supreme Court interpreted the Federal Rules as requiring the trial judge to make determinations about reliability, and to make an appraisal of scientific methodology, not by appeal to the scientific community, but in his own behalf. For what the *Daubert* Court has to offer by way of advice about how to make such determinations is—well, a little embarrassing.

Evidently the majority of the *Daubert* Court succumbed to the seduction of the honorific use of "scientific" and its cognates, looking for a simple criterion of demarcation that federal judges could use to determine whether proffered testimony was genuine science, not pseudo-science, and hence reliable; and hoped (apparently having read someone who read someone who read Popper) that Popper's criterion would do the trick. The justices were apparently unaware that Popper holds that no scientific claim or theory is ever shown to be true, reliable, or probable, but is at best "corroborated"; and (but for weak moments where he takes it back) that this is not equivalent to "confirmed."[64] If Popper were right, no scientific claim would be well warranted; in fact, it is hard to think of a philosophy of science less congenial than Popper's to the reliability approach (or to the admissibility of psychiatric evidence, but that is a whole other can of worms). And if the reference to Popper is a *faux pas*, running Popper together with Hempel—a pioneer of the logic of confirmation—is a *faux pas de deux*.

In and of itself, of course, the *Daubert* Court's mixing up its Hoppers and its Pempels is just a minor scholarly irritation. A more serious problem is that Hempel's philosophy of science will no more do the job the Court wants it to do than Popper's will. Though Hempel at least allows that scientific claims can be confirmed as well as disconfirmed, he offers nothing that would help a judge decide either whether evidence proffered is really scientific, or how reliable it is.

But the most fundamental problem is the *Daubert* Court's preoccupation with specifying what the method of inquiry is that distinguishes the scientific and reliable from the non-scientific and unreliable. There is no such method. There is only making informed conjectures and checking how well they stand up to evidence—which is common to every kind of empirical inquiry; and the many and various techniques used by scientists in this or that scientific field—which are neither universal across the sciences nor constitutive of real science.

The *Daubert* Court runs together (1) the tangled and distracting questions of demarcation and scientific method with (2) the question of the degree of warrant of specific scientific claims or theories and (3) the question of the reliability of specific scientific techniques or tests—which is different again, for the claim that this technique is unreliable may be well warranted, the claim that this other technique is reliable poorly warranted. Unlike determining whether a claim is falsifiable, however, determining whether a scientific theory (e.g., of the etiology of this kind of cancer) is well warranted, or whether a scientific test (e.g., for the presence of succinylcholine chloride) is reliable, requires substantive scientific knowledge. Justice Rehnquist is right: the reference to falsifiability is no help, and judges are indeed being asked to be amateur scientists.

Only four years later, in the *Joiner* ruling, some of *Daubert*'s epistemological chickens had already come home to roost: with the references to falsifiability gone and the distinction between methodology and conclusions dropped, it is starkly obvious that, as Judge Kozinski complained, federal judges are now obliged to determine substantive scientific questions. Moreover, with abuse of discretion as the standard of review, appeals courts may allow different courts' contradictory determinations of the very same scientific issues.[65]

Given the difficulties with the *Daubert* Court's efforts to specify what makes evidence genuinely scientific, perhaps the knots in which everyone ties themselves in *Joiner* (not to mention the absence from the ruling of any reference whatever to falsifiability, testability, Hepper, Pompel, etc.)[66] are not so surprising. What *is* surprising, to me at any rate, is that the *Joiner* Court should offer, as an interpretation of *Daubert*, a ruling that denies the legitimacy of a distinction *Daubert* presupposed. To be sure, a later ruling may make an earlier ruling determinate in respects in which it was formerly indeterminate (that is why the *Daubert* Court could *rule* that the *Frye* test is incompatible with the Federal Rules). But the idea that a later ruling which flatly denies a clear presupposition of an earlier ruling could qualify as an interpretation, rather than a revision, of it, still strikes me as very strange indeed.

But set that aside. What about the distinction between methodology and conclusions presupposed in *Daubert*, but repudiated in *Joiner*? In these cases the concept of methodology—never well defined in philosophy of science—has become an accordion concept,[67] expanded and contracted as the argument requires. Is the judge, in determining the validity of experts' "methodology," to decide whether the mouse studies on which Mr. Joiner's experts in part relied were well conducted, with proper controls and good records, using specially bred genetically uniform mice, etc., etc.; or what weight to give mouse studies with respect to questions about humans; or what weight to give those mouse studies in the context of other studies of the effects on humans of PCB and other contaminants; or what? There are so many ambiguities that everyone is right—and everyone is wrong.

Mr. Joiner's lawyers are right to suggest that drawing the reasonable conclusion from a conglomeration of disparate bits of information (mouse studies, epidemiological evidence, etc.) requires—well, weighing the evidence. But of course, it matters whether you weigh the evidence *properly*; and GE's lawyers are right, too, when they complain that Mr. Joiner's attorneys use "methodology" so loosely as to make *Daubert*'s requirements practically vacuous. But GE's accusation that Mr. Joiner's experts commit the "faggot fallacy" relies on an equivocation. There is an ambiguity in the reference to "pieces of evidence, each independently . . . suspect or weak": this may mean *either* "pieces of evidence each themselves poorly warranted" (which seems to be the interpretation intended by Skrabanek and McCormick, to whom the phrase "faggot fallacy" is due), *or* "pieces of evidence each by itself inadequate to warrant the claim in question" (which seems to be the interpretation most relevant to the case). True, if the reasons for a claim are themselves poorly warranted, this lowers the degree of warrant of the claim itself. But GE's brief offers no argument that the reasons based on the studies to which Mr. Joiner's experts refer are themselves poorly warranted. True again, none of those reasons by itself strongly warrants the claim that PCBs promoted Mr. Joiner's cancer. But GE's brief offers no argument that they don't do so jointly.

Sometimes bits of evidence which are individually weak are jointly strong; sometimes not—it depends what they are, and whether or not they reinforce each other (whether or not the crossword entries interlock). Neither party seriously addresses this question of interlocking, i.e., of explanatory integration. But in the very complex EPA guidelines to which Mr. Joiner's attorneys so casually refer, I find this: "Weight of evidence conclusions come from the *combined strength and coherence* of inferences appropriately drawn from all of the available evidence."[68]

Justice Stevens is right to insist that there is a difference between methodology and conclusions, as there is between ends and means; there is a difference, certainly, between a technique and its result, or between premises and conclusion. But on a more charitable interpretation, the majority's point is not that there is literally no distinction, but that (as Judge Becker had argued in *Paoli*) it is impossible to judge methodology without relying on some substantive scientific conclusions. And this is both true and important. To determine whether this evidence (e.g., of the results of mouse studies) is relevant to that claim (e.g., about the causes of Mr. Joiner's cancer) requires substantive knowledge (e.g., about the respects in which mouse physiology is like human physiology, about how similar or how different the etiologies are of small-cell lung cancer and alveologenic adenomas, etc.). And to determine the reliability of a scientific experiment, technique, or test, it is necessary to know what kinds of thing might interfere with the proper working of this apparatus, what the chemical theory is that underpins this analytical technique, what factors might lead to error in this kind of experiment and what precautions are called for, or to possess a sophisticated understanding of statistical techniques or of complex and controversial methods of meta-analysis pooling data from different studies. And so on. Which takes us back to that old worry of Justice Rehnquist's of which Justice Breyer's observation that judges are not scientists reminds us: judges are neither trained nor qualified to do this kind of thing.

In *Kumho*, the Supreme Court begins to shake off the honorific use of "science," to disentangle "reliable" from "scientific." Scientific testimony may be unreliable; non-scientific testimony may be reliable. So, yes, if what we care about is reliability, whether testimony is or isn't scientific isn't the point; and, yes, the *Daubert* factors may or may not be relevant to the reliability of proffered expert testimony. The problem, though, is that *Kumho* leaves judges with almost no guidance about how to determine whether such testimony is reliable.

A bit of scientific education for judges is at best a drop in the bucket. Not that teaching judges about DNA or whatever mightn't do some good; but a few hours in a science seminar will no more transform judges into scientists competent to make subtle and sophisticated scientific determinations than a few hours in a legal seminar would transform scientists into judges competent to make subtle and sophisticated legal determinations. ("This kind of thing takes a lot of training," as Mad Margaret sings in *Ruddigore*.) That *New York Times* report has me, for one, a little worried about the danger of giving judges a false impression that they *are* qualified to make those "subtle and sophisticated determinations."

"[N]either the difficulty of the task nor any comparative [*sic*] lack of expertise can excuse the judge from exercising the 'gatekeeper' duties that the Federal Rules impose," Justice Breyer avers.[69] More directly than the *Frye* test, calling on court-appointed panels of scientists turns part of the task over to those who are more equipped to do it. Isn't this a whole lot better than asking judges to be amateur scientists? Sometimes, probably, significantly better—the more so, the closer the work at issue is to black-letter science; not, however, quite as straightforwardly or unproblematically better as some hope.

As Judge Pointer's panel's report was made public, an optimistic headline in the *Washington Times* proclaimed "Benchmark Victory for Sound Science,"[70] and under the headline "An Unnatural Disaster," an editorial in the *Wall Street Journal* announced that "reason and evidence have finally won out."[71] ABCNEWS.com's "Health and Living" was considerably more cautious: under the headline "No Implant-Disease Link?" a sideline adds, "The panel found no definite links, but it also left the door open for more research."[72]

I should be quite surprised if it turned out that silicone implants do, in fact, cause the various diseases they have been alleged to. In June 1999, six months after the Pointer Panel report, a thirteen-member committee of the Institute of Medicine also concluded that "silicone breast implants do not cause chronic disease," though noting that "other complications are of concern."[73] Nor do I think it very likely that that $750 seriously affected Dr. Tugwell's opinion (though I must say that—even if this kind of thing is routine in funding applications, as for all I know it may be—his letter boasting of the applicants' influence with regulatory bodies leaves a bad taste in my mouth). I don't feel equally confident, however, that a really good way has yet been found to delegate part of the responsibility for appraising scientific evidence to scientists themselves. There is the worry about ensuring neutrality, and the appearance of neutrality—one of Judge Pointer's panelists, Dr. Diamond, remarks regretfully that she feels "extraordinarily naive" after realizing that, yes, she had discussed the issues with professional acquaintances who have connections with the defendants;[74] the worry about how much responsibility falls on how few shoulders—just four people, in the case of Judge Pointer's panel, all of whom combined this work with their regular full-time jobs, each of them in effect solely responsible for a whole scientific area; and the worry about what jurors will make of court-appointed experts' testimony. The history of the *Frye* test should warn us, also, of potential pitfalls in determining the relevant area(s) of specialization.

As for the revisions to the Federal Rules (at which, earlier, I poked a little gentle fun) the problem, as we now see, is that no legal form of words can ensure that only the reliable-enough is admitted. Adding "reliable," "reliably," "sufficient," to the Rules tells judges nothing substantive about what to exclude and what to admit; as Peirce might have said, it reaches only the second grade of clarity, not the third, pragmatic, or operational grade.

MORALS OF THE STORY

Hence the first moral: we shouldn't allow ourselves to be distracted by the futile hope that somehow, if only we could find the perfect legal formula, only reliable (how reliable?) scientific testimony would be admitted; there is no such formula. Nor should we allow ourselves to be sidetracked into debates about whether judges or juries are better equipped to determine the worth of scientific testimony; there is probably no general answer to be had.[75] But once we understand why scientific evidence provides so many opportunities for opportunism, we begin to see more clearly what kinds of legal arrangement can amplify the underlying problems, and where improvements may be feasible.

Justice Blackmun's valiant effort to articulate the differences between science and law that make their interactions so problematic gets it right in part, but only in part. Scientific inquiry aims to discover substantial, explanatory truth. Though it is fallible and fumbling, and progresses unevenly, it is nevertheless the best—the only—way of discovering the truth of the questions within its scope. But it takes—well, it takes the time it takes. The law, by contrast, aims to settle disputes in a timely manner. But of course it aims to settle cases not only promptly but rightly; "rightly," in this context, having two components, one focused on truth, and the other on due process. So the intersection of law and science is bound to be problematic; often, we are trying to arrive at justice on the basis of imperfect and imperfectly understood information, and, not so rarely, we are trying to create justice out of ignorance.

Moreover, in some ways the legal system amplifies the inevitable difficulties: the prospect of large contingency fees in the event of a major jury verdict for a client, for example, creates incentives for plaintiffs' attorneys to shop for experts willing to testify as they would like, which in turn creates incentives for scientists to move into the professional-witness business, or to devote themselves to research driven by the needs of litigators rather than by scientific interest. The more and less stringent approaches to admissibility in different jurisdictions creates incentives for forum shopping. And so on.

There are no easy answers; but there are, certainly, better questions and worse. Rather than worrying fruitlessly about the problem of demarcation or the distinction of methodology versus conclusions and all that, we would do better to turn our attention to questions of other kinds—keeping in mind that, though perfection is impossible, better is better than worse, and the cumulative effect of small improvements can be quite large; and that it may be unwise to restrict our attention too exclusively to issues about admissibility of testimony, or indeed to strategies internal to the legal system.

Some of the fruitful-looking questions focus on relatively modest changes, within the existing legal framework, that might enable better handling of scientific evidence. How might we ensure that forensic laboratories take the most rigorous precautions possible against known potential errors?[76] What could be done to ensure more competent representation of criminal defendants, and fewer convictions on the word of jailhouse informants or unreliable eyewitnesses? What could be done to help jurors deal better with scientific evidence: e.g., consistent with filtering out legally unacceptable questions, to allow them to ask for clarification when they can't follow an expert witness? What could scientists' professional associations do to help scientific witnesses communicate better with judges and juries, or to discourage those who abuse their expertise? Could the legal profession and legal educators do more to discourage unscrupulous witness-shopping and related abuses? What could courts do, when trying to determine whether a claim is generally accepted, to ensure that they are looking to a community of inquirers and not to a trade union or clique or mutual admiration society?[77]

What could we learn from the experience with Judge Pointer's panel about bridging some of the gaps between the folkways of science and of the legal system? What advice might best be given to court-appointed scientists about what connections should be disclosed, or what kinds of record-keeping will be expected of them? Might it be desirable to ask court-appointed scientists to provide details of the qualifications and affiliations of any assistants on whom they relied; of which studies they decided to look at in detail, and why; of which studies seemed most strongly to indicate the contrary conclusion to theirs, and why, in their opinion, those studies were flawed?

Other fruitful-looking questions focus on the interactions between the legal system and other social institutions. How could we make the legal system more responsive when new evidence comes in to the scientific community? How could we make the scientific community more responsive when legal disputes turn on scientific issues irresoluble by the presently available evidence? What, ideally, would be the role of tort litigation vis-à-vis other means of ensuring that, when there is a question about the safety of this or that product, it is carefully looked into, and appropriate action taken?—a question prompted in part by the singularly unfortunate interaction of the FDA and the tort litigation system in the silicone-implant affair; and in part by Justice Breyer's observations, in his concurring opinion in *Joiner*, about our ubiquitous dependence on synthetic substances, and the importance of ensuring that the "powerful engine" of tort litigation discourage the production only of the harmful stuff (though it spoils the effect somewhat that *Joiner* was about PCBs, so dangerous that they had been banned for decades).[78]

Then there are more policy-oriented questions. Is it appropriate for considerations about, for example, how to manage the risks inherent in our reliance on synthetic materials, chemicals, drugs, etc., also to determine what evidence is admissible in criminal cases? Is it appropriate for a uniform rule to allow that the same testimony be admitted by one court and excluded by another in the same jurisdiction? (What kinds of uniformity do we value in the legal system, and why?) How significant a gatekeeping role is it appropriate for judges to take? (What exactly do we value about trial by jury, and why?)

Given that mistakes are inevitable, should we be more willing to tolerate some kinds than others? Of course, in answering this question we shouldn't forget that scientific evidence plays a role both in civil and in criminal cases, and on both sides—as Paul Giannelli may have done when he wrote that "for me *Frye* functions much like a burden of proof. . . . If [in criminal cases] we are going to make mistakes in assessing the validity of a novel technique, they should be mistakes of excluding reliable rather than including unreliable evidence."[79]

And yet other fruitful-looking questions focus on those aspects of the U.S. legal system that amplify the problems of scientific testimony, such as trial by jury in civil cases, the contingency-fee system, those vast consolidated tort cases, and the opportunities for forum shopping. Here it may be helpful to ask whether and how matters are handled differently elsewhere, specifically in other scientifically and technologically advanced countries; and, if so, what the benefits, and what the drawbacks, are. A comparison of the United States with England or Canada,[80] for example, might be helpful with regard to issues about the contingency-fee system and the price—presumably, more limited access to the legal system for those without large resources—of changing it.

AND, IN CONCLUSION

Though certain aspects of this or that legal system may exacerbate them, the underlying epistemological difficulties in handling scientific testimony are the same everywhere. There is an interesting parallel with the relation of science to religion. Because of the constitutional provision against the establishment of religion, efforts to get religious teaching into schools under the rubric "creation science" have made the problem of the demarcation of science from non-science legally inescapable. But though creation-science cases such as *McLean* v. *Arkansas Board of Education* have a distinctively U.S. flavor, there is nothing local or parochial in the underlying epistemological questions about the relation of science and religion—to which I turn next.

CASES CITED

United States

Barefoot v. *Estelle,* 463 U.S. 880; 103 Sup. Ct. 3383 (1983).

Berry v. *City of Detroit,* 25 F. 3d 1342 (1994).

Bonnischen et al. v. *United States of America, Department of the Army, et al.,* Civil Judgment No. 96-1481-JE (2002).

Commonwealth v. *Lykus,* 327 N.E. 2d 671 (Mass. 1975).

Conti v. *Commissioner,* 39 F. 3d 658 (6th Cir. 1994).

Coppolino v. *State,* 223 So. 2d 68 (Fla. Dist. Ct. App. 1968).

Daubert v. *Merrell Dow Pharm., Inc.,* 509 U.S. 579; 113 Sup. Ct. 2786 (1993). ["*Daubert* I"]

Daubert v. *Merrell Dow Pharm., Inc.,* 43 F. 3d 1311 (9th Cir. 1995). ["*Daubert* II"]

Dow Chemical Co. v. *Mahlum,* 970 P. 2d 98 (Nev. 1998).

Flanagan v. *State,* 620 So. 2d 827 (Fla. 1993).

Frye v. *United States,* 293 F. 1013 (D.C. Cir. 1923).

General Electric Co. v. *Joiner,* 522 U.S. 136; 118 Sup. Ct. 512 (1997).

Hadden v. *State,* 690 So. 2d 573 (Fla. 1997).

Hall v. *Baxter Healthcare Corp.,* 947 F.Supp. 1387 (D.Or. 1996).

Kumho Tire Co., Ltd. v. *Carmichael,* 526 U.S. 137, 119 Sup. Ct. 1167 (1999).

McLean v. *Arkansas Board of Education,* 529 F.Supp. 1255 (1982).

Meister v. *Medical Engineering Corp.,* 267 F. 3d 1123 (D.C. Cir. 2001).

Minnesota Mining and Manufacturing Co. v. *Atterbury,* 978 S.W. 2d 183 (Tex. App. 1998).

Moore v. *Ashland Chemical, Inc.,* 151 F. 3d 269 (5th Cir. 1998).

Paoli R.R. Yard PCB Litig., In re, 35 F. 3d 745 (3d Cir. 1994).

People v. *Williams,* 164 Cal. App. 2d Supp. 858, 331 P. 2d 251 (Cal. App. Dep't Super. Ct. 1958).

Reed v. *State,* 391 A. 2d 364 (Md. 1978).

Rudd v. *General Motors Corp.,* 127 F.Supp. 2d 1330 (M.D. Ala. 2001).

United States v. *Addison,* 498 F. 2d 741, 744 (D.C. Cir. 1974).

United States v. *Black,* 831 F.Supp. 120 (E.D.N.Y. 1993).

United States v. *Bonds,* 12 F. 3d 540 (6th Cir. 1993).

United States v. *Chinchilly,* 30 F. 3d 1144 (9th Cir. 1994).

United States v. *Davis,* 40 F. 3d 1069 (10th Cir. 1994).

United States v. *Downing,* 753 F. 2d 1224 (1985).

United States v. *Lech,* 895 F.Supp. 582 (S.D.N.Y. 1995).

United States v. *Martinez,* 3 F. 3d 1191 (8th Cir. 1993).

United States v. *Posado,* 57 F. 3d 428 (5th Cir. 1995).

United States v. *Rodriguez,* 37 M.J. 448 (C.M.A. 1993).

United States v. *Sheffer,* 523 U.S. 303; 118 Sup. Ct. 1261 (1998).

United States v. *Starzecpyzel,* 880 F.Supp. 1027 (S.D.N.Y. 1995).

Canada

R. v. Bourguignon January 14, 1999 Doc. Ottawa, Flanigan J. (Ont. Gen. Div.). [unreported]
R. v. Mohan, [1994] 2 S.C.R.9.
R. v. J.-L.J., [2000] 2 S.C.R.600.

England

Alsop v. Bowtrell, Cro. Jac. 541 (1620).
Folkes v. Chadd, 3 Doug. 157 (1782).

NOTES

1. *Daubert* I, 509 U.S. 596–97, 113 Sup. Ct. 2798 (1993).
2. See Ruse, ed., *But Is It Science?*
3. See Bonnischen and Schneider, "Battle of the Bones"; Hunt, *The New Know-Nothings*, pp. 320–26; Chatters, *Ancient Encounters*. The legal isue was resolved in August 2002, when U.S. Magistrate Judge Jellerks ruled that the remains of Kennewick man, rather than being awarded to the Native American tribes that claimed them, could be studied by anthropologists.
4. Hansen, "The Great Detective."
5. See also chapter 11, p. 308.
6. *R. v. Bourguignon*. See also Koehler, Chia, and Lindsey, "The Random Match Probability in DNA Evidence."
7. Cohen, "Reasonable Doubt."
8. Huber, *Galileo's Revenge*; "Junk Science in the Courtroom"; Huber and Foster, eds., *Judging Science*.
9. Cheseboro, "Galileo's Retort."
10. Schmitt, "Witness Stand."
11. Hand, "Historical and Practical Considerations Regarding Expert Testimony," p. 54.
12. See Schwartz, "A 'Dogma of Empiricism' Revisited," (arguing in favor of *Frye* that feminist epistemology has established the social character of science); Notturno, "Popper, *Daubert*, and Kuhn," (arguing in favor of *Daubert* that Popper has established the fallibility of science).
13. Hand, "Historical and Practical Considerations Regarding Expert Testimony," pp. 40–49; the date (1620) is given on p. 45.
14. Landsman, "Of Witches, Madmen, and Product Liability," p. 141.
15. "The defendant in *Frye* was subsequently pardoned when someone else confessed to the crime," writes Paul Giannelli, in "The Admissibility of Novel Scientific Evidence," n. 42. Giannelli cites Wicker, "The Polygraphic Truth Test and the Law of Evi-

dence"; Wicker, he says, cites the *Fourteenth Annual Report of the Judicial Council of the State of New York* (1948), 265. But according to the most complete account I have been able to find of the many twists and turns of Mr. Frye's story—Starrs' "A Still-Life Water-Color"—none of this is true.

16. My source is Starrs, "A Still-Life Water-Color," p. 694; he refers to "Transcript on Appeal, File 3968, retired files, National Records Center, Suitland, MD."

17. From Judge Van Ordsel's opinion for the appellate court in *Frye*.

18. My source is Giannelli, "The Admissibility of Novel Scientific Evidence," p. 1207.

19. See Black, Ayala, and Saffran-Brinks, "Science and the Law in the Wake of *Daubert*," pp. 735 ff., listing *Reed* v. *State* and *United States* v. *Addison*, excluding voiceprint evidence under the *Frye* test; and *Commonwealth* v. *Lykus*, admitting voiceprint evidence under the *Frye* test. There is a useful summary of relevant cases in the Symposium on Science and Rules of Evidence in the *Federal Rules Decisions*, 1984.

20. Giannelli comments: "if the 'specialized field' is too narrow . . . the judgment of the scientific community becomes, in reality, the opinion of a few experts" ("The Admissibility of Novel Scientific Evidence," pp. 1209–10).

21. Again, my source is Giannelli, "The Admissibility of Novel Scientific Evidence," pp. 1222 ff.

22. McCormick, *McCormick on Evidence*, p. 364.

23. *Barefoot* v. *Estelle*, 463 U.S. 919; 103 Sup. Ct. 3408 (1983).

24. *Barefoot* v. *Estelle*, 463 U.S. 901; 103 Sup. Ct. 3398.

25. *Barefoot* v. *Estelle*, 463 U.S. 898; 103 Sup. Ct. 3397.

26. *Barefoot* v. *Estelle*, 463 U.S. 916; 103 Sup. Ct. 3407.

27. *Barefoot* v. *Estelle*, 463 U.S. 929; 103 Sup. Ct. 3413, and 463 U.S. 932, 103 Sup. Ct. 3415.

28. I rely on Giannelli, "The Admissibility of Scientific Evidence," pp. 1229–30. He mentions Saltzburg and Redden, *Federal Rules of Evidence Manual*, p. 426 as holding that the Federal Rules are compatible with the *Frye* test because they don't mention general acceptance; and Wright and Graham, *Federal Practice and Procedure*, p. 92 as holding that the Federal Rules are incompatible with the *Frye* test because they don't mention general acceptance.

29. Graham, *Federal Rules of Evidence* (1987 edition), p. 92.

30. My source is Gottesman, "From *Barefoot* to *Daubert* to *Joiner*," pp. 757–58.

31. *Daubert* I, 509 U.S. 598; 113 Sup. Ct. 2794.

32. *Daubert* I, 509 U.S. 580; 113 Sup. Ct. 2790.

33. *Daubert* I, 509 U.S. 593; 113 Sup. Ct. 2796.

34. Green, "Expert Witnesses and Sufficiency of Evidence in Toxic Substance Litigation," p. 645. A footnote (12) refers to Popper, but I can find no reference to Hempel.

35. *Daubert* I, 509 U.S. 600–601; 113 Sup. Ct. 2800.

36. *Daubert* II, 43 F. 3d 1316 (9th Cir. 1995).

37. Astara Moren, "Drug Revived to Fight Morning Sickness," *NurseWeek* [online], www.nurseweek.com/news/00-10/1011morn.asp [October 11, 2000].

38. As Gottesman points out in "From *Barefoot* to *Daubert* to *Joiner*."

39. According to Kesan, "An Autopsy of Scientific Evidence in a Post-*Daubert* World," p. 1992, before *Daubert*, 29 states followed *Frye*, and 20 some relevance and reliability standard. Six of the 13 states that considered whether to follow *Daubert* decided to retain *Frye*, and as of 1995, 22 states still followed *Frye*.

40. *United States* v. *Bonds*, 558, 560. See Kesan, "An Autopsy of Scientific Evidence in a Post-*Daubert* World," pp. 2030–31; Scheck, "DNA and *Daubert*," pp. 1991 ff.; the quotation is from p. 1993.

41. In *United States* v.. *Martinez* (1993) the court didn't consider the error rate or population substructure problems with the DNA evidence; in *United States* v. *Chinchilly* (1994) the court held that faults in the DNA profiling methodology went to the weight of the evidence, not its admissibility; in *United States* v. *Davis* (1994) the court held a *Frye* hearing. My source is Kesan, "An Autopsy of Scientific Evidence in a Post-*Daubert* World," pp. 2031–33.

42. In *United States* v. *Posado* polygraph evidence was admitted under *Daubert*; in *United States* v. *Black* it was excluded under *Daubert*; in *United States* v. *Lech* and *Conti* v. *Commissioner*, rather than a *Daubert* hearing being held, polygraph evidence was excluded under Rule 403; and in *United States* v. *Rodriguez* it was held that polygraph evidence can neither be excluded nor accepted out of hand. My source is Kesan, "An Autopsy of Scientific Evidence in a Post-*Daubert* World," pp. 2014–16.

43. According to David Faigman, there are about seven hundred federal cases a year involving *Daubert*.

44. *General Electric Co.* v. *Joiner*, 522 U.S. 136; 118 Sup. Ct. 514 (1997), quoting *Joiner* v. *Gen. Elec. Co.*, 864 F.Supp. 1310, 1326 (N.D. Ga. 1994), which in turn quotes *Daubert* I (where the phrase occurs three times), 509 U.S. 590, 597, 599; 113 Sup. Ct. 2786, 2795, 2800.

45. See *Gen. Elec. Co.* v. *Joiner*, 552 U.S. 140; 118 Sup. Ct. 516.

46. But the question with regard to furans and dioxins, according to the Supreme Court ruling, remained open.

47. Brief for Petitioners, *Gen. Elec. Co.* v. *Joiner*, p. 47.

48. Brief for Petitioners, *Gen. Elec. Co.* v. *Joiner*, p. 49, citing Skrabanek and McCormick, *Follies and Fallacies in Medicine*, p. 35, quoted in Huber and Foster, eds, *Judging Science*, p. 142. I notice that on the same page Skrabanek and McCormick refer to what they call the "weight of evidence fallacy"; this, they claim, is not scientific, because science, according to Popper, focuses on negative evidence (which can't be outweighed by confirming instances). While I am noting that GE's lawyers cite Peter Huber, I will also note that Kenneth Cheseboro was one of Mr. Joiner's lawyers.

49. Oral Argument of Michael H. Gottesman, *Gen. Elec. Co.* v. *Joiner*, pp. 43–44. Mr. Gottesman was also one of the attorneys for Mr. Daubert.

50. *Gen. Elec. Co.* v. *Joiner*, 522 U.S. 155; 118 Sup. Ct. 523 (emphasis added).

51. *Gen. Elec. Co.* v. *Joiner*, 522 U.S. 151; 118 Sup. Ct. 521 (Stevens, J., dissenting).

52. *Kumho Tire Co., Ltd.* v. *Carmichael*, 526 U.S. 138, 148; 119 Sup. Ct. 1169, 1174.

53. *Kumho Tire Co., Ltd.* v. *Carmichael*, 526 U.S. 151; 119 Sup. Ct. 1175.

54. *Kumho Tire Co., Ltd.* v. *Carmichael*, 526 U.S. 138–39; 119 Sup. Ct. 1170.

55. *Gen. Elec. Co.* v. *Joiner*, 522 U.S. 136, 148; 118 Sup. Ct. 512, 520.

56. For background and more details, see Hooper, Cecil, and Willging, "Assessing Causation in Breast Implant Litigation: The Role of Science Panels."

57. Submission of Rule 706 National Science Panel Report, p. 2, In re: Silicone Gel Breast Implant Products Liability Litigation (N.D. Ala. 1998) (No. CV 92-P-10000-S), Federal Judicial Center [online], www.fjc.gov/BREIMLIT/SCIENCE/report.htm.

58. "Only" not only because the sum is trivial relative to the compensation awarded in some implant cases, but also because it is really quite modest given the task undertaken. There was, however, the additional cost of special counsel to the panel: $1,157,594.74.

59. Submission of Rule 706 National Science Panel Report, p. 8.

60. "Pointer Rules Federal Science Panel Report Not Tainted by Payments to Panelist," p. 4. The discrepancy between reports about the sum of money involved—plaintiffs say $750, court $500—may be a matter of Canadian versus U.S. dollars. Plaintiffs also object that a colleague who assisted Dr. Tugwell in his work for the panel had received support from a company wholly owned by Bristol-Meyers Squbb.

61. See Walker and Monahan, "Scientific Authority: The Breast Implant Litigation and Beyond"; Hooper, Cecil, and Willging, "Assessing Causation in Breast Implant Litigation."

62. Goldberg, "Judges' Unanimous Verdict on DNA Lessons: Wow!"

63. Bandow, "Keeping Junk Science out of the Courtroom."

64. Black, Ayala, and Saffran-Brinks, in "Science and the Law in the Wake of *Daubert*," pp. 750 ff., seem to have confused corroboration, in Popper's sense, with confirmation. Green—who, incidentally, introduces Popper's philosophy of science in Kuhnian terms, as "the existing paradigm under which scientists work"!—acknowledges that Popper holds that "[t]heoretically . . . hypotheses are never affirmatively proved," but continues "of course, if a hypothesis repeatedly withstands falsification, one may tend to accept it, even if conditionally, as true" ("Expert Witnesses and Sufficiency of Evidence in Toxic Substance Litigation," pp. 645–46).

65. Faigman, "Annual Report on Science and Law" (2001).

66. And of any reassuring noises about jurors' ability to assess the weight of scientific evidence.

67. The term, and the idea, come from Sellars, "Scientific Realism or Irenic Instrumentalism?" p. 172.

68. 61 FR 19760-01: 17972 (1996), my italics.

69. *Gen. Elec. Co.* v. *Joiner*, 522 U.S. 148; 118 Sup. Ct. 520 (Breyer, J., concurring).

70. Peters, "Benchmark Victory for Sound Science."

71. "An Unnatural Disaster."

72. Jay Reeves, "No Implant-Disease Link?" ABC News [online], www.abcnews. go.com/sections/living/DailyNews/breastimplants981201.html [December 1, 1998].

73. "Silicone Breast Implants Do Not Cause Chronic Disease, but Other Complications Are of Concern," National Academies [online], www4.nationalacademies. org/news.nsf/isbn/0309065321?opendocument [June 21, 1999].

74. "Transcript of Rule 706 Panel Hearing 91," *Women in Health* [online],

womnhlth.home.mindspring.com/706/transcript_of_rule_706_panel_hea.htm [February 4, 1999].

75. I note, however, the discussion of this question in Gottesman, "From *Barefoot* to *Daubert* to *Joiner*," n. 39.

76. See Jonakait, "Forensic Science: The Need for Regulation."

77. See Schwartz, "A 'Dogma of Empiricism' Revisited," pp. 206 ff., for suggestions along these lines.

78. *Gen. Elec. Co.* v. *Joiner*, 522 U.S. 148; 118 Sup. Ct. 520.

79. In the Symposium on Science and Rules of Evidence, *Federal Rules Decisions*, 1984, p. 206. See also Schwartz, "A 'Dogma of Empiricism' Revisited," pp. 229 ff.

80. The seemingly relatively flexible standards of admissibility of scientific testimony adopted by the Canadian Supreme Court in *R.* v. *Mohan* (1994) were replaced in *R.* v. *J.-L.J.* (2000) by criteria closely akin to *Daubert* I.

10

POINT OF HONOR

On Science and Religion

Modern science kills God and takes his place on the vacant throne . . . as . . . the sole arbiter of all relevant truths.

—Vaclav Havel[1]

In the same way that each of us has had to grow up to resist the temptation of wishful thinking . . . , so our species has had to learn in growing up that we are not playing the starring role in any sort of grand cosmic drama.

—Steven Weinberg, *Dreams of a Final Theory*[2]

Havel and Weinberg agree that there is a real tension between religion and science. I believe they are right. Havel, however, thinks that science, pretending to be the sole source of truth, has blinded us to truths of a spiritual kind; while Weinberg thinks that intellectual maturity demands that we give up such wishful thinking about ourselves and our place in the universe, and takes it as a "point of honor" not to seek consolation by adjustment of our beliefs.[3] And here, I believe, Weinberg is right, and Havel wrong.

I say this more than a little diffidently, for the whole topic makes me somewhat queasy—I have, I realize, less the temperament of those village-pump atheists who relish speaking out against religion, than of those more retiring types for whom religious belief just isn't a live option. I have never felt moved to write a manifesto explaining "why I am not a Christian"; but now I can neither avoid the question of the relation of science to religion, nor duck the obligation to answer it honestly.

Of course, "the" question of the relation of religion and science isn't really one question, but a whole tangle. An initial complication is that religion may be construed quite narrowly, as a commitment to the existence of a personal god or gods interested in human beings' behavior, in our prayers and rituals, or very broadly, as with Einstein's Spinozistic conception of religious feeling as "rapturous amazement at the harmony of natural law."[4] Even with religion narrowly construed, there is the problem of the many competing religions. But I shall cut through these complications by focusing primarily (though not quite exclusively) on Christianity. The most important complication for present purposes is that religion and science differ from each other in a number of interrelated ways: in their conception of the essential character of the universe and our place in it; in the kinds of account they regard as genuinely explanatory; and not only in what they believe, but in how they believe it. This doesn't mean that science and religion are incommensurable, but it does mean that the most illuminating comparisons are neither so easy or so one-dimensional as is sometimes supposed.

Science is not primarily a body of belief, but a federation of kinds of inquiry. Scientific inquiry relies on experience and reasoning: the sciences have developed many ways to extend the senses and enhance our powers of reasoning, but they require no additional kinds of evidential resource beyond these, which are also the resources on which everyday empirical inquiry depends. Among other things, while in even the most ordinary of everyday inquiry we often depend on what others tell us, scientific inquiry has become the joint, ongoing effort of a vast inter-generational community.

The natural sciences seek explanations of natural, and the social sciences of social, phenomena and events. In the natural sciences, the explanations sought are in terms of physical forces and events. In intentional social science, as in history and detective work, the explanations sought are in terms of human beings' beliefs, goals, etc., and the actions they prompt. But both natural-scientific and social-scientific explanations are "natural" in the sense that they eschew appeal to any supernatural, other-worldly, spiritual forces.

Imaginative speculation is essential, but imaginative hypotheses have to stand up to evidence. In the scientific enterprise, respect for evidence, intellectual honesty, are prime epistemological (and ethical) virtues. At any time, there are new speculations as yet untried, and many contested issues, controversial claims, and competing theories or theory-fragments; the body of accepted claims and theories is far from complete, and it is fallible. Though much of it is by now firmly established, none is in principle beyond the possibility of revision in the light of new evidence. Parts of the presently accepted scientific account of the origin of the universe and of our place in it are well warranted, other parts less so; and many, many questions are as yet unresolved. But the main outlines, and many of the details, are pretty well warranted.

According to the best-warranted theories of modern science, the earth is just one small corner of a vast universe, a small corner which happened to be hospitable to life, and in which human beings evolved from earlier life-forms.

Religion, unlike science, is not primarily a kind of inquiry, but a body of belief—"creed" is the word that comes to mind. At the core of a religious worldview, as I shall understand it, is the idea that a purposeful spiritual being brought the universe into existence, and gave human beings a very special place. This spiritual being is concerned about how we humans behave and what we believe, and can be influenced by our prayers and rituals.[5]

Religious belief is supposed to be, not tentative or hedged, but a profound, and profoundly personal, commitment. To disbelieve, or to believe wrongly, is sinful, and faith, i.e., commitment in the absence of compelling evidence, often conceived as a virtue. (This is why we sometimes refer to religious people as "believers.") By contrast, although in their professional capacity scientists accept many propositions as true—some of them very confidently and firmly, and not a few pretty dogmatically—faith, in the religious sense, is alien to the scientific enterprise. (This is why it is sometimes said that belief has no place in science.) Not so incidentally, the different religions do not stand to each other as the different sciences do, but are rivals rather than complementary and interlocking parts of one enterprise.

Unlike religion, theology is a form of inquiry. Unlike scientific inquiry, however, theology welcomes—indeed, it seeks—supernatural explanations, explanations in terms of God's making things so. Usually, furthermore, it calls on evidential resources beyond sensory experience and reasoning, most importantly on religious experience and the authority of revealed texts. So, unlike scientific inquiry, theological inquiry is *dis*continuous with everyday empirical inquiry, both in the kinds of explanations in which it traffics and in the kinds of evidential resource on which it calls.

As I see it, religion and science really are profoundly at odds on all the dimensions I have distinguished; and science really is, on all those dimensions, far and away the more admirable enterprise. (I say "at odds," rather than "incompatible," because the vaguer term is appropriate to all three dimensions of comparison; and for the same reason I say "more admirable," rather than "true" or "better-warranted.")

Complicated as it is, however, this still abstracts from the even more complicated diachronic story of humankind as it has worked for millennia on that gigantic crossword puzzle. But only by looking at things historically can we see how the religious world-picture and the theological way of inquiring arose when early attempts to explain natural phenomena were prematurely inked in; how first one and then another aspect of the religious picture of the universe and our place in it has gradually been displaced as science advanced; and how the sciences have not only gradually arrived, entry by painstakingly worked-out interlocking entry, at a far better-warranted account of the world and of ourselves, but have also come to terms with the inevitability of ignorance and uncertainty, and developed modest and practicable ways, amplifying the resources of everyday empirical inquiry, of finding out how things are.

A BRIEF HISTORICAL EXCURSUS

Very early in the life of humankind, no doubt, people told stories about the origin of the world and the creatures in it; and, facing dangerous and uncontrollable phenomena—fire, flood, disease—hypothesized spirits or gods controlling natural events, gods who could be displeased by violations of their wishes, but might be appeased by rituals or sacrifices. No better explanations were available; and these religious explanations met an emotional as well as an intellectual need, suggesting that we are far from insignificant creatures, that how we humans behave is of concern to something superhuman.

Primitive religions shifted and changed, sometimes merged and blended. Most religions, and many religious ideas, were lost with the cultures that adopted them, but others took hold and spread, eventually becoming the central tenets of what we now think of as the great World Religions. Some religious ideas became firmly entrenched; some texts took on a special, sacred authority. New offices and institutions such as priesthoods and temples grew up, and came to be tightly interwoven into the fabric of society: in ceremonies to mark birth, marriage, and death, in the education of children, in government, in war, and in moral prescriptions and proscriptions—sometimes sanctioned by the promise of rewards and the threat of punishments in an afterlife.

As the Catholic Church, and later the Protestant churches, came into being, so did a new form of inquiry: theology, to which some quite remarkable minds were attracted. Certain entries in the crossword were by now indelibly inked in, and seemed to justify the presumption that, over and above the sensory experience and reasoning on which everyday empirical inquiry relies, there are other kinds of evidence: privileged persons' interactions with God, and the words of sacred scripture (although, to be sure, there were always difficulties, puzzles, and lacunae over which theologians debated, sometimes giving rise to heresies and schisms).

Probably there had always been those who had their doubts about the hocus-pocus of shamans and priests, and surely there were always the curious and ingenious ready to try this and test that. As soon as there were theological arguments, doubtless there were some who suspected that the Problem of Evil might be insoluble, or that the First Cause Argument might generate a regress. Well before Darwin, shrewd old David Hume had suggested that if the world is the way it is by design, it looks very much as if it must have been the design of a baby god just starting out as a creator, or perhaps of a whole squabbling committee of gods. And so on.

And as soon as science got into the business of explaining natural phenomena, there was potential for conflict with religion. Gradually—initially very gradually, but then faster and faster—evidence came in that seemed to threaten the key religious entries. Naturally and reasonably enough, at first people thought that the new evidence could be accommodated or explained away. But as more and more evidence came in, the long-standing, much-intersected old entries, and the efforts to keep them, looked more and more strained.

At least since Galileo's run-in with the Church, there has been felt to be a tension between science and religious belief-systems *qua* bodies of presumed knowledge; for the core idea of the religious conception has come increasingly under threat from what Weinberg aptly describes as a gradual but inexorable process of "demystification" first of the heavens and then of life. Copernicus suggested that the earth was not the center of the solar system; Galileo showed that he was right.[6] It was not until 1822, however, nearly three hundred years after Copernicus' *De revolutionibus orbium caelestium* was published, that the Catholic Church formally acknowledged the "new" astronomy.[7] And then, with the theory of evolution and Darwin's reply to William Paley's influential version of the Argument from Design[8]—the pièce de résistance of natural theology—the potential for conflict once again became acute.

Even before the publication of *The Origin of Species*, in response to the proto-evolutionism of Robert Chamber's *Vestiges of the Natural History of Creation*, Adam Sedgwick, the Anglican clergyman who had been Darwin's geology teacher, had protested that if evolution is true, "religion is a lie, human law is a mass of folly . . . , and morality is moonshine."[9] And in *Omphalos: An Attempt to Untie the Geological Knot*, published in 1857, Philip Gosse had tried to accommodate a strictly literal interpretation of the Genesis story to the evidence for a much older earth and a succession of fossil organisms: as God created

Adam with a belly-button—*omphalos* in Greek—so He made all creatures, and the earth itself, appear older than they actually are: trees with rings, animals with signs of earlier growth and wear and tear, even the skeleton of a Siberian mammoth at St. Petersburg "with lumps of flesh bearing the marks of wolves' teeth."[10] However, far from being the triumph for which he hoped, Gosse's book had proven an embarrassing failure, criticized not only as bad science, but also on the theological grounds that it postulated a deceiving God.

Darwin himself, as a young man, believed in the Bible as the word of God, and spent three years preparing for ordination in the Church of England—though rather half-heartedly; he was more interested in collecting beetles. Even before they were married, his Emma was imploring him to give up his habit of "believing nothing until it is proved."[11] He wrestled not only with the Argument from Design, but also with the Problem of Evil and the First Cause Argument. In 1837, he had written in his notebook: "how much grander [evolution is] than idea from cramped imagination that God created (warring against those very laws he established in all organic nature) the Rhinoceros of Java and Sumatra, that since the time of the Silurian, he has made a long succession of vile Molluscous animals—How beneath the dignity of him, who said let there be light and there was light."[12] But by 1860, the year after the publication of *On the Origin of Species*, Darwin wrote: "I am bewildered. I had no intention to write atheistically. But I own that I cannot see as plainly as others do . . . evidence of design and beneficence on all sides of us. . . . I cannot persuade myself that a beneficent and omnipotent God would have designedly created the Ichneumonidae with the express intention of their feeding within the living bodies of caterpillars";[13] although, the same year, apparently in hopes of conciliating angry clerics, in the second edition of *Origin*, Darwin modified its concluding sentence: "There is grandeur in this view of life, with its several powers, having been originally breathed into a few forms or into one," to: ". . . having been breathed *by the Creator* into a few forms or one."[14]

By the time of the candid autobiography he wrote for his family in 1876, Darwin suggested that the human mind may be incapable of answering religious questions, and described himself frankly as an agnostic. The word had only recently been coined—by Thomas Huxley (known as "Darwin's bulldog" for his tenacious defense of evolution), who had written in 1869 that whereas religious believers "were quite sure they had attained a certain 'gnosis'," he was quite sure he had not.[15]

When the *Origin* was published, the Bishop of Oxford, Samuel Wilberforce, had protested that Darwin was guilty of "a tendency to limit God's glory in creation," and that the theory of evolution "contradicts the revealed relations of creation to its Creator"; William Whewell, author of the *History of the Inductive Sciences*, had refused to allow Darwin's book in the library at Trinity College, Cambridge.[16] Huxley, however, wondered how he could have failed to see the extraordinary explanatory power of evolution for himself; and by 1863 Charles Kingsley could write that "Darwin is conquering like a flood, by the mere force of truth and fact."[17]

Nevertheless, when Darwin published *Descent of Man* in 1871, Pope Pius IX denounced it as "a system . . . repugnant at once to history, to the tradition of all peoples, to exact science, to observed facts, and even to Reason herself."[18] And Alfred Russel Wallace, the co-discoverer, with Darwin, of the theory of evolution by natural selection, eventually came to believe that man could not have evolved naturally; the human soul must have come from God. Well aware of the difficulty of believing that the Creator "should have any special interest in so pitiful a creature as man, an imperfectly developed inhabitant of one of the smaller planets attached to a second or third rate sun," in *Man's Place in the Universe* Wallace argued that the fact that the earth is uniquely suited for the existence of life shows that the universe was designed for man, that we humans really are the pinnacle of it all.[19]

Gertrude Himmelfarb writes that, as science advanced, so did theology; but one might better say that, as science advanced, theology retreated to higher ground. A reformed natural theology, rather than contesting the new biology, adapted to accommodate it. Kingsley wrote to Darwin that he had "gradually learnt to see that it is just as noble a conception of Deity, to believe that He created primal forms capable of self-development into all forms needful . . . , as to believe that He required a fresh act of intervention to supply the *lacunae* which he himself had made"[20] (the view Darwin himself had taken much earlier, before further reflection caused him such bewilderment and eventually forced him to the agnostic position of his autobiography). Henry Ward Beecher pronounced "design by wholesale" nobler than "design by retail." Baden Powell suggested that the uniformity, immutability, and sufficiency of the natural laws on which science depended was evidence of a higher purpose. Frederick Temple, observing that the author of the book of revelation is also the author of the book of science, argued that "the fixed laws of science can supply natural religion with numberless illustrations of the wisdom, the beneficence, the order, the beauty that characterize the workmanship of God." Henry Drummond derided those who "ceaselessly scan the fields of Nature and the book of Science in search of gaps—gaps which they will fill up with God"; God is to be found, not in gaps that science cannot fill, but in the entirety of nature.[21]

In his *History of the Warfare of Science with Theology in Christendom*,

Andrew Dickson White retreated to an even smaller patch of even higher ground. Tracing the gradual retreat of theological dogma in the face of scientific advance, White's book is an extraordinary compendium of the resistance of religious authorities, Catholic and Protestant, to every scientific step forward. But in the concluding section of the last chapter of the second of his two dense volumes, White urged that the new, scientific approach to the interpretation of the Bible had revealed an evolution in religious thought, a vision not of the Fall but of the Ascent of Man: from the tribal god of the Hebrews, to the Universal Father of the New Testament; from a moral code of cruelty and vengeance, to right for right's sake; from the idea of a "chosen people," to an ideal of the brotherhood of man.

It is hardly surprising that Darwin was bewildered, or that there was such confusion over whether the new scientific ideas could or couldn't be reconciled with the older religious world-picture. The scientific theories that threatened to displace religious ideas were not dogmatically certain or comprehensive, but uncertain, tentative, and incomplete; so far from being comforting psychologically, they were extremely disturbing; and they threatened the institutions, the social arrangements, and the moral codes that had been built on the foundations of the old entries.

What is more surprising is that plenty of people are still convinced that the old religious entries are correct. As in Darwin's day, some acknowledge the tensions between science and religion, while others try to save religious ideas by reconciling them with science. Some of those who acknowledge the tensions are unsophisticated folk clinging to the old texts, authorities, and certainties. Others are intellectually sophisticated types able to devise elaborate reasons for keeping the old entries, to exploit the incompleteness and uncertainty of scientific ideas, and to find lacunae which theistic ideas might fill. Some of those who seek a reconciliation are theologians claiming to be using the same methods of inquiry as the sciences; others are scientists who save religion in form, while stripping it of content. As we shall see, however, in practice the idea that the religious world-picture is better warranted than the scientific shades almost imperceptibly into the idea that it can fill lacunae in the scientific world-picture, and this into the idea that there is no real incompatibility.

CREATIONISM AND "INTELLIGENT DESIGN THEORY"

Among those who acknowledge the tensions between religion and science, the least sophisticated, and the most straightforward, are the creationists. However, "creationism" refers not to one view but to a whole family, the shared theme of which is that the Genesis account of how we and the world came into being is true, and the scientific account false. Young-earth creationists maintain that the Genesis story is strictly and literally true; God created the world by fiat and ex nihilo in six ordinary, 24-hour days at or around 4004 B.C., the date Archbishop Usher had calculated in 1650 on the basis of the Hebrew Scriptures.[22] (Some young-earth creationists, however, allow the possibility of a few thousand years more.) Old-earth creationists, by contrast, reinterpret the Genesis story so as to accommodate the standard geological chronology. The Day-Age Theory takes the "days" of the Genesis story to be long ages; the Gap Theory takes the six days of creation to be literal and recent, but preceded by the geological ages, after the original creation had been subverted by Satan.[23]

I shan't spend long on these epistemologically thin forms of creationism. But I will note for the record that just about every religion has its own creation story, of which, even in principle, at most one could be true; and that from the earliest times Christian (and Jewish) theologians have disagreed about the proper interpretation of the Old Testament creation story. There was, for example, grave puzzlement over the fact that the first Genesis account says that God created light, and the distinction between day and night, on the first day, but also that He brought the sun and the moon into being only on the fourth. And there was prolonged debate over the fact that the first of the two Genesis accounts has the operation extending over six days, while the second, speaking of "the day" in which God made the earth and heavens, suggests that God's creation was instantaneous: "He spake, and it was done; he commanded, and it stood fast." Aquinas, drawing on a distinction of Augustine's, maintained that God created the substance of things in an instant, but took six days to separate and form his creation; and this compromise held for centuries, even into the Reformation. Luther held that creation took six days, but was also, by a great miracle, instantaneous; Calvin, however, was a six-day man.[24]

For someone who accepts the Bible as the word of God, the conclusion Gosse drew long ago will seem inevitable: the evidence of the age of the earth and of evolution must be somehow misleading. But unless one acknowledges "evidences" in that old but not quite obsolete sense in which the word refers to especially significant biblical passages,[25] accepting the Genesis story, whether as-is or even stretched out by reconstruing days as "ages," really isn't a serious option.

Douglas Futuyma's *Science on Trial*, published in 1983 in response to a series of "creation science" cases in the courts, does a fine job both of summarizing the evidence for the scientific picture, and of demolishing creationist arguments. Understandably exasperated by the idea that the Genesis story is on a par with the results of the accumulated work of thousands of geologists, paleontologists, biologists, and biochemists, Futuyma is wonderfully scathing: "They must all have been in the ark—all 2 million individual animals, Australian kangaroos, South American boa constrictors, Arctic foxes, New Zealand Kiwis, and 250,000 species of beetles. Not to mention all their parasites. . . . I suppose all these species lived together in the Middle East, within easy reach of the ark, and that Noah was the best animal collector in the history of the world."[26] Quite so.

"Intelligent design theory" (aka "abrupt appearance theory," or "initial complexity theory") abandons biblical literalism, and eschews the term "creationism," which has become something of a liability. This variant seems far more intellectually respectable; so much so, indeed, that a recent article in the *Wall Street Journal* can describe intelligent design as "a sophisticated theory now being argued out in the nation's top universities."[27] A key theme, found in Michael Denton, D. S. Ulam, and Richard Goldschmidt, endorsed by Phillip Johnson in *Darwin on Trial*, and recently defended by Catholic biochemist Michael Behe in *Darwin's Black Box*, is that many structures, from mammalian hair to hemoglobin, are so complex that they could not have been produced by an accumulation of small mutations, but must be the creation of an intelligent designer.

Behe explains that when he calls a biological system "irreducibly complex" he means that it is "composed of several well matched, interacting parts that contribute to the basic function, wherein the removal of any of the parts causes the system to effectively cease functioning." According to Behe, numerous biological phenomena—cilia, flagella, blood-clotting, etc.—constitute such irreducibly complex systems; and these cannot have been produced by successive small modifications, because any precursor system, lacking some part, would be by definition non-functional.

This strand of Behe's argument, to the effect that the evolutionary explanation of such structures is false, and needs to be replaced by intelligent design, shades into another which suggests, rather, that the theory of evolution is incomplete, and needs to be supplemented by intelligent design. Darwinism cannot explain the origin of life itself; as Klaus Dose writes, "all discussions on principal theories and experiments in the field [of the origin of life] either end in stalemate or in confessions of ignorance."[28] Obviously, Behe concludes, the best explanation is God's design.

The same themes, and even a quotation from the same article of Dose's, had already formed part of the argument of Johnson's book, published a few years before. But Johnson, a professor of law at Berkeley who describes himself as a specialist in the analysis of arguments and a "philosophical theist and a Christian" who believes that a creator "plays an active role in worldly affairs," is far more ambitious philosophically than Behe; and he approaches the evidence for evolution as a prosecutor might, trying to discredit the exhibits produced by the defense. The defense attorney in this case is Futuyma, whose title Johnson echoes in his own.

Johnson concedes that peculiar circumstances can favor drug-resistant bacteria, or larger birds as opposed to smaller ones, or dark-colored moths as opposed to lighter-colored ones, etc.; but "none of these 'proofs' provides any persuasive reason for believing that natural selection can produce new species, new organs, or other major changes, or even minor changes that are permanent."[29] And Darwinists have not only based their defense on a few, inconclusive exhibits; they have also failed to produce the essential fossil evidence of intermediate species between fishes and amphibians, amphibians and reptiles, reptiles and mammals, apes and humans. Well, they did produce Piltdown Man, but he turned out to be a fake; and then there was Nebraska Man, but he was identified on the basis of a supposedly pre-human fossil tooth which turned out to be from a peccary (a kind of pig).

Darwinists, Johnson continues, fudge the distinction between fact and theory, telling us that evolution is a fact of nature as securely established as the revolution of the earth around the sun, conceding only that there is healthy disagreement about such matters of theoretical detail as the precise time-scale and mechanism of evolution. But really evolution is only a theory, a theory that scientists hold on faith and defend dogmatically.[30] Reconstruing Popper's criterion of falsifiability as a test of intellectual integrity rather than as demarcating science from non-science, Johnson argues that this is a test which defenders of Darwinism, who simply ignore or deny any phenomena that don't fit their theory, fail.[31] The root of the trouble, he continues, is Darwinists' commitment to "naturalism," a dogmatic denial that a supernatural being could influence natural events or communicate with natural creatures such as ourselves. "Scientific naturalism," he continues, makes the same point by assuming that science, which studies the natural exclusively, is the only reliable source of knowledge. Darwinism consequently treats science as equivalent to truth, non-science as equivalent to fantasy.[32]

The intelligent design people are more formidable than older-fashioned creationists, but not, in the end, much more convincing. Biologists object to Johnson's references to "Darwinism," fudging over the differences between Darwin's original account and evolutionary biology in its present, much adapted and modified form (you might as well call the NASA space program "New-tonism," Paul Gross complains).[33] Present evolutionary biology, after all, is anchored by a far larger range of observations and interwoven in a far broader and tighter fabric of biological and other theories than Darwin's account was when Kingsley described it as "rushing in like a flood by the . . . force of truth and fact."

Reading Johnson, you might well get the impression that the evidence Futuyma marshals amounts to a half-dozen examples of "observed phenomena that confirm the creative effectiveness of natural selection."[34] This is misleading, to put it mildly. Acknowledging that "we do not expect to see flies transformed into fleas in laboratory experiments,"[35] Futuyma describes evidence of mutations in every kind of characteristic, and different populations adapting to different environmental conditions; of continual gradations of difference among major groups of creatures such as insects; of the hierarchical arrangement of species evolution predicts; of fossil evidence in which, as predicted, groups which evolved relatively late are less widely distributed than groups which evolved earlier; and so on and so on. More recent textbooks, such as Mark Ridley's *Evolution*, spell out even more details of the role of evolution in developmental biology, ecology, genetics, molecular biology, phylogenetics, and paleobiology.

And yes, there are transitional fossils (Michael Shermer's *How We Believe* conveniently includes drawings of *Ambulocetus natans*, a transitional fossil between the land-based *Mesonychida* and the marine mammal *Archaeocetus*, the ancestor of modern whales).[36] Determining the exact line of descent of *Homo sapiens* has been difficult and remains highly controversial, because the fossil record of the great apes is meager, and it seems that during the past four million years there were long periods in which different forms of hominids were alive at the same time. According to John Maddox, however, "the human lineage plainly runs from the Australopithenes . . . to *Homo erectus* . . . the genetic source of several forms of hominids . . . as well as the forms of early man."[37]

As for those "irreducibly complex systems," the first thing to notice, as Robert Pennock points out,[38] is that Behe takes it to be true by definition that removing any component of an irreducibly complex system renders it no longer functional. But the claim Behe professes to have established is that complex organs and molecules cannot have evolved, because no precursor of such organs or molecules could have had any advantage to the creature. And this is not only not a tautology; it is not even true. A precursor to the eye—a light-sensitive spot, say—surely *could* convey an evolutionary advantage.

This mistake is hardly new. In *The Blind Watchmaker* Dawkins had already replied to another perpetrator, Hugh Montefiore (Bishop of Birmingham), who had in turn cited C. E. Raven on cuckoos: the cuckoo's parasitic lifestyle could not have arisen by chance, for each of its elements—female cuckoos' habit of laying eggs in other birds' nests, baby cuckoos' habit of throwing other nestlings out—would be useless by itself. But even a rudimentary eye or ear, or a half-baked scheme for sponging on other birds, certainly might be useful to the creature that has it; and anyway, the evolutionary alternative to divine design is not "chance," but cumulative selection. The odds on hitting on, say, hemoglobin "by chance" are indeed negligible. A hemoglobin molecule consists of four chains of amino acids twisted together, each 146 links long. Since there are 20 different kinds of amino acids, the number of possible 146-link chains is enormous—20 multiplied by itself 146 times. But this doesn't mean that hemoglobin couldn't have arisen through many steps of cumulative selection, in which each tiny improvement was the basis for future steps. Moreover, an omnipotent designer (unlike Hume's baby god) would presumably have designed perfectly. But what we actually find are useful but imperfect organs—such as the vertebrate eye, in which the light-sensitive cells point away from the light, and their connectors to the optic nerve stick out towards the side nearest the light. This is hardly what you would expect of a divine designer, but just what you might expect from cumulative selection.[39]

None of this is to deny that there are controversies and disagreements among evolutionary biologists, many anomalies as yet unexplained, and many, many details that are still controversial (Ridley helpfully indicates, in the notes on further reading after each chapter, which issues are the subject of ongoing controversy). Nor is it to say that it is impossible in principle that the theory of evolution should turn out to be mistaken after all. But when Johnson insists that evolution is only a "theory," not a "fact," he trades on a doubleness in the use of the word "theory." In ordinary speech, "that's just a theory" implies that whatever-it-is is highly conjectural, not well supported by evidence; and, as Johnson suggests, contrasts with "that's a fact," which implies that whatever-it-is is firmly established. In scientific contexts, however, the word "theory" is often entirely neutral with respect to whether the claims in question are well or poorly warranted. That evolution is a theory, in this usage, doesn't in the least imply that it is not well warranted.

In one way, Johnson is right about Popper's criterion of demarcation: falsifiability isn't a viable way to distinguish science from non-science; but willingness to take account of negative evidence is an important aspect of intellectual integrity. But in another way Johnson is wrong. Respect for evidence is a much subtler matter than he recognizes. The fit of theory and evidence is virtually always imperfect, predictions virtually never entirely and exactly borne out.[40] Intellectual honesty requires openness to revision, yes, but not a hair-trigger readiness to give up a promising idea in response to the slightest difficulty, which can be no less a mistake than hanging on to an idea too long when the evidence is strongly against it.

The question of the origin of life is, as Johnson and Behe say, unresolved—or, as I shall say, "Where did we come from?" falls in the category of What Remains to Be Discovered.[41] As early as 1868 Ernst Haeckel ("Darwin's German bulldog") had lectured on the spontaneous assembly of chemicals into primitive organisms. Darwin himself had at first thought the question was too difficult to consider, but by 1871 was ready to speculate about a "warm little pond, with all sorts of ammonia and phosphoric salts, heat, electricity, etc.," in which protein might have been formed.[42] Louis Pasteur, also, at one time thought the spontaneous generation of life impossible, but by 1878 he too had changed his mind. By 1924 A. I. Oparin speculated that simple organic compounds could have been formed from the chemicals in the atmosphere of the young earth. In 1956 Stanley W. Miller passed a high-voltage electrical discharge through gases of the suggested kind, and recovered a mixture of chemicals, including several amino acids, from the bottom of the reaction vessel.

But the assumption on which Oparin's argument depended, that the atmosphere of the young earth was "reducing," i.e., free of the oxygen which now sustains the life of animals, later came into doubt; and present calculations make the time available for life to arise shorter than was previously thought. The earth is now estimated to be 4,500 million years old, and life is estimated to have arisen more recently than 4,000 million years ago; there are fossil remains of living things from at least 3,500 million years ago, perhaps as long as 3,800 million years ago. This leaves only a relatively short period—somewhere between 500 and 200 million years—in which life could have arisen.

In *Chance and Necessity* Jacques Monod argued that a unique event such as the emergence of life on earth is in principle unanalyzable by science, which is inherently ill-equipped to explain unique historical events; and calculated just how very improbable it is that DNA should come to be assembled, at random, from its component parts. With his colleague Chandra Wickramasinghe, the astrophysicist Fred Hoyle, famously observing that it was as unlikely that DNA was randomly assembled from a "primordial soup" as that a tornado passing through a junkyard would randomly assemble a Boeing 747, fancifully suggested a creator sprinkling seeds of life; but also, less fancifully, suggested that life might have been brought to earth in an interstellar cloud of gas and dust. Francis Crick raised the possibility that life was brought to earth by extraterrestrial beings—"Panspermia."

Part of the problem, logically prior to *how* life arose, is exactly *what* the first kind of living, or pre-living, things or stuff might have been. The simplest organisms around today—slime molds, bacteria, and even viruses, which are so simple that some biologists question whether they are really "living"—are too complex to have been the first step;[43] and it is now thought that "the first living things did not have to wait on the random assembly of molecules comparable in complexity with those now found in modern organisms,"[44] but were preceded by a prebiotic phase in which there were not yet organisms, but molecules able to catalyze their own formation from raw materials then in the environment.

"We're working on it" is not, as Johnson and Behe seem to think, an admission of something badly wrong, but an acknowledgement of limitations which will hopefully one day be overcome. And even if the question were much more hopeless than it seems to be, their response would not be a reasonable one. If someone really had come up with a serious alternative to the theory of evolution, doubtless there would be just such ferment in the universities as Easterbrook's editorial in the *Wall Street Journal* suggested. So far as I am aware, however, the intellectual excitement created by this "sophisticated theory" does not extend to departments of biology—and for good reason: intelligent design hardly warrants the title "theory," let alone "sophisticated theory." It amounts to nothing more than a flat assertion that an intelligent designer created life and devised those "irreducibly complex systems"; it offers not the tiniest foothold from which to climb to a specific account of *how* God did it.

So in a way Johnson is right about naturalism too: supernatural explanations are alien to science. But again, in another way he is wrong. The commitment to naturalism is not merely the expression of a kind of scientific imperialism; for supernatural explanations are as alien to detective work and history or to our everyday explanations of spoiled food or delayed buses as they are to physics or biology. And the reason is not that supernatural explanations are alien to science; not that they appeal to the intentions of an agent; not that they rely on unobservable causes. The fundamental difficulty (familiar from the central mystery of Cartesian dualism, how mental substance could interact with physical substance) is rather that by appealing to the intentions of an agent which, being immaterial, cannot put Its intentions into action by any physical means, they fail to explain at all.

Others look not to biology but to the heady realm of cosmology for lacunae which religious hypotheses might fill. Recent theistic interpretations of the "anthropic principle," (a phrase introduced by physicist Robert H. Dicke in the mid-1960s) maintain, not that God's intervention was needed to create life, but that it can only have been God's design that brought about the very special circumstances in which life in general, and human life in particular, is possible. The universe seems "fine-tuned" for life, for if certain physical constants were even slightly different life would be impossible. And neither, it is argued, can physical cosmology explain why there is anything at all; ultimately, the answer to Leibniz's old question, "Why is there something rather than nothing?" can only be that God made it so.

Over the last half-century or so, physical cosmology has made considerable progress on the questions of the origin of the universe and the accretion of matter—but many issues are still controversial. The idea that the universe is expanding was established in 1929 by Edwin T. Hubble, who gave his name to the law that almost all galaxies within the observable universe are receding, at a speed increasing with distance. Hubble's Law fits neatly into the big bang theory proposed by George Gamow in 1947, according to which "until about 10 or 20 billion years ago, there was nothing, not even empty space. Then there sprang into being a tiny granule of space filled with such a huge amount of energy that it produced the 100 billion stars in the Milky Way, a comparable number of galaxies that lie beyond our own, the radiation that fills every corner of the universe and the momentum that even now sustains its expansion." Given enough energy, individual particles of radiation, photons, can generate matter out of empty space; the great energy of the big bang could create not only electrons, but also the protons and neutrons that are the ingredients of nuclear matter.[45]

In the 1970s, however, it was discovered that the universe is more uniform in temperature than, according to the big bang theory, it should be. Alan H. Guth's hypothesis that almost immediately after the big bang, and before matter of any kind appeared, the universe spontaneously expanded at great speed and accelerating pace, resolves the problem by smoothing out the temperature fluctuations expected in Gamow's model; however, it has the consequence that there must be a multitude of universes besides our own, each derived from a different speck of space-time.[46]

Still, theologians will ask, why is there something rather than nothing? What brought about the big bang(s)? Won't these questions remain no matter how controversial questions in physical cosmology are eventually settled? Not necessarily; for some cosmologists conjecture that the universe has no beginning and no end—Guth's "fractal universe"—and if this were so, the question to which theologians offer the answer "God did it" would not arise. If his program can be completed, Guth writes, "the laws of physics would imply the existence of the universe. . . . [P]erpetual 'nothing' [would be] impossible."[47] And Hawking: "If the universe is really self-contained, having no boundary or edge, it would have neither beginning nor end: it would simply be."[48]

Even if there was a beginning, the theistic answer leaves you wondering. Why is there God rather than nothing? And how did God bring about the big bang(s)? Jesuit theologian Michael Buckley concedes that "we really do not know how God 'pulls it off'";[49] but it would be hard to overstate how different this "concession" is from the "we're working on it" of serious scientists, or serious inquirers of any kind. Whereas physicists worked for decades to arrive at a detailed account of the accretion of the components of matter (and will doubtless work for many more to understand the origins of the universe), theologians feel able simply to dismiss the need to give any details.

Older and more anthropomorphic conceptions of God as, in White's phrase, a kind of giant toy-maker who

> . . . from his ample palm
> Launched forth the rolling planets into space[50]

have gradually given way to a more abstract conception of a disembodied God, timeless or perhaps eternal. Older conceptions of the physical, as simply material stuff, have also given way to an understanding of the equivalence of mass and energy. So perhaps it is not surprising if some are tempted to try resolving the difficulty about supernatural explanations by identifying God with whatever force brought the universe into being. Such a "resolution," however, would empty the concept of God of its essential content; for it makes no sense to imagine such a force listening to our prayers, performing the occasional miracle, or in any way taking an interest in our doings.

What about the necessary conditions for life? The matter left over from the first few minutes of the universe, Weinberg explains, was almost entirely hydrogen and helium, without the heavier elements such as carbon, nitrogen, and oxygen necessary for life. The chance of producing a carbon nucleus in its normal state (the state of lowest energy) in collisions of three helium nuclei is negligible; but appreciable amounts would be produced if the carbon nucleus could exist in a radioactive state with an energy of roughly 7 MeV (million electron volts) above the normal state, and then there would be no problem producing ordinary carbon. And, as it turns out, the carbon nucleus has just such a radioactive state, 7.65 MeV above the normal state. Isn't this evidence of divine "fine-tuning"? Hardly, Weinberg replies: the formation of carbon in stars is a two-stage process, and the crucial thing is not the 7.65 MeV energy of the radioactive state of carbon above its normal state, but the 0.25 MeV energy of the radioactive state, an unstable combination of a beryllium-8 nucleus and a helium nucleus, above the energy of those nuclei at rest. And this misses being too high for the production of carbon, not by a tiny margin suggesting fine-tuning, but by 20 percent.

Similarly, Weinberg continues, if the cosmological constant—the energy density of empty space—were larger, it would either, if positive, have prevented matter from clumping together in the early universe, or, if negative, have caused the universe to recollapse. Isn't this evidence of divine fine-tuning? Not necessarily, if our universe is only one fragment of a much larger universe in which big bangs go off all the time, each with different values for the fundamental constants.[51]

Life is, as Maddox puts it, in a sense an aberration, requiring a constant supply of chemicals from the environment and a way of producing energy from them.[52] But this doesn't mean that the universe was designed for us; indeed, as Clarence Darrow pointed out long ago, if it was designed, it looks to have been for insect rather than for human life.[53] Nor is the fact that we are here a source of wonder; where else would we be but in parts of the universe which allow the possibility of life? John Earman asks us to imagine "the wonderment of a species of mudworms who discover that if the constant of thermometric conductivity of mud were different by a small percentage they would not be able to survive."[54] Exactly.

"Can science explain everything?" theists ask, in a tone suggesting that a negative answer clinches the matter. Indeed, there are many questions within the competence of science that it can't answer yet (not to mention many questions outside its competence altogether). But it doesn't follow, and it isn't true, that religion can fill the gaps.

ATTEMPTS AT RECONCILIATION

When Judge Overton ruled in *McLean* v. *Arkansas Board of Education* that creationism is not science, he also objected to the creationist claim that belief in a creator and belief in evolution are mutually exclusive, an idea he described as "offensive to the religious views of many." Frank Press, then president of the National Academy of Sciences, wrote in a 1984 NAS pamphlet on *Science and Creation* that it is "false . . . that the theory of evolution represents an irreconcilable conflict between religion and science."[55] But William Provine wrote in response to the NAS pamphlet that such "rationalizations are politic but intellectually dishonest."[56] I suspect not so much intellectual dishonesty as a kind of polite reluctance to offend religious believers; as when Paul Gross writes that "there are endless ways in which evolution could have happened in a 'created' world. . . . Scores of evolutionists are Christians. Their pleas are that creationism hurts, rather than helps."[57] But I agree with Provine that the hope of reconciliation is ill founded.

It is true, of course, that the sciences say nothing about God; and it might seem to follow that there can be no real incompatibility. But this doesn't follow, any more than it follows from the fact that the modern scientific account of combustion doesn't mention phlogiston that it can't be incompatible with the phlogiston theory (or than it follows from the fact that my vote not to make an appointment at this time doesn't list all those whom we should not appoint, that it can't be incompatible with a unanimous vote to appoint X).[58] The kind of theistic evolutionism with which Darwin toyed and which Kingsley accepted disguises, but does not resolve, irreconcilable differences between the scientific and the religious world-pictures. As Darwin came to realize, it fails to acknowledge the deep tension between the idea of an omnipotent and perfectly benevolent God and the many imperfections and cruelties of nature; as I suggested earlier, it fails to acknowledge the deep tension between a scientific picture of a vast and uncaring universe, and the religious idea of a caring and involved God. As Stephen Hawking once put it, "We are such insignificant creatures on a minor planet of a very average star in the outer suburbs of one of a hundred thousand million galaxies. So it is difficult to believe in a God that would care about us."[59]

In *The Mind of God*, Paul Davies, also a physicist, but a believer (and winner of the million-dollar Templeton Prize "for progress in religion") concludes that "belief in God is largely a matter of taste, to be judged by its explanatory value rather than logical compulsion. Personally I feel more comfortable with a deeper level of explanation than the laws of physics. Whether the use of 'God' for that deeper level is appropriate is, of course, a matter of debate."[60] This, from the idea that explanatoriness is just a matter of taste, through the play on "deeper," to the insouciance about the meaning of "God," sounds to me like—well, a million-dollar muddle. I shall focus instead on a book apparently written in part to counteract the atheistical tendencies of Hawking in physics as well as of Dawkins in biology, in which Richard Swinburne explains that rather than challenging science or postulating a "God of the gaps" to explain what science cannot explain, he postulates God "to explain why science explains." In a late-twentieth-century version of the reformed natural theology envisaged in the 1860s by Powell, Temple, Beecher, and Drummond, Swinburne offers a kind of theistic foam that will penetrate and solidify in the frame and all the joints of the scientific model.[61]

Treating knowledge as a huge interconnected web, acknowledging that there is a vast and heterogeneous array of evidence to be accommodated, and noting the connections among the concepts of explanation, kinds, and laws, Swinburne argues that, by the same criteria for theory-choice that all empirical inquirers use, the hypothesis that there is a God best "explains *everything* . . . the fact that there is a universe at all, that scientific laws operate within it, that it contains conscious animals and humans with very complex intricately organized bodies, . . . that humans report miracles and have religious experiences." The very same criteria that scientists use to reach their theories lead us "to a creator God who sustains everything in existence."[62]

Like myself, Swinburne sees continuities between scientific inquiry and the investigations undertaken by historians and detectives; unlike myself, he adds theologians to the list. But despite his efforts to minimize them, the obvious discontinuities in how theology conceives of explanation, of evidence, and of belief, cannot be disguised. "The simple hypothesis of theism," according to Swinburne, is to explain why there is anything at all—and why there is *so much of it*;[63] why all electrons behave alike; why tigers behave alike; how humans can understand enough of how the world works to find food, keep warm, etc.—in short, "*everything* that we observe."[64] Evidently the (very elastic and ambiguous) concept of simplicity is going to carry quite a burden; for according to Swinburne, whereas in other kinds of inquiry a good explanation must not only be simple but also fit with the rest of our knowledge, the latter requirement does not apply to theological explanations.[65] This suggests that the connection of Swinburne's theological account with the sciences is likely to be tenuous at best.

As, indeed, it is. Apparently Swinburne wouldn't be fazed if there were fifty-seven varieties of tiger, for it doesn't trouble him that there are umpteen varieties of beans, beetles, bacteria, etc. Again, apparently Swinburne wouldn't be fazed if the universe were more orderly than it is, or less, for it doesn't trouble him that quantum mechanics is indeterministic. An understandably exasperated Adolf Grünbaum protests that it apparently doesn't matter at all how the world is; Swinburne would offer the same explanation *whatever* laws scientists discovered.[66]

The most that God's existence is really to explain, apparently, is why there is a universe at all, and why there are kinds and laws. (The contrast with the "God of the gaps" of theistic cosmology begins to blur somewhat.) But Swinburne sidesteps some difficulties by holding that God is not timeless but eternal; and rather than shrugging his shoulders like Buckley, he does try to explain the nature of divine agency: it is analogous to human agency, except that "as we may need to postulate unobservable planets and atoms to explain phenomena, so we may need to posit non-embodied persons."[67] But intentional human action, as Swinburne acknowledges, involves physical movements; the problem with non-embodied agents is not that they are unobservable, but that they are non-embodied. What is needed is nothing like regular agent causation, and nothing like the operation of unseen bodies; as Grünbaum says, it is *magical* causation.[68] Swinburne likes to speculate about God's motives—why He is both self-revealing and self-concealing, why He sustains the laws of nature but occasionally miraculously suspends them, etc.; but no amount of such speculation can overcome the essential difficulty.

Swinburne treats religious experience as the operation of a kind of sixth sense, analogous to sight and hearing. Defining religious experiences very weakly, as "experiences which seem to the subject to be experiences of God,"[69] he relies on "the principle of credulity," that we ought to believe that things are as they seem unless and until we have evidence that we are mistaken,[70] to bridge the gap between religious experience in his non-committal sense and Religious Experience in the sense of genuine interaction with God.[71] Swinburne grants that not everyone has religious experiences, though "millions of people" do; but some people are color-blind, too. Some of those who have religious experiences have ingested drugs, or subjected themselves to such disciplines as fasting; but it could be that drugs or fasting open people's eyes to God. Some drugs are known hallucinogens; but most religious experiences are not "made" [*sic*] under the influence of such drugs. Almost all those who have religious experiences have previously been exposed to religious belief (actually, given Swinburne's definition of religious experience, the "almost" is surprising); but only those who already know what a telephone is report seeing what seems to be a telephone, too.

But Religious Experience would be quite unlike sight or hearing insofar as it involves interaction with something non-embodied. In fact, there are far more important differences than similarities. Virtually everyone who learns the word "telephone" does so in the first instance by hearing it used in the visible or audible presence of telephones, whereas no one learns the word "God" ostensively. Virtually everyone who is shown what a telephone is and is subsequently in the presence of a telephones sees it, whereas by no means all of those told about God subsequently report "sensing" him. The vast majority of people can see colors; and we understand the physiological defect that causes color-blindness. Most people, however, don't "sense" God; and we also understand that exhaustion, starvation, drugs, suggestion, etc., can cause illusions and delusions.

For most people, if religious experience is to warrant belief in God, it will have to be other peoples' religious experience.[72] Many of us have rely on the assurances of others that they have seen black swans; but in the case of religious experience, most of us will have to rely on a relatively few others who claim to have had experiences of a kind radically different from any that we—the unprivileged majority—have ever had. What do we have to go on?

Religious experience, in Swinburne's weak sense of the phrase, is a real phenomenon, all right, but a very complex one: not only because, for just about every religion, there are people who claim to have had experiences which authenticate it, but also because "religious experience" covers everything from actual voices and visions, through every kind of illusion and self-deception, to the vaguest sense of edification, of being taken out of oneself. Shermer describes Martin Luther's report of waking suddenly to find Satan disputing with him; and provides photographs of the appearance of the Virgin Mary at the Ugly Duck car rental firm in Clearwater, Florida—an image on the showroom window produced when oil from palm trees was sprayed on the window by water sprinklers—and of the sad people who showed up in wheelchairs and leaning on canes, hoping for a miracle.

Recent work in neurophysiology has begun to investigate what happens in the brain during religious experiences. One conjecture is that religious experiences are evoked by electrical disturbances in the temporal lobes, which can be triggered by anxiety, lack of oxygen, low blood sugar, or fatigue.[73] In and of itself, of course, such a neurophysiological story cannot tell us whether the sense of being in contact with some higher being is an artifact of brain activity, or a bona fide interaction with God. But if Swinburne's "Principle of Credulity" is not to degenerate into a Principle of Credulousness, we would need, at the least, to be told how to distinguish the genuine article, Religious Experience, from other experiences mistakenly taken for interactions with God, and what evidence there is that some religious experiences *are* Religious Experiences.

In what seems to be an allusion to the Bayesian arguments on which he had relied in an earlier book,[74] Swinburne argues that when evidence of religious experience and miracles is taken into account along with the fact that there is a universe at all, that there are kinds and laws in the universe, etc., it "tops up" the probability that God exists sufficiently to justify belief. He never suggests, in the manner of a priori theology, that God's existence is certain; but the reason he gives for not expecting certainty is quintessentially theological: "If God's existence, justice and intentions became items of common knowledge, then man's freedom to choose [to believe or disbelieve] would in effect be vastly curtailed."[75] Also quintessentially theological is the idea that once the probability of God's existence is high enough (51 percent? 65 percent? 77 percent?), the appropriate response is not tentative acceptance pending further investigation, but "an unlimited commitment."[76] It is as I said earlier: the differences between religion and science extend not only to what is believed, but also to how it is believed.

Contrary to his intent, Swinburne's book highlights not only the differences between scientific and theological inquiry, but more importantly the discontinuities between theological inquiry and our everyday empirical investigations. Whereas the sciences have amplified and refined the evidential resources on which we all rely daily, theology relies on additional evidential resources the authenticity of which depends on dubious old crossword entries—rather as if a whole new kind of inquiry had sprung from the phlogiston theory, and somehow survived its demise as an account of combustion.

In an encyclical of 1950, Pope Pius XII had written that "[s]ome imprudently and indiscreetly hold that evolution . . . explains the origin of all things"; "[c]ommunists," he continued, "gladly subscribe to this opinion." The theory of evolution should not be accepted as proven, and "the Catholic faith obliges us to hold that souls are immediately created by God."[77] But he conceded the possibility of "research and discussion" between scientists and theologians on the evolution of the human body. In a 1996 address entitled "Truth Cannot Contradict Truth," Pope John Paul II acknowledged that "new knowledge has led to the recognition of the theory of evolution as more than a hypothesis."[78] He maintained, however, as Wallace had done much earlier, that "if the human body takes its origin from pre-existent living matter, the spiritual soul is immediately created by God."[79]

Recently, claiming Pius XII and John Paul II as allies, Stephen Jay Gould has also maintained that science and religion have distinct and non-overlapping domains or "magisteria" (hence his acronym "NOMA").[80] Each domain of inquiry "frames its own rules and admissible questions, and sets its own criteria for judgment and resolution."[81] The role of religion is not to fill gaps in the scientific picture, but to answer its own distinctive questions. But Gould's view is not, as Pope John Paul's is, that science is concerned with the natural world, and religion with the human soul, with spiritual matters; rather, he takes science to be concerned with the realm of fact and religion with the realm of values, with meaning and morals—something like White, who, however, was far more candid about the long history of bitter conflict between religion and science than he; and something like Einstein, who, however, had made it clear that he conceived of the religious attitude simply as an aspiration to "superpersonal" values, such as knowledge of the truth.[82]

Gould's position may seem attractively irenic; but its appeal derives from its vagueness and ambiguity, and evanesces under closer scrutiny. Setting aside the objection that religions often make quite specific historical/factual claims,[83] and postponing the question, whether science has anything to tell us about values,[84] I want to focus for the moment on the thesis that the magisterium of religion is the realm of values. I don't suppose Gould thinks religion concerns itself with aesthetic or epistemic values. Nor does it seem helpful to construe his talk of "meaning and morals" as gesturing towards the way religious people sometimes invest events which from a scientific point of view are mere accidents, coincidences, with a deeper significance. Why was *my* cow struck by lightning, and not yours? Or—a real example, this, from a disturbing television program in the aftermath of Hurricane Andrew—why did the hurricane destroy *their* church, but leave ours untouched? This would reduce religion to a kind of vulgar superstition; which I don't believe is Gould's intent.

But Gould *has* kidnapped the term "religion," and put it to a quite non-standard use. This is most explicitly confirmed when, observing that Thomas Huxley rejected the Christian doctrine that a belief in resurrection serves as a necessary prod for decent behavior during our earthly life, "on a higher principle he regards as essentially religious in nature," Gould tells us that he will construe as fundamentally religious "all moral discourse on principles that might activate the idea of universal fellowship among people."[85]

Huxley would not have been pleased to be co-opted along with Popes Pius and John Paul; for, though he imagined a kind of secular "religion" that might meet some of the same emotional needs without untenable theistic doctrine, he seems to have thought of the relation of religious belief and the scientific world-picture much as I do. Our ancestors believed that "the earth is the centre of the visible universe, [and] man the cynosure of things terrestrial"; and that "the course of nature . . . could be altered by the agency of . . . spiritual beings," he wrote; but by his time, "[i]t is very certain that the earth is not the chief body in the material universe, and that the world is not subordinated to man's use. It is even more certain that nature is the expression of a definite order with which nothing interferes."[86]

In any case, Gould's conception of religion is both too narrow and too broad, for morals are neither the only concern of religion nor the concern only of religion. Set aside the problem that a moral code that distinguishes Jew or Mormon from Gentile is certainly religious, even though positively discouraging to "the idea of universal fellowship among people." Gould's account of religious discourse is clearly too narrow in another way as well: religions have at their core a body of doctrine about God, or the gods, and their relations with human beings and role in their fate. Moral precepts usually follow, or are appended, the point generally being that following this code will please God, or appease the gods; they may even come to occupy a central place in religious discourse of an uplifting rather than theological style. But they are not the heart of the matter. At the same time, Gould's account of religious discourse is also too broad: not only the religious, but also agnostics and atheists, acknowledge moral constraints.[87] Not only religious belief, furthermore, but many other things—family, personal relationships, love of music or of animals, commitment to humanitarian goals, etc.—can give meaning to life.[88] In short, Gould is guilty of the confusion for which H. L. Mencken criticized Alfred North Whitehead, of "the religious impulse [with] common decency."[89]

Wondering why "various alarmed and conciliatory scientists"—he mentions Robert Millikan and Arthur Eddington—were so anxious to reconcile science and religion, Mencken suggested two motives: a kind of cowardice, a disinclination to provoke unscrupulous antagonists too far; and inner doubts, an inability entirely to throw off the pieties in which they were brought up.[90] More recently, Dawkins writes in exasperation of "a cowardly flabbiness of the intellect."[91] I shan't speculate about Gould's motives, but I will repeat that in my opinion the reconciliation he proposes works only by stripping religion of its essential content.

RELIGION, MORALITY, AND THE "WILL TO BELIEVE"

Still, even if it isn't true, isn't religious belief a Noble Lie, perhaps even a necessary one? Aren't "its dreadful threats and solemn sanctions . . . needed to keep frail mankind in order"?[92] Weinberg, reminding us of the long and gory history of "crusades, pogroms, and jihads," writes that "on balance the moral influence of religion has been awful."[93] Drawing up a balance sheet doesn't strike me as easy; but at the very least it is far from obvious that the bottom line would leave religions, morally speaking, in the black.

On the positive side, one might put the psychological comfort religion offers to believers; the social solidarity religions encourage, and their contributions to social reform; the extraordinary heritage of religious poetry, art, and music; and—doubtless this is part of what Havel had in mind—the usefulness of religion as a bulwark against the excesses of godless communist regimes.

But a too-easy confidence that religion is reliably an influence for the good is fostered by our tendency to select the moral precepts of which we approve, and ignore the rest.[94] A correspondent points out in the *Wall Street Journal* that you can find in the Old Testament not only the proscription against homosexuality cited in an article earlier in the week, but also proscriptions against touching the skin of a pig, against eating shrimp, and, in the "exquisitely numbered" Leviticus 20:20, against allowing people with imperfect eyesight to approach the altar.[95] Many religious prescriptions and proscriptions are morally objectionable, even barbaric: the Hindu practice of suttee, the ritual suicide of a widow on her husband's funeral pyre; the prescription in Islamic law of amputation of a thief's right hand; death by stoning as the Old Testament punishment for adulteresses. True, I have chosen somewhat melodramatic examples; but there are plenty of others, less melodramatic but equally apropos—especially, perhaps, in the realm of sexual morality.

Theological responses to the Problem of Evil can be downright scary. Here is Swinburne: "I am fortunate if the natural possibility of my suffering if you choose to hurt me is the vehicle that makes your choice really matter. . . . [M]y good fortune is that the suffering is not . . . pointless."[96] As it happened, the same day I read this I also read an article by a woman who had been raped, sodomized, terrorized, beaten, and left for dead by a gang of thugs.[97] Can you accept that she was benefited by having been the vehicle that made her attackers' choice really matter? I can't.

And religions have certainly done harm as well as good. The other face of the power of religion to unite people is its power to divide them; and so my list of harms would begin, like Weinberg's, with religious and sectarian wars. It might continue: the Inquisition; the persecution of supposed witches; tyrannical theocratic governments; human sacrifice; the effect of Hindu fatalism in inhibiting social reform; the role of the Dutch Reformed Church in supporting apartheid; etc., etc.

We are selecting among the evidence, again, when we notice the role religious people have sometimes taken in social reform, and forget that not all religious people are reformers, and not all reformers religious. Freeman Dyson stresses the role of religious belief in the suppression of slavery; Weinberg notes that while devout Christians such as William Wilberforce were influential in opposing slavery,[98] plenty of devout Christians supported it, and humanists like Adam Smith and Jeremy Bentham fought against it.[99]

We humans are creatures capable both of great cruelty and of great kindness—both with the help of religion, and without it. Sometimes people are kind and generous, sometimes cruel and heartless, from religious motives; and sometimes people are kind and generous, sometimes cruel and heartless, without the help of religion.

Life is often difficult, disappointing, discouraging, and worse; sometimes much worse. Hence James Beattie's protest to the skeptical Hume: "there is many an honest and tender heart pining with an incurable anguish . . . whom nothing but trust in Providence, and the hope of a future retribution, could preserve from the agonies of despair. And do [you], with sacrilegious hands, attempt to violate this last refuge of the miserable, and to rob them of the only comfort that had survived the ravages of misfortune, malice and tyranny!"[100]

Beattie is eloquent, even touching; but for the classic argument that religious belief is legitimated by its enabling the believer to live better, one turns to William James's "The Will to Believe."[101] In *The Ethics of Belief*, W. K. Clifford had maintained that "it is wrong always, everywhere, and for anyone to believe anything on insufficient evidence." James responds by defending "our right to adopt a believing attitude in religious matters, in spite of the fact that our merely logical intellect may not have been coerced." When a hypothesis cannot by its nature be decided on the basis of evidence; when it is live for us, appeals to us as a real possibility; when the choice between believing it and disbelieving it is forced, unavoidable, and momentous for our lives, then *"our passional nature ... lawfully may decide."*[102] Religious belief, though it cannot in principle be verified or falsified, is legitimated by its salutary effect on the believer's life.

One complication is that, as a Pragmatist, James is officially committed to holding that a hypothesis which cannot by its nature be decided by evidence is meaningless. But I shall set such scholarly complications aside for now,[103] to concentrate on the problems caused by the fuzziness of James's conception of the content of religious belief: "the best things are the most eternal things"—and the "salutary effect" it is supposed to have on a believer's life. You might argue like this: if the salutary effect is to open the believer's heart not only to the beauty of the natural world, to creativity, to passionate concern, and to loving relationships with others, but also to an appreciation of the divine meaning in the prosaic events of life, James's thesis presupposes the truth of theism, and is question-begging. If, on the other hand, the salutary effect is to enable the believer to overcome otherwise insuperable obstacles and bear otherwise intolerable burdens, James's thesis really does begin to sound, as critics objected, like a defense of wishful thinking.

But this is still something of an over-simplification. First of all, there is an ambiguity in "faith in God," which may mean either *belief that God exists*, or else—quite differently—*trust in God*; an ambiguity which, if it goes unnoticed, may hide what is otherwise obvious—that trust in God presupposes His existence; and that although, as James sees, placing your trust in someone may itself contribute to bringing a trusting relationship into existence, it certainly can't bring the other person into existence. Second of all, even if it is true that believing that there is an interested, caring God enables you to bear those otherwise intolerable burdens, that it does so is not evidence of God's existence.

The year after *The Will to Believe* was published, Peirce, to whom it was dedicated, is found defending the "scientific attitude," the "Will to Learn." A little later, he writes that, though he would not condemn a man who persuaded himself that he would be reunited with his late wife in the hereafter, if that was the only way to save his reason, for himself, he "would not adopt a hypothesis, and would not even take it on probation, simply because the idea was pleasing to him."[104] Here, implicitly, Peirce distinguishes the epistemological and the prudential aspects that James implicitly runs together. From an epistemological point of view, wishful thinking is always a defect. That we wish things were thus and so is irrelevant to whether it is true that they are thus and so—which is why wishful thinking doesn't work directly, but by distorting the evidence to enable you to give too much credence to propositions you would like to be true, too little credence to propositions you would prefer to be untrue.

Whether and to what degree wishful thinking is imprudent, however, depends on what the wishful believer has to lose if his wishful belief is false. If, having purchased my single ticket, I manage to persuade myself that I will win the lottery, and give away my house, car, and savings, and commit myself to paying for the mansion, the Rolls, etc., that I shall easily be able to afford when I win, you would rightly think me grossly imprudent. But if, stricken with a cancer for which there is no known cure, I manage to persuade myself that visualizing the cancer cells in retreat will save me, or that I shall shortly go to a Better Place, however misguided you thought me, you might be willing to concede that I was better off than if I gave up hope and turned my face to the wall.

Human beings, I said earlier, are capable of both cruelty and kindness; now it is time to emphasize that we are divided creatures intellectually as well as morally. So far as we know, we humans are the smartest creatures around (rather a scary thought, as David Stove somewhere remarks!). We have the capacity to investigate, to discover something of how the world is; and in consequence we can sometimes gain control over natural events and phenomena that affect us—cure a disease, dam a river, improve the harvest. And at least some of us, at least some of the time, actually enjoy figuring things out. However, finding out how things are can be painfully hard work; the results, furthermore, are often uncertain, and sometimes unpalatable. We dislike not knowing; we dislike uncertainty; and we naturally prefer to avoid facing unpalatable truths—sometimes, indeed, we are better off not facing them.

No wonder, then, that our penchant for wishful thinking survives our knowledge that it is an epistemic defect. No wonder, either, that while we are awed by the work of the greatest scientific minds, we are also a little resentful, a little abashed, at the difficulty we have in understanding it. The other side of our ambivalently resentful admiration for extraordinary scientific work is our yearning for religious certainties, and our fondness for religious mysteries, the contemplation of which makes us feel elevated, and where a failure of comprehension is no reproach. We can even, some of us, persuade ourselves that it is a virtue to accept on faith where evidence fails us, that Tertullian's *Credo, quia absurdum est*[105]—I believe because it is absurd—is not foolishness, but morally meritorious. For others, this is almost as hard to swallow as the old idea that, since the abasement of man contributes to the glory of God, filthiness is next to godliness.[106]

Even those of us who, like myself, find the idea that there is moral merit in faith epistemologically repugnant may feel a twinge of something like envy when we read the touching story of little Eva, dying of consumption, telling her Papa not to be too sad: "I had rather be in heaven; though, only for my friends' sake, I would be willing to live."[107] But this is rather nostalgia for the lost innocence of childhood than a true wish that we too had such faith. Uncle Tom's faith is no less steadfast; but we don't find it so unambiguously admirable—and the word that comes to mind is "childlike." As Dawkins points out, for good biological reasons small children are naturally credulous—they are "information caterpillars," sucking in information as caterpillars suck in cabbage leaves. Believing what grown-ups tell them has survival value; but since grown-ups sometimes give contradictory advice something like an override mechanism is needed, to encourage persistence of belief in what you were told first. "Those old Jesuits," Dawkins concludes, "knew what they were about."[108] Indeed.

Religion is no less quintessentially human an enterprise than science; it is much older, and its roots in our psychological makeup perhaps deeper. But its fundamental appeal is to the side of the human creature that craves certainty, likes to be elevated by mysteries, dislikes disagreeable truths, and clings to the flattering idea that we are not just remarkable animals, but the chosen creatures.

AND, IN CONCLUSION

There will be those—doubtless they will be many—who object that what I have said here reveals a narrow, scientistic rationalism. That is seriously untrue, for I have maintained all along that the sciences are not the only legitimate kinds of inquiry, not the only source of truth. But it *is* true that, when Mencken writes, celebrating the triumphs over nature that humankind has achieved through the exercise of its intelligence, "[e]verything that we are we owe to Satan and his bootleg apples,"[109] though I pause to note the literal falsity of his words, I am moved by their spirit. I see a nobility in the human capacity to puzzle, dream, calculate, check, and test, to work to find out how things are, to refuse false comfort, to try to find ways to make life better. I make no apology for reserving my greatest admiration for those who "delight to exercise the mind, no matter which way it takes [them]," as it delights others to exercise their muscles,[110] those for whom doing their damnedest with their mind, no holds barred, is "a point of honor."

NOTES

1. Quoted in Vadislav, ed., *Vaclav Havel, or Living in the Truth*, pp. 138–39, cited in Holton, *Science and Anti-Science*, pp. 177, 188.

2. Weinberg, *Dreams of a Final Theory*, p. 260.

3. Ibid.

4. Einstein, "The Religious Spirit of Science," p. 40.

5. This probably excludes Buddhism as not, as I am using the term, really a religion—a consequence I am willing to accept, especially because the issue of conflict between science and religion seems to arise primarily with respect to Christianity, Judaism, and Islam.

6. Weinberg, *Dreams of a Final Theory*, pp. 245–46.

7. My source is Mencken, *Treatise on the Gods*, p. 259.

8. Paley, *Natural Theology*.

9. My source is McIver, *Anti-Evolution*, p. 245.

10. McIver, *Anti-Evolution*, p. 91; the quotation about the mammoth skeleton is from White, *A History of the Warfare of Science with Theology in Christendom* 1:242.

11. Browne, *Charles Darwin, Voyaging*, cited in Gardner, "The Religious Views of Stephen Gould and Charles Darwin," p. 8.

12. Darwin, *On Evolution*, p. 73.

13. Darwin, letter to Asa Gray, quoted in Gardner, "The Religious Views of Stephen Gould and Charles Darwin," p. 9.

14. For the amended sentence in the second edition, see Gardner, *Great Essays in Science*, p. 13. Italics are mine.

15. Huxley, *Collected Essays*, 5:237–38.

16. White, *A History of the Warfare Between Science and Theology in Christendom*, 1:70 (Wilberforce), 84 (Whewell).

17. Darwin, *Autobiography and Letters*, p. 267.

18. My source is Futuyma, *Science on Trial*, p. 24.

19. McIver, *Anti-Evolution*, p. 290.

20. Darwin, *Autobiography and Letters*, p. 242.

21. Himmelfarb, *Darwin and the Darwinian Revolution*, pp. 392–93.

22. White, *The History of the Warfare Between Science and Theology in Christendom*, 1:253.

23. McIver, *Anti-Evolution*, p. xiv.

24. White, *A History of the Warfare of Science with Theology in Christendom*, 1:5 ff.

25. I learned of this usage from Pennock, *Tower of Babel*, to which I refer readers wanting a detailed discussion of the various forms of creationism.

26. Futuyma, *Science on Trial*, p. 205. Perhaps believers will reply that God sent the animals to Noah; but I rest my case.

27. Easterbrook, "The New Fundamentalism."

28. Behe, *Darwin's Black Box*, p. 168, citing Dose, "The Origins of Life: More Questions Than Answers," p. 348.

29. Johnson, *Darwin on Trial*, p. 27.

30. Ibid., pp. 11–12.

31. Ibid., pp. 145–54.

32. Ibid., pp. 114–16.

33. Gross, "Politicizing Science Education," p. 9.

34. Johnson, *Darwin on Trial*, pp. 25–26.

35. Futuyma, *Science on Trial*, p. 204.

36. Shermer, *How We Believe*, p. 114.

37. Maddox, *What Remains to be Discovered*, pp. 166–67. For a summary account of scientific discoveries and controversies since the publication of Maddox's book, see Begley, "The New Old Man," and "Bickering over Old Bones."

38. Pennock, *Tower of Babel*, pp. 268 ff. Elsewhere in his book there are detailed critiques of other arguments in Behe.

39. I have relied on Dawkins, *The Blind Watchmaker*, pp. 39–41, 45, and 91–92. He cites Montefiore, *The Probability of God* and, on imperfection as evidence of evolution, Gould, *The Panda's Thumb*.

40. See also Weinberg, *Dreams of a Final Theory*, pp. 48–50, on why he wouldn't bother testing certain claims, e.g., about telekinesis.

41. I am alluding to the sources on which I have drawn here: Pigliucci, "Where Did We Come From?" (which includes a useful bibliography of historical and contemporary source material); Maddox, *What Remains to Be Discovered*, part 2.

42. Darwin to Hooker, quoted in Maddox, *What Remains to Be Discovered*, p. 127.

43. Pigliucci, "Where Do We Come From?"

44. Maddox, *What Remains to Be Discovered*, p. 133.

45. Ibid., pp. 25 ff. (the quotation is from p. 25).

46. I continue to rely on Maddox, *What Remains to be Discovered*, this time on pp. 52–59.

47. Guth, *The Inflationary Universe*, p. 276; my source is Shermer, *How We Believe*, p. 108.

48. Hawking, *A Brief History of Time*, p. 175; my source is Shermer, *How We Believe*, p. 103.

49. Buckley, "Religion and Science: Paul Davies and John Paul II," cited in Grünbaum, "A New Critique of Theological Interpretations of Physical Cosmology," p. 26.

50. White, *A History of the Warfare of Science and Theology in Christendom*, 1:2; the quotation is from ancient Chaldean religious writings that inspired the Old Testament.

51. Weinberg, "A Designer Universe?"

52. Maddox, *What Remains to Be Discovered*, p. 151.

53. I rely on Grünbaum, "A New Critique of Theological Interpretations of Physical Cosmology," p. 30.

54. Earman, "The SAP [Strong Anthropic Principle] Also Rises," p. 134; my source is Grünbaum, "A New Critique of Theological Interpretations of Physical Cosmology," p. 30.

55. My source is Johnson, *Darwin on Trial*, p. 123.

56. My source is Johnson, *Darwin on Trial*, p. 124.

57. Gross, "Politicizing Science Education," p. 10.

58. No, I am not making this up.

59. The quotation comes from a BBC television program called "Master of the Universe"; my source is Shermer, *How We Believe*, p. 102. At the end of *A Brief History of Time* Hawking had written that if we knew the ultimate laws of nature, "we would know the mind of God"; but it seems clear that this was a figure of speech rather than an affirmation of belief in a personal God.

60. Davies, *The Mind of God*, p. 189; my source is Shermer, *How We Believe*, p. 103.

61. My discussion will be relatively brisk. There is a much lengthier and more detailed discussion in Olding, *Modern Biology and Natural Theology* (and a lengthier and more detailed discussion of Behe—more sympathetic than mine—in Olding's "Maker of Heaven and Microbiology").

62. Swinburne, *Is There a God?* pp. 68, 2.

63. Ibid., p. 49.

64. Ibid., p. 2.

65. Ibid., p. 41.

66. Grünbaum, "A New Critique of Theological Interpretations of Physical Cosmology," p. 21.

67. Swinburne, *Is There a God?* p. 37.

68. Grünbaum, "A New Critique of Theological Interpretations of Physical Cosmology," p. 23.

69. Swinburne, *Is There a God?* p. 130.

70. Ibid., p. 134.

71. The typographical distinction is not Swinburne's but mine.

72. As Weinberg also notes (*Dreams of a Final Theory*, p. 254).

73. Shermer, *How We Believe*, pp. 34–35 (the Virgin Mary at the Ugly Duck car rental) and 65–69 (neurophysiological evidence; Luther is on p. 68). See also Begley, "Religion and the Brain."

74. Swinburne, *The Existence of God*; Swinburne's Bayesian arguments are criticized in detail by Grünbaum in "A New Critique of Theological Interpretations of Physical Cosmology."

75. Swinburne, *The Existence of God*, p. 244.

76. Swinburne, *Is There a God?* p. 141.

77. Pope Pius XII, *Humani Generis*; my source is Gould, *Rocks of Ages*, pp. 77–80.

78. Pope John Paul II, "Message on Evolution to the Pontifical Academy of Sciences"; my source is Dawkins, "When Religion Steps on Science's Turf," p. 19.

79. Again, my source is Dawkins, "When Religion Steps on Science's Turf," p. 19.

80. Gould, *Rocks of Ages*.

81. Ibid., pp. 52–53.

82. Einstein, "Science and Religion," pp. 42, 44–45.

83. See also Dawkins, "When Religion Steps on Science's Turf," p. 18.

84. See chapter 11.

85. Gould, *Rocks of Ages*, pp. 60–62 (the quotation is from p. 62).

86. Huxley, "Science and Culture," pp. 135, 138.

87. See for example Robinson, *An Atheist's Values*, and the essays collected in Kurtz, *Moral Problems in Contemporary Society*.

88. See Haack, "Worthwhile Lives."

89. Mencken, *Treatise on the Gods*, p. 269.

90. Ibid., pp. 244 ff.

91. Dawkins, "When Religion Steps on Science's Turf," p. 18.

92. Mencken again: *Treatise on the Gods*, p. 267.

93. Weinberg, "A Designer Universe?".

94. Again, see Dawkins, "When Religion Steps on Science's Turf," p. 18.

95. Smith, "Leviticus Contains Many Proscriptions."

96. Swinburne, *Is There a God?* p. 103.

97. Ruder, "Opening the Drawer One More Time."

98. Father of Samuel Wilberforce, the Anglican bishop who attacked Darwin so bitterly.

99. Weinberg, "A Designer Universe?" referring to Dyson, *Imagined Worlds*.

100. James Beattie, *An Essay on the Nature and Immutability of Truth*, part 2, chapter 1. My source is Stove, "D'Holbach's Dream," p. 87.

101. From which, ironically enough, the stirring words of James's quoted at the beginning of this book about the "magnificent edifice" of science are also taken.

102. James, *The Will to Believe*, pp. 12, 11.

103. But see Haack, "'The Ethics of Belief' Reconsidered," and "American Pragmatism."

104. Peirce, *Collected Papers*, 5:583, 5:598.

105. Mencken, *Treatise on the Gods*, p. 230.

106. White, *A History of the Warfare of Science with Theology in Christendom*, 2:69.

107. Stowe, *Uncle Tom's Cabin*, p. 275.

108. Dawkins, *Unweaving the Rainbow*, pp. 138–44; the quotation is from p. 144.

109. Mencken, *Treatise on the Gods*, p. 266.

110. Ibid., p. 292.

11

WHAT MAN CAN ACHIEVE WHEN HE REALLY PUTS HIS MIND TO IT

The Value, and the Values, of Science

> [I]f, as our race approaches its maturity, it discovers, as I believe it will, that there is but one kind of knowledge and one method of acquiring it; then we, who are still children, may justly feel it our highest duty to recognise the advisability of improving natural knowledge, and so to aid ourselves and our successors towards the noble goal which lies before mankind.
>
> —Thomas Huxley, "On the Advisableness of Improving Natural Knowledge"[1]

Perhaps, to a modern ear, these lines sound a little overwrought; nevertheless, I confess, I wish I'd written them. For Huxley packs more than one insight into this single, subtle sentence: the continuity of science with everyday empirical inquiry; the difference between the credulity of children, and of the child in each of us, and the rigors of really trying to find things out; the enterprise of inquiry as a characteristically human undertaking; the epistemological, and the ethical, significance of this enterprise. Much of what I shall have to say here will be, in effect, an amplification and defense of precisely these ideas.

People often talk about "the question of science and values"; but this, like "the" question of the relation of science and religion, isn't really one question, but a whole tangle of questions. What kinds of value are relevant to the appraisal of science? How do epistemological values relate to aesthetic values, or to the ethical? Can the sciences themselves tell us anything about values, or are they competent only within the realm of fact? And what is the value of science itself?—which, in turn, is really a whole cluster of questions: what is the value of science *qua* intellectual enterprise, of science *qua* social institution, of this specific branch of scientific inquiry, of that specific piece of scientific work, or of a specific item of scientific knowledge?—questions which may be asked from an epistemological, a practical, an ethical, an aesthetic, a political, or an economic point of view. Needless to say, I can't hope to offer even superficial answers to all these questions; I shall have to be selective. The core of my argument will concern the value of science *qua* intellectual enterprise, and will keep the epistemological dimension in central focus.

The first phase will be to the effect that what is epistemologically valuable about the natural sciences is not simply the vast body of knowledge they have accumulated about the world and how it works, but also the way they have expanded and refined human cognitive capacities, overcome human cognitive limitations, and amplified our capacity to inquire effectively. They represent a remarkable manifestation of man's cognitive potential, of what human beings can achieve when they do their damnedest with their minds.

The second phase of my argument will be that, though epistemological and ethical values are distinct, intellectual integrity is a moral as well as an epistemological virtue. Like Huxley, I see the amplification of the "method of experience and reasoning" achieved by the natural sciences as a step of ethical as well as epistemological significance.

Again like Huxley, I see scientific inquiry as continuous with everyday empirical inquiry—only, I would add, "more so." The sciences are not epistemologically privileged, but they are epistemologically distinguished; i.e., by the standards by which we judge evidence better or worse, inquiry well or poorly conducted, they have succeeded, though very far from perfectly, remarkably well. The third phase of my argument will be that, though we cannot look to the sciences for an account of core epistemological concepts, cognitive psychology has a contributory relevance to our understanding of epistemological standards, and history of science to our understanding of the remarkable successes of the sciences in meeting them.

Some disparage science because they believe it destroys the wonder and mystery of the natural world; others because they believe it is an inherently white, male, Western enterprise. The fourth phase of my argument will be that science is an expression of capacities which are not so much white, Western, or male, as human; and that while, to be sure, scientific understanding dispels mystery, the discoveries of science reveal the world to be far more wonderful than we could have imagined.

Some fear science because they believe the technological developments it has spawned are dangerous and damaging. The last phase of my argument will respond to, and where appropriate acknowledge the elements of truth in, this criticism; so that finally I shall be able to offer, by way of conclusion, a diagnosis of the deep ambivalence that many people obviously feel about science.

THE EPISTEMOLOGICAL VALUE OF SCIENCE

The natural sciences have accumulated a vast and ever-growing body of knowledge about the natural world and how it works, patchy and incomplete to be sure, but increasingly well anchored in experience and increasingly integrated internally. Of course, the sciences are far from perfect; of course, there have been remarkable achievements in history, detective work, literary scholarship, etc. Moreover, the body of well-warranted theories the sciences have thus far produced is the result of a long process in which most of what was proposed has been, or sooner or later will be, discarded as untenable. Furthermore, plenty of the knowledge scientific work has produced is trivial, boring, unimportant even from a scientific point of view. Nevertheless, some of the knowledge the natural sciences have produced is stunning; and the way it all fits together is extraordinary. In fact, the body of scientific knowledge accumulated even thus far is so extensive, and its internal interconnections so dense and complex, that I am a little at a loss to know how to illustrate the point at modest length. One way, perhaps, is to look at Huxley's remarkable lecture "On a Piece of Chalk," and Weinberg's equally remarkable riff on the same theme.

Huxley's lecture, delivered in 1868 to the working men of Norwich, presented evidence of the vast age of the earth—older by far than Archbishop Usher's estimate—and of the evolution of new living species. The "piece of chalk" of his title symbolizes the chalk layer, in places more than a thousand feet thick, extending under much of southeast England, Europe, and the Levant, forming familiar geographical features from the white cliffs of Dover and the Needles in the Isle of Wight to the mountain ranges of Lebanon. "A great chapter of the history of the world is written," in this chalk layer, which was once the mud of an ancient sea bed. It is composed of the compressed, fossilized shells of the myriad simple organisms that drifted to the bottom of the ancient seas that once covered Europe. Caught in the chalk are the fossils of larger sea creatures, including a whole series of proto-crocodiles, and of land animals from periods when geological upheaval brought the chalk above the sea. The further down they are, the more different these fossilized creatures look from their modern counterparts; these species were evolving during the millions of years in which the chalk deposits were laid down.[2]

I like to imagine Huxley enjoying this sentence from a recent best-selling detective story: "[t]he floors were polished limestone," reports private investigator Kinsey Millhone; "I could see ancient marine creatures pressed into the surface, a tiny fossil museum at my feet."[3] But I digress.

Borrowing Huxley's title, Weinberg approaches that piece of chalk from a physicist's point of view. He explains first why chalk is white (light reflected from chalk has about the same distribution of visible wavelengths as the light that illuminates it); why some substances and not others absorb visible light at particular wavelengths (they are composed of molecules which do not happen to have any state that is easy to jump to by absorbing photons of any color of visible light); why atoms and molecules come in discrete states, each with a definite energy (the wave functions for the particles in an atom or molecule can appear only in certain quantum states, each with its own energy, and the molecules of calcium carbonate—unlike, say, the copper atom—do not happen to have any loose electrons that absorb photons of a specific wavelength); and finally why the quantum-mechanical equations that govern the particles in atoms are what they are (ordinary matter is composed of electrons, protons, and neutrons because all the other massive particles are violently unstable). And why does the world consist of just the fields of quarks, electrons, photons, etc.? "Sorry," Weinberg replies, "these questions are still unanswered."[4]

"Scientific discoveries are not independent isolated facts," he comments, stressing the way "one scientific generalization finds its explanation in another, which is itself explained by yet another."[5] But, comparing his and Huxley's reflections on that piece of chalk, we see that subsumption of less under more general laws is only one aspect of the enormously complex intermeshing of the system of scientific knowledge.

But it is not only the knowledge produced by the sciences that is epistemologically valuable; it is also the character of the scientific enterprise itself. Growing out of our everyday inquiries into things, extending human beings' unaided evidential reach, stretching their unaided imaginations, stiffening their unaided respect for evidence, refining their judgment of evidence by means of statistical techniques, controlled and double-blinded experiments, etc., science has been like the remarkable child of ordinary-seeming parents; or—given that its reliance on the cooperative and competitive work of a vast inter-generational mesh of sub-communities has been one crucial way in which science has extended the powers of even the most talented individuals—rather like the remarkable sports teams or orchestras or ballet companies where later generations build on and go beyond the achievements of the first. Science has not only produced, at its best, some of the most remarkable intellectual work of which human beings are capable; it also represents, at its best, a remarkable amplification and refinement of human cognitive capacities.

Not every remarkable manifestation of a characteristically human capacity is admirable; the genocidal cruelties of a Pol Pot or a Hitler, unfortunately, are no less remarkable manifestations of a characteristically human capacity than the intellectual achievements of a Newton or an Einstein, or the musical achievements of a Mozart or a Beethoven. (A controversial advertisement for a Holocaust exhibit at the Imperial War Museum in London urged patrons to "Come and See What Man Can Achieve When He Really Puts His Mind to It.")[6] But though it can be put to bad uses, human beings' capacity to figure things out is clearly a talent, and not, like our capacity for cruelty, a defect.

Peirce writes that science, conceived not as a body of knowledge but as "diligent inquiry into truth for truth's sake . . . from an impulse to penetrate into the reason of things," embodies "the epitome of man's intellectual development."[7] He is using the word "science," as he often does, to refer to genuine inquiry generally, to all bona fide efforts to discover the truth of some question. Still, it is no accident that "science" is the word he uses. Similarly, after observing that the scientist sees intellectual integrity as "the last flowering of the genius of humanity," Bridgman hastens to add that not only scientists but serious inquirers of every kind respond emotionally to this ideal. Still, it is no accident that he has "the scientific worker" in central focus; for, he argues, in science "intellectual honesty is the price of even a mediocre degree of success."[8] He means, I take it, that in the sciences reputation and real achievement are well enough correlated that success in the worldly, professional sense, as well as success in investigation, requires respect for evidence.

You might object that some scientists have achieved considerably more than a mediocre degree of professional reputation despite being guilty of self-deception or worse. In 1903, the distinguished French physicist René Blondlot thought he had found evidence of a new kind of emanation from the X-ray source, in the increased brightness of an electric spark between two wires. Another colleague at the University of Nancy found that these emanations were also emitted by the nervous system of the human body; and "N-rays" were soon found in gases, magnetic fields, and chemicals. Between 1903 and 1906 the effects of N-rays were observed by at least forty people and analyzed in some three hundred scientific papers; and in 1904 Blondlot was awarded the Leconte prize by the French Academy of Sciences. But when American physicist R. W. Wood, observing an experiment of Blondlot's in which N-rays were supposedly separated into different wavelengths after passing through a prism, surreptitiously removed the prism and slipped it into his pocket, Blondot got the expected results just the same. There are no N-rays; the whole affair seems to have been the result of collective self-deception.[9]

But it isn't hard to imagine how Bridgman would respond: Blondlot got away with it temporarily, but what we remember him for now is those non-existent N-rays. Scientists certainly can get away temporarily with self-deception, or outright fraud, or any of the many shades of intellectual dishonesty in between; they may even get away with it permanently, if their work is of no lasting interest. But in scientific work of any significance, fraud and self-deception are likely to be discovered sooner or later—perhaps as a result of someone else's relying on the shaky results, and finding things don't turn out as expected, or perhaps because historians of science take an interest in pioneering work.[10] Not all scientists are intellectually honest; not all scientific dishonesty, whether knowing or self-deceptive, is sure to be unmasked. Nevertheless, a tendency to discourage dishonesty—a fallible and imperfect tendency, as always—is built into the cooperative, competitive, and, especially, the cumulative character of the scientific enterprise. Robert J. Oppenheimer, acknowledging that "science is not all of the life of reason," but only part of it, and that what matters is "not only what the scientist finds out but how he finds it," writes of the intellectual craftsmanship of science, its character as a collective effort—and the discipline of knowing that errors will eventually be found out.[11]

THE EPISTEMOLOGICAL AND THE ETHICAL

Huxley's remark about our "duty" to advance the "noble goal" of improving natural knowledge, like Bridgman's reflections on intellectual integrity, hint that there is something morally as well as epistemologically admirable about the achievements of the sciences.[12] When we describe a claim as well or poorly warranted, evidence as strong or flimsy or partial, an investigation as thorough or sloppy or only a cover-up, an inquirer as brilliant or scrupulous or slapdash, we are appraising them from an epistemological point of view; but many of the terms we use in this context have other uses as well: "simple," for example, plays as important an aesthetic as an epistemological role; and "honest," though important epistemologically, is primarily a term of ethical appraisal. So what is the relation of epistemological evaluation to the ethical?

Epistemological and moral appraisals of an investigation, or an investigator, can and do come apart. An investigation may be ethically impeccable, conducted with scrupulous honesty in obtaining funds, reporting results, and explaining the benefits and drawbacks of participation to those involved, creating no environmental hazards, causing no pain to laboratory animals, and so forth, and yet be poorly conducted from an epistemological point of view, e.g., failing to control for potentially significant interfering factors. Conversely, an investigation may be conducted in a way which is okay or even admirable from an epistemological point of view, but morally dubious.

After the famous blow-up in which James Watson thought that "in her hot anger" Rosalind Franklin might strike him, he found Maurice Wilkins more willing to confide in him; so much so that he showed him Franklin's X-ray diffraction photograph of the B form of DNA—yes, *that* photograph, the one of which Watson wrote that as soon as he saw it, "my mouth fell open and my pulse began to race."[13] From the point of view of efficiency in figuring out the structure of DNA, Watson's impatience to see the photograph can't be faulted; but from an ethical point of view it would have been desirable to get Franklin's permission first.

Or—a fresher and more interesting example: according to recent press reports, a five-year study among prostitutes in Africa and Thailand of a vaginal spermicide called Advantage-S clearly falsified the expectation raised by laboratory tests that the cream, containing the microbicide nonoxynol-9, would be effective in reducing HIV infection rates.[14] The tests would certainly have been more informative—epistemologically better—if the women used no other protection; but given that infection would be deadly, researchers felt they must—morally must—also give the women condoms, and advise them to use them.[15] Whether or not you think they did the best thing, this illustrates not only the potential for divergence between epistemological and ethical considerations, but also the possibility that ethical considerations might override epistemological concerns.

And whether or not, as some philosophers have held, moral considerations are always, perhaps by definition, overriding,[16] some ways of obtaining information, however desirable that information may be from an epistemological point of view, may be ruled out on ethical grounds. A study or experiment may be informative but cruel; and the moral defect of cruelty may override the epistemological merit of informativeness. Epistemological and ethical values are distinct, and need not coincide.

Nevertheless, there is a significant overlap. Intellectual honesty is an ethical as well as an epistemological virtue, and negligence in inquiry an ethical as well as an epistemological fault. This is not to say that intellectual honesty is sufficient to make you a good person; far from it. But then none of the virtues is. That a person has one virtue by no means guarantees that he has them all; and even those he has he may manifest in some contexts but not in all. A person may be courageous but unkind; similarly, he may be intellectually honest but inconsiderate of his wife's feelings or domineering with his students. A person might be courageous in speaking out against injustice, but timid in the face of physical danger; similarly, he might be intellectually honest, but cheat on his tax returns. And, as courage can be put to bad use (as with the soldier who fights bravely in an unjust war), so can intellectual honesty (as with the scientist who scrupulously investigates the most efficient way to infect the enemy population with a deadly disease).

We describe a person as honest who is truthful in his dealings with others; we describe a person as intellectually honest who is truthful in his dealings with himself. It is no accident that another term for intellectual honesty, the phrase Bridgman uses, is "intellectual integrity," with its suggestion of wholeness, of harmony between the will and the intellect; for the characteristic failing of intellectual dishonesty is allowing your wishes, hopes, or fears to distort your relation to evidence—trying to stay out of the way of evidence that may be unfavorable to what you would prefer to believe, or to evade acknowledging the relevance or the significance of the unfavorable evidence you can't avoid. Intellectual honesty is an epistemological virtue because it stiffens respect for evidence; intellectual dishonesty is an epistemological vice because it *un*stiffens it.

And why is intellectual honesty also a moral virtue, and intellectual dishonesty a moral vice? In part at least because someone who lies to himself, avoiding or twisting evidence as it suits him, is as untrustworthy, as unreliable, as someone who lies to others. The value of trustworthiness in science, where progress depends in part on each scientist's being able to rely on others' work, is obvious.

MODEST NATURALISM

Neither expresses it quite overtly, but both Bridgman and Huxley hint that the sciences have something to teach the rest of us about evidence and inquiry. The successes of the sciences have fed back into our everyday inquiries, making us more aware of the need to rule out potential interfering factors, and so on; the sciences can provide us with raw materials for epistemological theorizing, in the form of examples of well-, and of poorly, conducted investigations, and of strongly, and of weakly, warranted claims; and when they reflect on their own enterprise, scientists sometimes come up with real epistemological insights: Bridgman on scientific method, Einstein on the differences between science and fiction, Delbrück on the differences between scientific and literary writing, Weinberg on beautiful theories. But what, if anything, could the sciences, *qua* disciplines or branches of inquiry, have to tell us about epistemological values?

This question is often posed in terms of a contrast between "naturalistic" approaches, which look to the sciences themselves for an understanding of evidence, warrant, etc., and rival approaches looking to a priori epistemological principles (in this context "naturalism" contrasts, not with supernaturalism, but with apriorism). The truth, however, lies somewhere in between: it is not within the competence of the sciences to articulate core epistemological concepts and values; nevertheless, they do have an epistemological contribution to make.

But let me begin at the beginning. Some scientific claims and theories are strongly warranted, some weakly; some scientific investigations are well conducted, some poorly; some scientists are imaginative, scrupulous, passionate in the pursuit of truth, some hidebound, dishonest, passionate only about how many times they are cited or how large a grant they can finagle. Of course, the same goes for other kinds of inquiry, and other kinds of inquirer, as well. In fact, as Samuel Butler well knew, self-deception, sham inquiry, and hypocrisy, "the homage vice pays to virtue," really are *The Way of All Flesh*.

A key turn in Butler's semi-autobiographical novel comes when as an eager young curate Ernest Pontifex tries to convert his neighbor, the freethinking tinker Mr. Shaw, to Christianity. Ernest is reduced to stammering embarrassment as Mr. Shaw exposes his ignorance of what the Gospels really say, and sends him away to re-read the different accounts of the Resurrection. As Ernest tries "to find out, not that [these accounts] were all accurate, but whether they were all accurate or no"[17] (as in real life Huxley had done with the different accounts of the miracle of the loaves and fishes),[18] he begins to grasp for the first time the difference between sham inquiry and the genuine article.

In another epistemologically intriguing novel, Dorothy Sayers' *Gaudy Night*, Miss de Vine, the new history don at Sayers' imaginary Oxford women's college, has exposed the dishonesty of a professorial candidate who, when he found an old letter that undermined his thesis, purloined and hid the evidence. The exposure costs him his career and, as he turns to drink and falls into despair, his life. His widow, Annie Wilson, takes a post as a scout at Shrewsbury College, where she vents her rage at Miss de Vine, and her resentment of women scholars generally, in vandalism, poison-pen letters, and even attempted murder. Significant among Annie's acts of vandalism is the destruction of the college library's copy of C. P. Snow's novel, *The Search*, in which a young man starting out in science is tempted to destroy the X-ray photograph that undermines his beautiful theory, but resists the temptation. Later, however, just as he is about to be appointed to an important post, he finds he has made a careless mistake in his work, the discovery of which costs him the position—after which he decides he doesn't really want to be a scientist after all.

As the examples illustrate, the core epistemological values, intellectual honesty, concern for truth, respect for evidence among them, are not peculiar to the sciences, but are relevant to inquiry of every kind, whether it be biblical scholarship, history, crystallography, or whatever. It isn't realistic to expect the psychology of cognition or the history of science to tell us what makes evidence stronger or weaker, whether and if so why true predictions confirm the truth of a theory, why an intellectually honest inquirer is, other things being equal, a better inquirer, what the connection is between warrant and likely truth, or whether there is a distinctive scientific method. It is entirely realistic, however, to expect the psychology of cognition to tell us something about human cognitive capacities, limitations, and weaknesses, and about the circumstances in which our weaknesses tend to be most manifest.

Psychologists as well as epistemologists are concerned with how people learn, perceive, and inquire, how they react to inconsistencies in their beliefs, how they manage to deceive themselves, and so forth. Their concern, furthermore, is not always strictly descriptive; they are interested in the causes and conditions of misperception, faulty inference, etc. One psychologist, for example, has gone back over decades of raw data to check how often he had mistranscribed numbers (too often for comfort), and in what proportion of cases his mistranscriptions favored the hypothesis for which he was arguing (significantly more often than not). Another interesting study suggests that jurors who are told not only the probability of a random match of this DNA sample with the defendant, but also the frequency with which DNA testing laboratories mislabel samples or make other kinds of mistake, tend seriously to misjudge the probability of a random match given this total evidence.[19] In these and many other ways the psychology of cognition can help us understand human cognitive capacities and weaknesses: how large is the risk of experimenter bias, when can additional information actually impair people's judgment of evidence?

It is also entirely realistic to expect the history of science (construed broadly, to include very recent history) to trace the constantly evolving array of ways in which scientific inquirers have gradually managed to extend those natural human cognitive capacities and overcome those natural human cognitive limitations and weaknesses: think of the earliest fumblings towards a recognition that scientists need assurance of the reliability of others' experiments,[20] the long and continuing history of the development of scientific instrumentation,[21] or recent and still controversial techniques of meta-analysis.

So, unlike those who think it is beyond the competence of the sciences to tell us anything about epistemology, I take the psychology of cognition and the history of science to have epistemological contributions to make; but, unlike those who think epistemology is internal to the sciences, I see their relevance as only contributory, not exhaustive. This is a modest naturalism, in which the concept of contributory relevance is key.

Philip Kitcher writes that the hallmark of naturalistic epistemology is its "[p]lacing the knowing subject firmly back into the discussion of epistemological problems."[22] Of course, I agree that the knowing subject (and interaction among knowing subjects) is crucial, and consequential.[23] But if granting an important place to the knowing subject were the criterion of naturalism, Descartes would qualify as a naturalized epistemologist—a curious consequence, to put it mildly. In any case, this is not the "hallmark" of my naturalism, which is, rather, acknowledgment of the contributory relevance of the sciences to epistemology.

My modest naturalism differs from Alvin Goldman's more ambitious naturalism as my epistemology differs from his.[24] Apparently eschewing the concept of evidence, Goldman hopes that psychology will tell us what cognitive processes are reliable, and hence what beliefs are justified. But in my view the concept of evidence is epistemologically essential; moreover, it clearly does not fall within the competence of psychology to tell us, e.g., to what degree, given this evidence, the proposition that the defendant did it is warranted. Nor, for that matter, can psychology tell us what inferences or other cognitive processes are reliable; that study of the effect of additional information on jurors' judgments, for example, *presupposes* that the correct probability was such-and-such (a probability arrived at, not by psychological research, but by mathematical calculation). Mine is the weaker thesis that, insofar as it can advance our understanding of human beings' cognitive capacities and limitations, psychology has a contributory relevance to the epistemology of the scientific enterprise.

Similarly, my modest naturalism differs from the more ambitious style favored by Larry Laudan as my philosophy of science differs from his.[25] Laudan thinks it falls to the study of scientific practice to track an evolving "scientific method." But, as I see it, rather than a distinctive "scientific method," there is only, on the one hand, the core epistemological values common to all inquiry, and, on the other, the myriad local and evolving techniques and helps that make scientific inquiry "more so." Nevertheless, insofar as it can advance our understanding of the evolution of these helps, the history of science, also, certainly has a contributory relevance to the epistemology of the scientific enterprise.

And, when Ronald Giere writes of *scientific judgment without rationality* (apparently meaning, like Goldman, to eschew such concepts as evidence and warrant, which are central to my approach), it appears that my modest naturalism differs from his more ambitious naturalism as my Innocent Realism differs from his constructivist metaphysics. But, as usual, Giere's headline is more melodramatic than his text, where he acknowledges a role for an instrumental, means/end conception of rationality. If the end in question might be truth—as seems possible when Giere backs away from another melodramatic headline, *realism without truth*—then perhaps we might after all find some measure of agreement.[26]

REPLIES TO SOME RESERVATIONS ABOUT THE WORTH OF SCIENCE

The argument thus far has been not only modestly naturalistic, but also modestly optimistic. Many, I'm afraid, will object that it has been much *too* optimistic. Some fear that science destroys the wonder and mystery of the natural world, "unweaving the rainbow," in the words of John Keats, who once proposed a toast to "Newton's health, and confusion to mathematics," on the grounds that Newton had "destroyed the poetry of the rainbow by reducing it to a prism":

> Do not all charms fly
> At the mere touch of cold philosophy?
> There was an awful rainbow once in heaven:
> We know her woof, her texture; she is given
> In the dull catalogue of common things.
> Philosophy will clip an angel's wings,
> Conquer all mysteries by rule and line,
> Empty the haunted air, and gnomed mine—
> Unweave a rainbow.

("Lamia," 1820)

It is true, of course, that when science explains a natural phenomenon—the mockingbird's musical versatility, the elegance of the orchids,[27] the beauty of a rainbow, or the ferocity of a storm—that phenomenon is no longer mysterious (the word derives from the Greek *mystos*, keeping silence, from *myein*, to be closed, as of eyes or lips). And it is also true that scientific explanations of natural phenomena displace supernatural explanations; God didn't, after all, design the mockingbird, the orchid, and the rainbow for our pleasure, or send the hurricane because He was angry with us sinners. But none of this makes the mockingbird less charming, the orchid less elegant, the rainbow less beautiful, or the hurricane less terrible.

"Wonder and mystery": the phrase came naturally to mind, but now it is helpful to pull the two concepts apart. The wonderful is "a source of amazement or admiration." If the rainbow is beautiful, the explanation of how a rainbow comes about is, in precisely this sense, wonderful. Here is a very brief summary, from Dawkins' *Unweaving the Rainbow*: Sunlight enters a raindrop; at the boundary of air and water it is refracted, the different wavelengths that make it up bent through different angles as in Newton's prism. Hitting the concave far wall of the raindrop, the colors are reflected, leave the raindrop, and are refracted for a second time. Suppose it is early morning or late afternoon on a rainy day, and the sun is behind and slightly above you. A complete spectrum leaves every raindrop; and though you see only a small part of the spectrum from each, you see a complete rainbow, because a band of thousands of raindrops is giving you green light (and simultaneously blue light to anyone suitably placed above you and red light to anyone suitably placed below you), another band of thousands of raindrops is giving you red light, and so on.[28]

The simplicity-in-complexity of this explanation is pretty amazing. And the vast edifice of knowledge the sciences have thus far accumulated—even though it is far from complete, even though parts of it are bound to be torn down and rebuilt, perhaps over and over—is far more wonderful even than those medieval cathedrals to which Popper compares it. And, by my lights, the enormous extension of our unaided cognitive powers that the sciences have achieved, pooling evidential resources within and across generations, discarding earlier efforts when they proved mistaken, developing human intellectual muscles, is no less wonderful.

Still, I am struck by the way our capacity to explain and to take pleasure in understanding sits side by side with our capacity for obfuscation and our sneaking enjoyment of the "unexplained"—usually, after all, purportedly explained by spooks, aliens, angelic intervention, or whatever. The fun of a good detective story is that it exercises the part of us that likes to figure things out. But the pleasure of junk food for the mind, such as those dreadful television programs about Mysterious Phenomena, Miracles, and all that, is a kind of guilty indulgence of that other side of the human creature, the side that likes to be elevated by mysteries, and enjoys a boost to its uneasy sense of self-importance.[29] We humans, as Carl Sagan once wrote, are "significance junkies."

Sometimes it is said that, in our society, science has taken the place of religion. This seems on its face a considerable exaggeration. Churches still thrive, after all, and ministers, priests, bishops, popes, mullahs, ayatollahs, etc., are still respected, even revered. Alongside its first report on that meteorite possibly indicative of early life on Mars, *Newsweek* published an article by theologians happy to learn that We Are Not Alone; its features on Jesus, Miracles, etc., are warmly received by readers. Television "news" programs offer advice on how to improve your spirituality, as well as on how to lose weight or prevent colon cancer. Books like Behe's promoting the Intelligent Design idea, and Johnson's criticizing "Darwinism," sell briskly. According to recent press reports, in New York a new Pentecostal church opens every three weeks;[30] across Africa, twelve hundred new churches spring up every month.[31] In the formerly godless Communist countries of Eastern Europe, the Orthodox and Catholic churches are once again forces to reckon with; conservative Hindus are an important political force in India, as Christian conservatives are in the United States. And now it is hardly necessary to speak of the potential global consequences of fundamentalist Islamic fanaticism.

Sometimes, again, it is said that science has become a religion. Perhaps this is suggested by a superficial reading of Einstein's observation that "you will hardly find one among the profounder sort of scientific minds without a religious feeling of his own"; but when Einstein goes on to describe the religious feeling of the scientist, his "rapturous amazement at the harmony of natural law," as "closely akin to that which has possessed the religious geniuses of all ages,"[32] it is clear that he means less to make a religion of science than to stress the kinship between the self-transcendence required of the scientist and that required of the religious man. Einstein also writes that it is "a matter of faith" for the scientist that nature is comprehensible to creatures like us.[33] And, indeed, the existence of a world independent of how we believe it to be, and yet susceptible to investigation by us humans, is a presupposition of scientific inquiry. But this doesn't mean that science, like religious belief, requires a peculiar leap of faith: for everyday inquiry of the most ordinary kind rests on just the same presuppositions.

But perhaps all that is meant, when it is said that science has become a religion, is that people now attribute the kind of authority to the claims of science that they once attributed to the tenets of religion, or that they give the same deference to scientists that they once gave to figures of religious authority. Indeed, well-known scientists *are* often seen by an admiring public as authorities even on matters far from their areas of expertise; and that ubiquitous epistemologically honorific use of "science," "scientific," and "scientifically" *is* uncomfortably reminiscent of the honorific use in which "gospel" means "guaranteed true." Lay people are often in no position to make a truly judicious estimate of the strength of evidence with respect to a scientific claim or theory; and then they have no choice but to rely on what scientists in the relevant field tell them. In this sense, deference to scientists' expertise is unavoidable, and reasonable. However, those who defer to science, or take the pronouncements of scientists, as "gospel," miss the essential epistemological point: science is precisely *un*like religion in its permanent commitment to the possibility of revision of even the most established claims, should new evidence demand it. As forms of credulousness go, this kind of deferential scientism is less dangerous than falling for those cheap-and-nasty "wonders" I have been complaining about; but it is a form of credulousness, all the same.

Some disparage science because they see it as peculiarly Western, white, and male. It is true that modern science arose in Western Europe, and has been for most of its history the work mainly of white men. It doesn't follow, and it isn't true, that it is inherently a white male enterprise. There were many anticipations of modern science in the civilizations of ancient China, the Indus Valley, and Babylon, as well as in ancient Greece. If these anticipations didn't quite make it to critical mass, they nevertheless left contributions of enduring value; ancient Chinese records of astronomical events, for example, have over the last thirty years proven indispensable in the interpretation of the exploding stars called supernovae.[34] There is, furthermore, no reason to doubt that even the most primitive peoples have the mentality that makes science possible; as James Michener observes, anyone who could devise a tool as intricate as the triple-jointed poison arrows with which the bushmen of the Kalahari desert hunt their game could, given the right circumstances, design an airplane.[35]

And by now there are many, many capable scientists of just about every race. In 1947, Oppenheimer wrote that "of all intellectual activities, science alone has turned out to have the kind of universality among men which the times require";[36] almost a half-century later, Sheldon Glashow—referring to Tycho Brahe, Copernicus, Kepler, Galileo, and Newton—writes that "[s]cience is the unique truly international forum and has been for at least five centuries. . . . [F]ive men from five nations taught us our place in the universe."[37]

Science is a *human* thing. Not that every human being has the capacity to be much of a scientist—but then, not every human being has the capacity to be much of an athlete, or much of a musician, either. Nevertheless, science is a manifestation of a capacity which (so far as we know) is distinctively human, but which has come to its fullest fruition thus far only, in Bridgman's words, as "the culmination of a long cultural history."[38]

As modern science has spread around the world, it has, not surprisingly, proven disruptive to previously settled traditions and cultures. Even the simplest technology can have unforeseen and disturbing effects. Katharine Milton describes the introduction of steel axes to the Panare Indians of Venezuela, who traditionally had worked together with stone axes to clear trees for a garden; with the new steel axes, one man could do the work alone, and the old cooperative ways were abandoned.[39] Understandably enough, the labor-saving new axes were welcomed, all the same.

Some, misconceiving science as an expression of specifically European "epistemic practices," criticize its introduction into non-Western cultures as disrespectful, an objectionable colonialism. But introducing modern scientific ideas to a pre-scientific culture really isn't on a par with, say, trying to impose English cooking on the gloriously subtle and versatile cuisine of India, or to displace their foolish superstitions by our True Religion. Science is, in a way, a culturally specific phenomenon; but it is also universal. It arose in a particular time and place, and the circumstances of that time and place are relevant to its having arisen there and not elsewhere, at that time and not another. Nevertheless, it is a manifestation of human cognitive capacities, and continuous with the everyday empirical inquiry in which people of every culture engage.

In 1962 the first prime minister of India, Jawaharlal Nehru, spoke of science as India's best hope to "solve the problems of hunger and poverty, of insanitation and illiteracy, of superstition and deadening custom of tradition, of vast resources running to waste."[40] By 1990, however, American anthropologist Frédérique Apffel Marglin, endorsed by prominent Indian social scientists, could write that the eradication of smallpox using modern cowpox vaccine was "an affront to the local custom of variolation," inoculation using human smallpox matter accompanied by prayers to the goddess of smallpox[41]—though she acknowledges that the traditional method is ten times likelier to cause smallpox in the patient. While it is just barely conceivable that people might prefer to keep the old ways despite the medical superiority of the cowpox method, to grant its superiority in that respect is absolutely not a mere expression of a cultural preference for modern science over traditional medical practices. (And it would be worse than an "affront" to parents who want to protect their children from a horrible disease to offer them, as Meera Nanda puts it, "epistemic charity," rather than the best information possible.)

None of this is to deny that scientists have sometimes been too quick to scoff at traditional medical practices, which sometimes turn out to have surprising merits. Moxibustion, which uses heat from burning herbs at pregnant women's little toes to stimulate acupuncture points, can apparently help move a fetus into the proper position for a head-first delivery;[42] and an old Chinese herbal remedy (a potion of "Baikal skullcap, rabdosia, mum, and Dyer's woad," which sounds remarkably like something the witches in *Macbeth* might cook up!) is proving effective in the treatment of prostate cancer.[43]

If such remedies really are effective, I take it, there must be some scientific explanation of how they work. No, I haven't succumbed to the honorific use of "scientific"; I mean only that, whatever the true explanation is, it must interlock somehow with the truths about the world which the sciences have already discovered, must fit somehow with the rest of the crossword. For all their imperfections, the sciences really have managed to find out far more about the natural world and how it works than anything else the human race has ever tried; nonetheless, scientists can sometimes learn from older efforts to understand and control natural phenomena.

"For all their imperfections": insofar as there are races, doubtless there are differences among them (the Zulu really are, on average, taller than the Japanese, and so on); but sometimes the medical and human sciences, carelessly endorsing prejudices prevalent in the wider society, have accepted unwarranted racial claims. Between 1830 and 1851, for example, the prominent Philadelphia physician Samuel G. Morton collected more than a thousand human skulls, and, taking skull volume as a measure of intelligence, ranked the races: whites the most intelligent, and whites of Western European extraction above Jews; blacks the least intelligent; and American Indians in between (apparently, where the "inferior" races were concerned, he didn't bother to discriminate the larger, male skulls from the smaller, female skulls).[44]

Of course, unwarranted scientific claims about the races—whether the result of honest but mistaken efforts to find out, or simply an expression of prejudice—never excuse mistreating people. But the sciences have accepted unwarranted claims about many things; and it is overly hasty to conclude that this particular failure reveals the inherent racism of the scientific enterprise.

Among the many other things about which the sciences have sometimes made unwarranted claims are the differences between the sexes, and as a result they have sometimes, also, seemed to excuse the mistreatment of women. That, like the unwarranted claims about race, is greatly to be regretted. As post-colonialist critics of science sometimes object that scientific values are Western, so feminist critics of science sometimes object that scientific values are masculine, or masculinist; but, again, it is overly hasty to conclude that scientists' unwarranted endorsement of mistaken ideas about women reveals the inherent sexism of the scientific enterprise.

In its simplest form, the idea of science as a masculinist enterprise relies on a crude conception of the values of science, along the lines of "objectivity," "rationality," "linear reasoning," etc.—the sneer quotes hold these supposedly masculine concepts at arms' length—and a crude contrast with such supposedly feminine values as emotion, connectedness, holism; as if a unified account of apparently diverse phenomena weren't a goal of scientific inquiry, and a passionate desire to understand a motive for undertaking it. It also relies on old stereotypes about women as emotional, illogical creatures; as if there were no cool-headed and reasonable women, and no over-emotional and unreasonable men.

Though the differences between the sexes are less superficial than the differences among the races, nonetheless, as Sayers writes in one of her insightful feminist essays, women really are "more like men than anything else on earth."[45] Rather than disparaging a stereotyped science misperceived as embodying stereotypically masculine values, we need to rid ourselves of stereotypes, both of women and of science. Insofar as racial and sexual stereotypes misrepresent people, the antidote is more evidence, or more careful attention to the evidence already available. So those post-colonialist or feminist critics of science who are tempted to argue from the true premise that science has sometimes endorsed racial or sexual stereotypes as established truth, known fact, or well warranted by evidence, to the false conclusion that the ideas of truth, known fact, warrant, evidence, etc., are racist or sexist humbug, commit the Passes-for Fallacy; and in the process cut themselves off from the only way to overcome stereotypical thinking.

In a somewhat subtler form, the idea of science as a masculinist enterprise relies on the idea that the values science exemplifies are *perceived as* masculine, and that it neglects values *perceived as* feminine. This avoids endorsing stereotypes of the masculine and the feminine, which is certainly all to the good; but it still relies on an over-simplified conception of the scientific enterprise, blind to the way the sciences combine passion for a beautiful theory with respect for evidence, concern for detail with a sense of potential connections, evidential rigor with imaginative sweep (I might almost say—but of course I won't!—"yin with yang").

Other feminist critics, holding that science is unavoidably a political enterprise, and thus far informed by masculinist values, advocate its transformation by the infusion of more progressive, feminist values; as if we hadn't learned from the awful examples of Nazi and Soviet science that to put pressure on scientists to arrive at politically desired conclusions is to court disaster. "The truth—whatever that is!—will not set us free," writes Sandra Harding, calling for "politically adequate research and scholarship." The thought is chilling. It is also incoherent; for we can't improve the position of women unless we can figure out what women's true interests are, and what social changes would truly advance them.

None of this is to deny that, though some remarkable women managed to overcome them, in the past there have been many obstacles to discourage women from science (it's no wonder if Rosalind Franklin was a bit prickly—excluded from the shop-talk in the staff common room at King's College, London, in which women weren't allowed, condescended to by Maurice Wilkins, treated by

Watson as if she were Wilkins' research assistant, and so on). Nor is it to deny that, though they are surely fewer and less severe than they once were, obstacles still remain. Some of these might be properly described as discrimination, while others arise from the greater complications of women's lives: pressure to give priority to a husband's career, subtle discouragements on the part of parents, teachers, and peers, and the disinclination to take intellectually serious women quite fully seriously that still lingers like an unpleasant odor in the academy, and doubtless outside it as well.[46]

Those subtle discouragements, and that lingering disinclination, are in part the legacy of the old sexist stereotypes; but now, just as they might finally have been put behind us, they are being reinforced by the unfortunate revival of those stereotypes by some feminist critics of science—a grave disservice to women, and to science. The problem as I see it is by no means so melodramatic as those who repudiate science as a sexist enterprise suppose; but it is not nothing. There are still talents lost to science, and opportunities lost to women.

It is legitimate to wonder how science might be different if there were more women scientists. Not only men are influenced by sexist stereotypes; but perhaps it would be at least a bit less likely that those stereotypes would be endorsed as established facts (as, in recent decades, our understanding of the social life of the apes seems to have benefited from the fresh perspectives brought by a small influx of women into primatology). It is also possible to envisage a continuation of the trend in medical research away from an emphasis on male subjects and male diseases (though I'm not sure, after reading an article about how a drug developed to treat breast cancer turned out to be much more effective for prostate cancer, and watching a television program about the under-diagnosis of osteoporosis in men, that a simple shift of emphasis to "women's diseases" would be an ideal upshot). In any case, I expect that serious female scientists, just like serious male scientists, will continue to be most attracted to questions they think are hard enough to be really important, but not so hard as to be beyond their capacity to solve—the crossword entries with enough letters filled in already to give them a start, and interlocking with enough others to promise further payoffs down the line.

COSTS OF SCIENCE, RISKS OF TECHNOLOGY

When I wrote, just now, of "payoffs," it was, to borrow Bacon's terminology, light rather than fruit that was foremost on my mind. But of course I don't mean to sidestep issues about the dangers of technology.

We are accustomed to thinking of technology simply as the fruit of theoretical developments in science; but this is something of an over-simplification. Barzun lists "the steam engine, the spindle frame and power loom, the locomotive, the cotton gin, the metal industries, the camera and plate, anaesthesia, the telegraph and telephone, the phonograph and electric light," as all invented "by men whose grasp of scientific ideas was slight, or at best empirical."[47] Huxley noted how the advancement of science depends on technological developments such as the easy availability of alcohol, glass, etc. Oppenheimer takes up the same theme in a twentieth-century key when he writes that the technology of ultra-high-energy accelerators has been "the repayment technology makes to basic science, in providing means whereby our physical experience can be extended and enriched."[48] My present concern, however, is with the value of technology *qua* spin-off from science.

The classic expression of the Enlightenment hope that science would overcome human suffering is Baron D'Holbach's: "The source of man's misery is his ignorance of nature." David Stove replies: "This must be one of the silliest things ever said."[49] And to be sure, if D'Holbach had meant that by explaining Newton's theories to him, you could cheer up someone who has lost his wife and children to the plague and is now dying an agonizing death himself, it would be very silly indeed. Interpreted, however, as saying that the life of humankind would be far happier if we knew more about how to prevent or control the diseases and natural forces that cause so much misery, it doesn't seem so silly after all.

The truths science discovers, Francis Bacon wrote, "will carry whole troops of works along with them." For "Nature to be commanded must be obeyed," so "those twin objects, human knowledge and human power, really do meet in one."[50] Thinking back imaginatively to the time at which Bacon wrote, you begin to appreciate the grandeur of his vision of how science might improve the human condition. In the essay from which the quotation that opens this chapter is taken, Huxley writes of the plague year, 1664, as described by Defoe: "death, with every accompaniment of pain and terror, stalking through the narrow streets of old London . . . a silence broken only by the wailing of the mourners of fifty thousand dead [and] the woful denunciations and mad prayers of fanatics." At the time, he continues, people "submitted to the plague in humility and penitence, for they believed it to be the judgment of God." But by the time he wrote—1866— people had "learned somewhat of nature, and partly obey her,"[51] and had defeated the plague, though not typhoid or cholera.

As Bacon anticipated, technology *has* lengthened and improved our lives. Again, it is hard to convey the range or the extent of this in a few words; but Max Perutz's essay, "Is Science Necessary?" spells out benefits Bacon couldn't have dreamt of, the myriad ways in which technology has revolutionized agriculture, improved public health, reduced our dependence on fossil fuels, etc., etc. Bacon, however, seems to have been more vividly conscious of the benefits that knowledge of nature might bring than of the possibility that this power might threaten human welfare. But the fact is, as Perutz is well aware, that like any great increase in human power, science brings real pitfalls and real dangers as well as real benefits.

Some of those who fear the dangers of technology also disparage the epistemological *bona fides* of science. "[I]t is simply too dangerous to placate the scientists by granting them honorary objectivity. Inevitably, the scientists will use their objectivist credentials to advance the subjective beliefs that all along attended their ostensibly objective knowledge. . . . Those who falsely claim objectivity inevitably smuggle their values into social decisions," writes William Dean.[52] It is true that, however knowledgeable they may be about their own field, scientists are not necessarily more knowledgeable about the social consequences of technology, nor morally or politically wiser, than other people. But it is incoherent to deny, as Dean does, that the sciences have made objectively true discoveries, while warning of the dangers of the technological developments scientific discoveries have spawned; for those technological developments are evidence of the truth of the science concerned.

But no such objection applies to those who, acknowledging the light that science has brought, fear its fruit: Kurt Vonnegut Jr., for example, who writes: "I used to think science would save us, and science certainly tried. But we can't stand any more tremendous explosions, either for or against democracy. . . . If you want to become a friend of civilization, then become an enemy of truth and a fanatic for harmless balderdash";[53] or Michael Dummett, who avers that if only it were feasible, "there are overwhelming reasons for bringing *all* scientific research to a halt."[54] I disagree; but I certainly acknowledge that there are costs, limits, and dangers.

Though some scientific discoveries are still made with tiny resources, the days in which major breakthroughs can be made with the help of a candle and a piece of string are mostly behind us; and the most expensive work generally has to be supported either by governments or by large industrial concerns. Where public money is concerned, there are knotty questions about priorities. It is hardly surprising if people are troubled by the expense of scientific work, especially where the issues at stake are, as they often are with the most expensive research, esoteric in the extreme, though of the profoundest scientific importance; nor that they are troubled when they read that their taxes have paid for scientific blundering and waste, or discover that scientists have, wittingly or unwittingly, succumbed to false optimism about what practical benefits are likely to result, or how soon.

In 1999 the $125 million Mars Climate Orbiter burned up in the Martian atmosphere; engineers had confused pounds and kilograms in their calculations. In 2000 the $165 million Polar Lander, which was to have recovered samples from Mars, crashed because a simple software glitch shut down the engines too soon before landing. The chair of the relevant House committee, observing that everyone is entitled to one stupid mistake, but not to two, warned that NASA shouldn't look for more funding from Congress.

Government and businesses have other interests besides arriving at the truth. Private concerns may want to keep potentially profitable truths from their competitors, putting pressure on the free exchange of ideas required for the scientific enterprise to flourish. And both governments and private concerns may want unpalatable truths covered up. The manufacturers of Vioxx and Celebrex, for example, apparently sought to downplay research suggesting that these hot new arthritis drugs—on which more advertising dollars are being spent than on Coca-Cola—might carry an increased risk of heart attacks.[55]

And the pervasive dependence of medical research in universities on financial support from the drug companies raises legitimate concern both that researchers may be tempted to skew their results to favor the interests of the companies that support their work (see cartoon, right), and that they may be obliged to keep potentially valuable information secret.[56] In an editorial in the May 2000 issue of the *New England Journal of Medicine*, Marcia Angell asked, "Is Scientific Medicine for Sale?" Since 1984 the journal has required authors to disclose financial ties to companies with an interest in their research; but disclosure statements are often so long they have to be on the journal's Web page rather than in the journal itself (a footnote to an article on anti-depressants disclosed more than 350 ties to pharmaceutical companies manufacturing those drugs). In the same issue of the journal as Angell's editorial, a study concluded that "researchers with ties to drug companies are indeed more likely to report results that are favorable to the products of these companies than researchers without such ties." The same year, Immune Response took legal action to try to prevent publication of a university study showing that its AIDS drug, Remune, was ineffective; when this failed, and the study was published, the company sued for millions of dollars in damages. In 2002, another study in the *New England Journal* concluded that "a re-evaluation of the process of contracting for clinical research is urgently needed."[57]

The questions answers to which are most urgent from a practical point of view are not always or necessarily questions answers to which are scientifically

"We want you to do some pure disinterested fundamental research into something immensely profitable."

Reprinted from Watson and Tooze (eds.),
The DNA Story, p. 282. Originally published in
The New Scientist (London).

most feasible. Sometimes we desperately want solutions to entries in areas of the crossword puzzle which are, thus far, pretty much blank. At best, we must accept that setting to work on those parts of the puzzle may be very expensive in time, effort, and money; at worst, we may have to accept that, for now, solutions are simply not possible. According to Paul Gross and Norman Levitt, for example, an understanding of the nature and etiology of AIDS wasn't feasible before the late 1970s, the first point at which enough knowledge of cellular immunology, molecular biology, and virology was in place.[58] In a grim sort of way, you might say, we were lucky.

Even when scientific answers are available, they may fail to yield practicable solutions to the larger problems we want fixed. Medical science has by now found ways to control AIDS—for some; but the costly and elaborate drug regimes involved are beyond the reach of many, and in particular of the vast number of sufferers in sub-Saharan Africa. The most that scientific medicine could do is find a vaccine to prevent and/or drugs to cure or block transmission of the disease. It is not within its competence to solve knotty political problems, to resolve issues about international drug patents and pricing, to figure out what regimes of treatment or prevention are most feasible in conditions where even running water is unavailable, nor to overcome the stigma carried by AIDS, or the complacency of those in the more affluent parts of the world to whom the disaster seems distant and unreal (though the social sciences *might* be able to shed some light on such problems).

Even the most obviously beneficial technological developments sometimes turn out to have undesired side-effects. John Snow, the doctor who noticed, during the London cholera epidemic of 1854, that the disease was prevalent among those whose water came from a well supplied by one water company but not among those who drank from another, helped establish what caused the disease, and how to prevent its spread (though, sad to say, the story that he stopped the epidemic by having the handle of the offending pump removed turns out not to be true.)[59] However, the outbreak of polio in the early twentieth century was apparently in part the unanticipated result of making clean drinking water widely available; children were more susceptible to polio because they no longer had the immunities that previous generations had gradually built up from drinking tainted water.[60]

Scientific work may present dangers to those involved, such as the earliest workers on X-rays, who suffered horrendous burns; and—though the fear now seems exaggerated that genetically modified organisms might escape the laboratory and cause who knows what kind of plague (see box, right)—it may also present dangers to the public. As this reveals, scientists can make only informed guesses about what discoveries will prove most dangerous; and they quite often

FEARS ABOUT RECOMBINANT DNA RESEARCH

In the early 1970s Herbert Boyer and Stanley Cohen devised procedures to produce recombinant DNA molecules from two different parental DNA molecules, such as mouse and *E. coli* DNA, and Paul Berg's laboratory was working to clone the animal tumor virus SV40 in *E. coli*. On a summer course at Cold Spring Harbor, one of Berg's graduate students described this work to other molecular biologists, who became concerned about the potential hazards.

In 1973, a first meeting of concerned molecular biologists at Asilomar, California, produced the "Moratorium Letter," published in *Science*, temporarily suspending certain experiments. In 1975, a second Asilomar meeting produced a set of guidelines voluntarily restricting further cloning of DNA to organisms that had been genetically disabled so they wouldn't grow well outside a test-tube.

In 1976, the National Institutes of Health imposed regulations on recombinant DNA research, delimiting four levels of risk requiring corresponding levels of biocontainment: from elementary laboratory hygiene at level P1, to the highest level of containment, for anthrax bacilli, smallpox and lassa fever viruses, etc., at P4.

When molecular biologists at Harvard proposed to convert several rooms to level P3, the mayor of Cambridge took the issue to the city council, and the debate soon became a major political, legal, and social tangle.

By 1977, scientists in the field were mostly convinced the dangers had been grossly exaggerated. Cohen concluded: "Our earlier concerns were greatly overstated; the work has proceeded without adverse consequences."

In 1979, the NIH guidelines were relaxed.

In 1980, the Supreme Court approved a disputed patent submitted by Cohen and others on "biologically functioning molecular chimeras." ("Chimera," originally referring to a mythological creature with a lion's head, a goat's body, and a serpent's tail, had been adopted by biologists for their real, recombinant organisms.) The same year, construction work began on the first industrial plant to make insulin using recombinant DNA procedures.

In 1981, the first recombinant DNA company, Genentech, went public.

guess wrong. In 1930, just fifteen years before Hiroshima and Nagasaki, commenting on the fear expressed by fellow Nobelist Frederick Soddy that atomic energy might be used to make weapons of unprecedented power, Robert Millikan averred that this, "like most of the hobgoblins that crowd in on the mind of ignorance—[is] a myth."[61]

Freedom of inquiry is politically as well as epistemologically desirable; but it doesn't follow, and I have been careful not to say, that absolutely unfettered freedom to pursue scientific inquiry is always or necessarily best. Scientific progress is enabled by freedom of inquiry, and hampered by obstacles to the free exchange of ideas, whether in the form of government regulation of hazardous experiments, financial support conditional on agreements not to disclose results to competitors, or the threat of incarceration in a mental institution or a concentration camp for such thoughtcrimes as "bourgeois genetics" or "Jewish physics." But that science is valuable doesn't mean that there aren't other values, or that the advancement of science is always overriding. Abuses of the Nazi and Soviet variety are never justifiable; but regulation of potentially hazardous research may be, when the benefits outweigh the costs (even non-disclosure agreements might be, if they make discoveries possible that would otherwise not be made at all).

More difficult even than the issues raised in the previous paragraph are questions about the abuse of scientific discoveries by governments themselves. It isn't scientists' fault if thugs put their discoveries to evil uses, Bridgman writes; and, indeed, scientists can't be blamed for, say, the high-tech terrorism of the religious fanatics of Aum Shinrikyo.[62] But what no one anticipated, but we all now fear, is that nuclear weapons in the hands of a rogue nation or an individual lunatic might cause irreparable damage. The danger (somewhat analogous to that posed by a retarded adolescent with the mind of a child but the body and sexual appetites of a grown man) is that the power over nature made possible by the work of some of the intellectually most sophisticated among us, which inevitably includes power for harm as well as for good, may fall into the hands of the emotionally least stable among us. And while it is true that scientists can't be blamed for the actions of such a nation or such an individual, and also true that the only means of preventing such a catastrophe are political, scientists, who know most about what the possible consequences are, and the rest of us, who may have to live—or die—with them, must inevitably share the responsibility for trying to ensure that scientific discoveries are used for good and not for harm.

AND, IN CONCLUSION

Fallible and imperfect as it is, science is a manifestation of the human mind at its cognitive best. But it is no wonder that it evokes fear and resentment as well as admiration. It imposes an intellectual discipline unwelcome to many of us, and threatens some of our cherished illusions about ourselves. We are disappointed when it fails to produce, on demand, the fruits we most want, and fearful when we realize that its fruits may have unpleasant, or in some cases, disastrous side-effects, especially if abused. We feel threatened, I think, both by the successes of science and by its failures; not surprisingly, perhaps, since it, and we, are only human.

NOTES

1. Huxley, "On the Advisableness of Improving Natural Knowledge," *Collected Essays*, volume I, p. 41. His title is an allusion to the Royal Society for the Improvement of Natural Knowledge.

2. Huxley, *On a Piece of Chalk* (I have relied in part on Weinberg's summary); the quotation is from Huxley, p. 27.

3. Grafton, *O is for Outlaw*, p. 64.

4. Weinberg, "On a Piece of Chalk," in *Dreams of a Final Theory*, pp. 20–50; the quotation is from p. 25.

5. Weinberg, *Dreams of a Final Theory*, p. 19.

6. Ellison, "Ads for a Holocaust Exhibit in London Cause a Stir."

7. Peirce, *Collected Papers*, 1:44, 7:49.

8. Bridgman, "The Struggle for Intellectual Integrity," pp. 365–66.

9. My source is Broad and Wade, *Betrayers of the Truth*, pp. 112ff.

10. See chapter 7, pp. 196–201 on fraud in science.

11. Oppenheimer, "Physics in the Modern World," pp. 198, 196–97.

12. In a letter to F. Dyster dated January 30, 1959, Huxley is much bolder and more explicit, averring that "thoroughly good work in science cannot be done by a man who is deficient in high moral qualities" (my source is Ruse, *From Monad to Man*, p. 217, drawn to my attention by Andrew Reynolds).

13. Watson, *The Double Helix*, p. 98.

14. It is now known that N–9 weakens the cell lining of the vagina, making infection more likely. As a result, several makers and vendors have now withdrawn N–9 spermicides from the market; see Zimmerman, "Wary of HIV, Some Vendors Stop Selling N–9 Spermicide."

15. Brannigan and Carrns, "Surprise Failure Dashes Hopes for HIV Product"; Kalb, "We Have to Save Our People."

16. See Hare, *The Language of Morals* (moral values as overriding); Chisholm, "Firth and the Ethics of Belief."

17. Butler, *The Way of All Flesh*, pp. 90, 281.

18. Huxley, "Possibilities and Impossibilities," *Collected Essays*, 5:202–204.

19. Koehler et al., "The Random Match Probability in DNA Evidence."

20. The theme (despite the unfortunate title) of Shapin's *A Social History of Truth*.

21. See Wilson, "Instruments and Ideologies: The Social Construction of Knowledge and Its Critics."

22. Kitcher, *The Advancement of Science*, p. 9.

23. As I had argued as early as 1979, in "Epistemology *with* a Knowing Subject."

24. Goldman, *Epistemology and Cognition*; see also Haack, *Evidence and Inquiry*, chapter 7.

25. Laudan, "Progress or Rationality? The Prospects for Normative Naturalism."

26. Giere, *Science Without Laws*, p. 7.

27. See Darwin's *The Various Contrivances by Which Orchids Are Fertilised by Insects*, excerpted in Glick and Kohn, *Charles Darwin on Evolution*, chapter 11.

28. Dawkins, *Unweaving the Rainbow*, pp. 45–47; Dawkins refers to Whelan, *The Book of Rainbows*.

29. See chapter 10, pp. 265, 293.

30. Carnes, "The Pentecostal City."

31. Woodward, "The Changing Face of the Church."

32. Einstein, "The Religious Spirit of Science," p. 40.

33. Einstein, "Physics and Reality," p. 295.

34. Maddox, *What Remains to Be Discovered*, p. 2.

35. Michener, *The Covenant*, p. 18.

36. Oppenheimer, "Physics in the Modern World," p. 198.

37. Glashow, "The Death of Science!?" p. 23.

38. Bridgman, "The Struggle for Intellectual Integrity," p. 366.

39. Milton, "Civilization and Its Discontents"; my source is Dennett, "Faith in the Truth," p. 12.

40. My source is Dawkins, *Unweaving the Rainbow*, p. 30.

41. Nanda, "The Epistemic Charity of Social Constructivist Critics of Science," p. 291, citing Marglin, "Smallpox in Two Systems of Knowledge."

42. Cowley and Underwood, "What's 'Alternative'?"

43. Burton, "In Trials, Potion of Herbs Slows Prostate Cancer" (the quotation is from p. B1).

44. My source is Broad and Wade, *Betrayers of the Truth*, p. 194, citing Gould, *The Mismeasure of Man*.

45. Sayers, "The Human-Not-Quite-Human," p. 142 (Sayers herself eschews the word "feminist," which she felt was being used to refer to ideas contrary to women's interests—as it still is). Cf. Haack, "After My Own Heart."

46. See Sonnert and Holton, *Gender Differences in Science Careers*, especially chapter 4, and Hanson, *Lost Talent*.

47. Barzun, *Science: The Glorious Entertainment*, p. 20.

48. Oppenheimer, "Physics in the Modern World," p. 191.

49. Baron D'Holbach, *The System of Nature*, first page; cited in Stove, "D'Holbach's Dream," p. 81.

50. Bacon, *Great Instauration*, pp. 80, 84; *New Organon*, aphorism 3.

51. Huxley, "On the Advisableness of Improving Natural Knowledge," *Collected Essays*, 1:19, 27.

52. Dean, "Foreword" to Elvee, *The End of Science?* p. viii.

53. Vonnegut, *Wampeters, Foma & Granfaloons*, pp. 163–65 (my source is Holton, *Science and Anti-Science*, pp. 177 and 188).

54. Dummett, "Ought Research to be Unrestricted?" p. 3.

55. Burton and Harris, "Note of Caution: Study Raises Specter of Cardiovascular Risk for Hot Arthritis Pills."

56. Dickson identifies a $23 million, ten-year agreement signed in 1975 between Monsanto and the Harvard Medical School members as pointing the way to a succession of industry-university arrangements; see *The New Politics of Science*, p. 66.

57. Schulman, "A National Survey of Provisions in Clinical-Trial Agreements between Medical Schools and Industry Sponsors"; for a summary, see Zimmerman, "Medical Schools' Research Pacts Are Criticized."

58. Gross and Levitt, *Higher Superstition*, p. 181.

59. According to Bailar, "The Powerful Placebo and the Wizard of Oz," the cholera epidemic had already begun to wane before Dr. Snow had the pump disabled.

60. Sager, "A Profession in Crisis, a World Still in Misery."

61. Millikan, "Alleged Sins of Science." My source is Hunt, *The New Know-Nothings*, pp. 344, 386; he does not give a page reference.

62. The cultists who nerve-gassed the Tokyo subway in 1995.

12

NOT TILL IT'S OVER

Reflections on the End of Science

Exact science must presently fall upon its own keen sword. . . .
An orgy of two centuries of exact scientific-ness brings satiety.
—Oswald Spengler, *The Decline of the West*[1]

Science is nearer its beginning than its end.
—Sheldon Lee Glashow, "The Death of Science?!"[2]

The problems that remain unsolved are gargantuan. They
will occupy our children and their children and on and on for
centuries to come, perhaps even for the rest of time.
—John Maddox, *What Remains to Be Discovered*[3]

The idea that science is nearing its end has recently become fashionable—
again; far from being novel, it is, as historians of science will point out, a
startlingly familiar idea. It is also, however, far from being one single,
simple idea, but comes in numerous different and even incompatible forms.

In an article entitled "How to Think about the End of Science" Gerald Holton
contrasts those, like Einstein, who take a linear view of science as progressive and,
albeit unevenly, cumulative, with those, like Oswald Spengler, who take a cyclical
view of science as, like other cultural institutions, rising, growing, decaying, and

eventually dying away.[4] As Holton's title hints, it is a serious question how best to approach questions about the end of science. But Einstein's position is quite subtle and complex, and Spengler's is an extraordinary amalgam of cultural-historical, philosophical, and analogical elements. Any number of logically independent contrasts are implicit in the apparently simple dichotomy of "linearists" and "cyclicists":[5] between objectivists and cultural relativists; between those who see science as essentially cumulative and those who believe it undergoes an unending series of revolutions; between those who think the sciences will eventually discover the ultimate laws of the universe and those who believe they will eventually run up against insuperable human limitations; and so on.

In fact, it's such a tangle that the best strategy may be first to sketch some influential variants of the "end of science" idea, and in the process to identify some main themes—demotion, annihilation, culmination, diminishing returns, completion—which can then be scrutinized more carefully.

IDEAS OF THE "END OF SCIENCE": A BRIEF HISTORY AND ANALYSIS

In 1872 the German physiologist Emil du Bois-Reymond published a celebrated lecture entitled "On the Limits of Scientific Knowledge," and in 1880 an even more influential lecture entitled "The Seven Riddles of the Universe." Taking our knowledge of how the world works to be limited to purely mechanical principles, du Bois-Reymond argued that some of the most fundamental questions—the nature of matter and force, the ultimate source of motion, the origin of sensation and consciousness—are insoluble. *Ignoramus et ignorabimus* was his slogan: we don't know, and we shall never know. In *The Riddle of the Universe*, published in 1889, Ernst Haeckel replied that, so far from being in principle intractable, the questions du Bois-Reymond thought unanswerable had mostly been answered; and those that had not, such as the "problem of substance," were really metaphysical, not scientific.[6] As Nicholas Rescher observes, the sharply controversial exchange between du Bois-Reymond and Haeckel disguises their agreement, at a deeper level, that "science was coming pretty near the end of the road."[7] Indeed, in the latter part of the nineteenth century Haeckel's idea that science was close to completion was almost a commonplace. When Max Planck entered the University of Munich in 1875, the chair of physics advised him against studying science, on the grounds that there was nothing left to be discovered. Robert Andrews Milliken encountered the same attitude when he began as a graduate student at Columbia in 1894,[8] the year of Albert Michelson's now-notorious prediction that "[w]hile it is never safe to say that the future of Physical Science has no marvels even more astonishing than those of the past, it seems probable that most of the grand underlying principles have been firmly established and that further advances are to be sought chiefly in the rigorous application of those principles."[9]

Very soon after Michelson's pessimistic prediction, however, "complacency gave way to excitement," an excitement which, by 1902, Michelson himself had come to share. For a physicist, as Steven Weinberg observes, the twentieth century begins in 1895 with Wilhelm Roentgen's discovery of X-rays, continues with Henri Becquerel's discovery of radioactivity in 1896, J. J. Thompson's discovery of the electron in 1897, and then in the next decades with the work of Rutherford, Planck, Einstein, and Bohr.[10]

But in *The Decline of the West*, published in 1918, Oswald Spengler decried this new physics as undermining the hope of certainty and exactitude, and a sure sign of the inevitable and imminent decay of science. Written mostly by candlelight during the First World War "by an obscure and impoverished German high school teacher . . . with a doctorate in Greek mathematics and an encyclopedic ambition," this extraordinary opus offered "a grand Teutonic theory . . . interspersed with dramatic predictions [and] a good share of absurd-sounding speculations."[11] It is not easy to convey its character or tone in a few sentences; as Holton observes, even its English title falls short of the full sense of exhaustion, of enervation, conveyed by the German, *Der Untergang des Abendlandes* (etymologically, "the going-under of the evening-land"), the sinking away of Western civilization. Two dominant themes work in concert: the epistemological demotion of science to the status of cultural construct, comparable to styles of music or painting, and the imminent annihilation of science with the decay and death of Western civilization. When Spengler observes that soon there would be no more great scientists, only "gleaners," a third theme briefly appears: diminishing returns.

"That which physics—which exists only in the waking consciousness of the Culture-man—thinks it finds in its methods and its results was already there, underlying, and implicit in, the choice and manner of its search," Spengler avers; the formulae of physics are meaningless until they are interpreted, and their interpretation can only be an expression of cultural presuppositions. The "facts" which the scientist imagines are really thus and so are the creation of his own culturally conditioned subjectivity. "Polarized light-rays, errant ions, flying and colliding gas-particles, magnetic fields, electric currents and waves—are they not one and all Faustian visions," Spengler asks, "closely akin to Romanesque ornamentation, the upthrust of Gothic architecture, the Viking's voyaging into unknown seas, the longings of Columbus and Copernicus?"[12]

So the profoundest history of science tracks, not its theoretical results, but its symbolism, its style, as it shifts and changes through inherent historical necessity: from the seventeenth-century conception of force, akin to the then-dominant style of oil painting, through the more abstract eighteenth-century conception, in unison with Bach, to the nineteenth and early twentieth century, when, with "culture at its end," it becomes "pure analysis." With the "ruthlessly cynical" theory of relativity, Spengler thought, and with the concept of entropy, the "symbol of decline," physics was drawing near to the limit of its possibilities. Its "mission as a historical phenomenon has been to transform the Faustian Nature-feeling into an intellectual knowledge, the faith-forms of springtime into the machine-forms of exact science"; but "[f]rom our standpoint of today, the gently sloping route of decline is clearly visible."[13]

Every major civilization has taken the same course, from burgeoning spring to decline and death in winter; and this will be the fate of our civilization, the sciences included. Spengler even predicted precisely when Western science would come to an end[14]—2000 (the very year in which we imagined new blades of grass coming up bleen!). The sciences would fail as, overreaching, they tried to apply their methods of reason and cognition to history, where intuitive perception is needed; and eventually they would exhaust themselves even within their own sphere, as their essentially subjective character became unmistakable.

Also in 1918, in the address celebrating Max Planck's sixtieth birthday to which Holton refers, Einstein asks: "What place does the theoretical physicist's picture of the world occupy among all [the] possible pictures" made by scientists, artists, poets, and philosophers? Theoretical physics demands the highest possible mathematical rigor, and in consequence must restrict itself to the very simplest events. But for all that, Einstein continues, it is entitled to call itself a theory of the universe, for the laws it seeks are valid for any natural phenomenon whatever; and only the limited capacity of the human intellect stands in the way of a complete, and completed, explanation of every natural process.[15] So the contrast is not that Spengler predicts the end of science, and Einstein thinks this inconceivable; it is rather that Spengler predicts the annihilation of science, while Einstein envisions its culmination and even, in principle, its completion.

Much of Gunther Stent's *The Coming of the Golden Age: A View of the End of Progress* (1969) is devoted to tracing the history of genetics and molecular biology: from the "Classic Period" of the rediscovery of Mendel's laws at the turn of the twentieth century, through the "Romantic Period" of the Phage group and the identification of DNA as the genetic material, and the "Dogmatic Period" in which the structure of DNA was solved and the Central Dogma of molecular biology formulated, to the "Academic Period," in which the details of the coding problem were worked out. Initially, Stent argued, success built on success; but before long biology would have solved its major problems, leaving only the "stamp-collecting" work of filling in details; and eventually biology would wind down as diminishing returns set in. Physics, Stent conceded, is apparently more open-ended; but as it became more and more expensive and less and less comprehensible, society would be less and less willing to continue to support it.

But these are only the first steps in Stent's much more ambitious argument that the "end of progress" is at hand. The idea of a theoretical culmination followed by the speedy onset of diminishing returns is soon intertwined with the idea of annihilation. The end of progress, Stent argued, would soon come not only in science but also in music, the arts, and literature; for though at first progress picks up speed as it builds on itself, eventually it is self-limiting: "the dizzy rate at which progress is now proceeding makes it seem very likely that progress must come to a stop soon, perhaps in our lifetime, perhaps in a generation or two."[16] With increased economic security, Spengler's Faustian Man, "the prime creative mover of history," would soon fade from the scene as the adaptive value of his boundless Will to Power diminished, to be displaced by the inner-directed personality. And so we would shortly enter a "New Polynesia"— the New Age anticipated by the beatniks and celebrated by the hippies—in which the urge to find out has exhausted itself.

Reprinted, by permission, from James Thurber, *Men, Women and Dogs* (New York: Dodd, Mead, 1975), p. 88.

By now, of course, the first of Spengler's themes, the demotion of science to mere cultural construct, is a familiar motif in the New Cynicism. An especially striking example is the letter of invitation to a conference at Gustavus Adolphus College in 1989—the culmination of a series, endorsed by the Nobel Foundation, beginning with "Genetics and the Future of Man," and including meetings devoted to physics, chemistry, biology, economics, creativity, peace, the place of women in the life of the mind: an "intellectual festival," attended by some four thousand people, on The End of Science. "In recent times," according to the conference announcement, "science has come to be understood as . . . more subjectivist and relativistic . . . the product of such things as paradigmatic focuses, ideological struggles and the basic instrument of power." Whereas "previously science was thought to be 'tainted' if it was under the direction of government/culture/society," and Stalinist intervention in Soviet biology was seen as something to be deplored, the new understanding is that "science as science is coming to an end . . . , no longer a fortress of objectivity."[17]

Though Spengler is truly inimitable, in his introduction to the proceedings of the Gustavus Adolphus conference Richard Q. Elvee makes a creditable attempt: "science is a series of interpretations of nature; however, the interpretations are creations of interpreters of nature, and to a large extent science is an account of the imagination of the interpreters. . . . This means not only that the history of the intellectual endeavor that for 400 years has been called science is, to a large extent, a creature of the scientist's subjectivity but also that those objective regularities that the scientists interpret themselves are composed of observer responses to the world." "Is this new understanding," he asks rhetorically, "forecasting THE END OF SCIENCE?"[18]

The lead speaker, Sandra Harding, turns in a remarkable performance in which she seems simultaneously to endorse the organizers' desire to politicize science, and to reassure readers that she is not "down on physics," not unfriendly to the sciences. The "anti-democratic" elements of science, she explains, "deteriorate [its] ability to provide objective, empirically defensible descriptions and explanations" of nature and society.[19] Science is a "ruling-class activity," and as such "can have only a partial and distorted understanding of nature and social relations."[20] The model for good science should be "research programs that have been explicitly directed by liberatory political goals." This doesn't *guarantee* good empirical results, but "since sexism, racism, imperialism and bourgeois beliefs have been among the most powerful influences on the production of false scientific belief," it is the best way to such results. To acknowledge this, Harding maintains, is not to succumb to relativism or subjectivism, but to adopt a new, "strong" conception of objectivity, referring to the fit of hypotheses with "prevailing cultural interests and values."[21]

If this kind of stuff makes you a little giddy, the next speaker, Sheldon Glashow, will steady you as he brilliantly parodies his letter of invitation, asking: "Whatever does it all mean, and why did I agree to participate in this farce?" Observing that "The Big Questions here are: Is Science simply a grown-up version of a boys-only game of 'Let's Pretend'? And can a new kinder and gentler Science be invented with Force replaced by Love, and Power by Tenderness?" he answers: "No, and no again." Science "is not sick and it is not dead."[22]

But in 1996, with John Horgan's *The End of Science*—a book which, apparently catching a current of end-of-the-millennium portentousness, seems to have been briefly quite influential—the death of science was announced yet again. Horgan's subtitle, *Facing the Limits of Knowledge in the Twilight of the Scientific Age*, echoes Spengler; and his book opens with a discussion of Stent's "astonishingly prescient" book ("astonishingly prescient," of course, because it anticipates Horgan's own ideas!). One of Horgan's main themes, familiar from du Bois-Reymond, is that soon all the answerable scientific questions will have been answered; his conclusion, however, is not exactly that science is coming to an end, but that future science will be "ironic," open and unsettled, more like literary interpretation than present-day science. Adapting Harold Bloom's conception of "strong poets" who, under the "anxiety of influence," assert their individuality by struggling to define themselves relative to Shakespeare or Dante,[23] Horgan predicts that when the answerable questions are all answered, "strong scientists" will speculate about unanswerable questions, asserting their individuality by defining themselves relative to their great predecessors.

Incomparably more intellectually serious than Horgan, Weinberg (like Einstein) dreams of a "final theory": a culmination of the dream that began with the Milesian philosophers, of explaining all natural phenomena in terms of a fundamental constituent of matter; a dream continued by Newton, but which really began to take shape in the mid-1920s with the discovery of quantum mechanics and the recognition that chemical phenomena can be explained by the electrical interactions of electrons and atomic nuclei. Already by 1902 even the formerly pessimistic Michelson had begun to hope that, perhaps very soon, "the converging lines from many apparently remote regions of thought will meet on . . . common ground . . . marshalled into a single and compact body of scientific knowledge."[24] And in 1992 Weinberg writes: "Though present theories are of only limited validity, still tentative and incomplete, behind them we catch now and then glimpses of a final theory . . . of unlimited validity," of the fundamental principles to which all the arrows of explanation in the independent sciences seem to point, in which a reconciliation of relativity with quantum mechanics will be a large part.[25]

The fundamental idea here is the idea of the fundamental. The final theory Weinberg envisages would constitute a comprehensive and integrated account of the laws governing the fundamental constituents of the world, of matter/energy. It would be, in a sense, comprehensive; nevertheless, Weinberg argues (and here he departs from Einstein),[26] this "final theory" would not enable us, even in principle, to deduce all the truths of biology, or even of physics; for biology, and even physics itself, partly depends on historical contingencies. Nevertheless, the final theory of which Weinberg dreams would constitute the ultimate explanation of why physical things are the way they are.

If particle physicists were eventually to grasp and articulate the final theory, and perhaps in the not-too-distant future, this would constitute a kind of culmination of the scientific enterprise. In his inaugural lecture as Lucasian Chair at Cambridge, Stephen Hawking had suggested that the then-fashionable extended supergravity theories might provide the basis for such a final theory. But Weinberg doesn't suggest that a final theory is just around the corner, only that tantalizing glimpses of such a theory are occasionally to be had. And he is very clear that "[a] final theory will be final in only one sense—it will bring to an end a certain sort of science, the ancient search for those principles which cannot be explained in terms of deeper principles." Furthermore, the work Weinberg sees as remaining for science once the final theory has been discovered would be nothing like the "stamp-collecting" of which Stent speaks: "Of course a final theory would not end scientific research, not even pure scientific research, not even pure research in physics. Wonderful phenomena, from turbulence to thought, will still need explanation whatever final theory is discovered."[27]

"The" idea of the end of science, in short, encompasses both pessimistic visions of the ultimate failure of science and optimistic visions of its ultimate success, and accommodates themes as disparate as annihilation and completion, demotion and culmination—disparate themes, however, often intertwined and sometimes blurred together. Those who react strongly against the thesis that the pretensions of the sciences to discover truths about the natural world are delusory sometimes react equally strongly against any and every version of the "end of science" idea. But it is essential to disentangle serious issues about the external factors that might bring science to an end, and serious questions about whether it makes sense to suppose that physics might eventually discover the ultimate laws of nature, or that the work of the sciences might eventually be completed in every last detail, from sociological confusions and rhetorical posturing about objectivity.

Beginning with the theme of demotion, my first task will be to get to the bottom of the muddles about the place of science in culture, about objectivity and politics, and about the relation of science to art and literature, variously committed by Spengler, Stent, the Gustavus Adolphus cynics, Harding, and Horgan. Turning to the theme of annihilation, my next task will be to understand what conditions might be necessary for science to flourish, and whether we can expect them to continue, or predict that or when they might come to an end. Then, turning to questions about the limits of science, my final task will be to investigate whether there could, or must, be ultimate laws of nature, whether scientific inquiry might, or must, eventually produce diminishing returns, and whether science really is the "endless frontier"—or only an enormous crossword in which every last entry might eventually be completed.

THE THEME OF DEMOTION

When I write of "demotion," it is as shorthand for "epistemological demotion," referring to the idea of science as mere cultural construct.

As I acknowledged earlier, science is, in a sense, culturally specific: modern science arose at a particular time and in a particular part of the world; and for science to flourish, or even to survive, requires the right kind of cultural environment. Nevertheless, as I also argued earlier, science really isn't, in the sense Spengler imagined and the New Cynics have recently once again made fashionable, merely one cultural phenomenon among many. For science is also universal—in more senses than one: it is a manifestation and amplification of the capacity for inquiry possessed in some measure by all human beings;[28] moreover, permanently important scientific discoveries may arise out of local, temporary cultural preoccupations.

Here is a striking example, borrowed from Holton: Johannes Kepler was trained in the old quadrivium—arithmetic, geometry, music, and astronomy—and had particularly studied music theory. His *Harmonices mundi* of 1619 was concerned primarily with harmony in music. During their orbital motions, Kepler found, Earth and Venus sometimes sing out a minor and a major sixth; he even identified a heavenly six-part chord generated by the orbital velocities of all the six then-known planets. Kepler's third law—for every planet, the square of its period of revolution around the sun divided by the third power of its mean distance from the sun is a constant—is presented in just one sentence in the fifth section of this book. Yet this law supported the Copernican heliocentric astronomy, contributed to Newton's dynamic conception of the solar system, and survives and is useful to this day.[29]

The example suggests that Spengler gets things exactly backwards when he insists that the deepest history of science lies, not in its results, but in its cultural symbolism and style. As Holton puts it, "it is precisely the strength of science that, in the long run, it throws off the various cultural scaffoldings, which in the nascent phase helped some individuals in the building of the ever-unfinished Temple of Isis. The concepts and theories of mature science, as used in daily work and publications, can do without any residue of their various individual human origins."[30]

The example also illustrates historically what I have argued philosophically: that natural-scientific inquiry really is, in an important though subtle sense, cumulative. Not that, at every step, a new truth is added, nor that progress is always a matter of the accumulation of new truths; but, as scientific inquiry proceeds in its ragged and uneven way, it finds new truths, better instruments, better vocabulary, etc., and ways to build on them; so that over the centuries the sciences have built a great edifice of well-warranted claims and theories (even though, to be sure, the trash-heap of discarded concepts and theories is larger by far).[31]

Science really isn't, in the sense Stent borrows from Spengler and Horgan from Stent, an enterprise of just the same kind as art or literature. Like science, traditions of art, music, and literature arise and flourish in specific cultural contexts, and yet are universal insofar as they manifest capacities and tendencies common to all human beings. In art, music, and literature, as in science, traditions grow up and develop, later generations build on the work of previous generations, and so on. But art, music, and (as I argued earlier) literature are not, like the sciences, kinds of inquiry; and are not cumulative in the special, epistemological sense that science is.

Nor is the speculative science of today, as Horgan suggests, essentially different from the speculative science of the past. At the frontier scientists have always wondered if this or that newly thought-up possible kind of thing or stuff is really real, whether, in Glashow's words, it will "go the way of phlogiston or of quarks."[32] Far from being, as Horgan suggests, a sign that science is about to go the way of literary theory, this is a mark of the vitality of science.

As the organizers of the Gustavus Adolphus conference point out, in recent times the claims of science to arrive at objective knowledge have (again) been challenged. However, when they say that science "has come to be understood" as relative or subjective, they are relying on the implication of truth carried by "understood" to convey the impression that the New Cynics' challenge to the epistemic pretensions of science succeeds. It doesn't. It is, rather, the Passes-for Fallacy writ large. I will say it one last time: Scientific knowledge is not the only knowledge, nor scientific inquiry the only legitimate kind of inquiry; the rationality of the scientific enterprise cannot be completely captured by any narrowly logical model;

scientific claims and theories are fallible and revisable; scientists are seldom if ever wholly objective, impartial inquirers; and the social context in which scientific work is conducted can affect not only what questions are thought worthy of investigation, and not only what solutions come to mind, but even, sometimes, what results are arrived at. But it doesn't follow that there is no such thing as objectively stronger or weaker evidence, objectively better- and worse-conducted inquiry, objectively true or false scientific claims and theories. And it isn't true.

And allow me to say one last time, also: a claim is true if things are as it says, false if they are not; evidence is better or worse depending on how supportive it is, how independently secure, and how comprehensive; an investigation is better or worse conducted depending on how scrupulous, how honest, how imaginative, how thorough it is. These concepts are, all of them, objective; in each case one could add, "irrespective of whether anyone, or everyone, believes that this is (or is not) strong evidence, or a well-conducted investigation, or a true claim." They are concepts taken for granted not only by scientists but by anyone who makes any claim, or inquires into any question; and presupposed in pointing out—as of course is true—that scientific investigations have sometimes been poorly or even dishonestly conducted, and that many claims accepted by scientists have been ill-founded or false.

Granted, developments in science and technology have political consequences; what scientific, and especially what technological, work gets done is often influenced by political considerations; access to scientific careers, the applications of scientific results, the availability of technology, are all issues of legitimate political concern; and sometimes (usually but not quite exclusively in the human and social sciences) the engagement of people of differing political views, or of women as well as men, or of non-white as well as white scientists, has made it possible to overcome formerly entrenched misconceptions. But it doesn't follow that science ought to be reinvented as an overtly politicized enterprise. And it isn't true.

As I stressed earlier,[33] there is a real distinction between inquiry and advocacy. Inquiry, which is an attempt to discover the truth of some question, regardless of what that truth may be, requires the inquirer to seek out and assess the worth of evidence. Advocacy, which is an attempt to make a case for some proposition or propositions determined in advance, requires the advocate to select and emphasize whatever evidence favors the proposition(s) in question, and to ignore or play down the rest. C. S. Peirce once wrote: "if a man occupy himself with investigating . . . for some ulterior purpose, such as to make money, or amend his life, or to benefit his fellows, he may be ever so much better than a scientific man . . . but he is not a scientific man."[34] Though many scientists, doubtless, have mixed motives, Peirce is right to stress that inquiry, truth-seeking, is the heart of science.

To blur the distinction between inquiry and advocacy, moreover, is to threaten the intellectual honesty that serious scientific work, like serious inquiry of every kind, demands; and at the same time to undermine the possibility of genuinely beneficial social change, which requires knowing what the present situation is, and what the consequences of this or that change would be. Here is Peirce again: "I must confess that I belong to that class of scallawags who purpose . . . to look the truth in the face, whether doing so is conducive to the interests of society or not. Moreover, if I should ever tackle that excessively difficult problem, 'What is for the true interest of society?' I should feel that I stood in need of much help from the science of legitimate inference."[35]

The organizers of the Gustavus Adolphus conference suggest that it is naive and old-fashioned to imagine that the politicized science of the Nazi and Soviet regimes was tainted—or, as they say, scare-quotes "tainted." Doubtless, given the powerful effect on many academics of the fear of being thought naive, this is a shrewd rhetorical strategy. But I shall risk the accusation of naiveté, and say plainly that Nazi science *was* tainted. Nobel Prize winners Philip Lenard and Johannes Stark, maintaining that the race of researchers determined their physics, admired "German physics" as in direct contact with nature, but deplored "Jewish physics" as guilty of "abstraction." To the idea that science is international, Lenard replied: "It is false. Science, like every other human product, is racial and conditioned by blood." Adolf Hitler announced: "I don't want there to be any intellectual education. . . . [W]e stand at the end of the Age of Reason. . . . A new era of the magical explanation of the world is rising, an explanation based on will rather than knowledge. There is no truth, in either the moral or the scientific sense. Science is a social phenomenon, and like all those, is limited by the usefulness or harm it causes."[36]

Nazi science was tainted epistemologically, because its goal was not to seek truth, but to promote a political agenda. It was also, of course, tainted politically: the reinvented, "Aryan" science promoted by Nazi ideologues was to be *völkisch*, i.e., to reflect the perspective of the desirable races. But it is not the hideousness of the political agenda that Nazi science was to promote, but the fact that it was to promote a political agenda at all, that is most to the present purpose. The same goes, mutatis mutandis, for Stalinist science.

Harding urges a science directed by "liberatory political goals"; but it is not entirely clear whether what she proposes is that science be reinvented as a tool for promoting the interests of women and other oppressed people, or as a forum for negotiation among competing interest groups. It is clear enough, however, that an essential premise of her argument is that the chief source of error, in the natural as well as the social sciences, is the covert operation of race, gender, and class interests. Harding tells us that "some critics have detected social values in contemporary studies of slime mold, and even in such abstract theories as relativity theory and formal semantics."[37] In a variant of a now-familiar rhetorical maneuver, she relies on the connotation of success carried by "detect," and offers only large, vaguely Spenglerian observations to the effect that the interpretation of the mathematical symbolism of physics is social, and that science relies on social metaphors,[38] instead of evidence that these critics' claims to have detected the covert influence of sexism in theories of slime mold, semantics, etc., are correct. I don't believe they are.

What is the chief source of error in the sciences? The work is hard; and scientists are fallible. To be sure, one of the many respects in which scientists are fallible is that they sometimes have prejudices, i.e., entrenched and unexamined commitments to poorly warranted background beliefs; to be sure, one of the many kinds of prejudice is stereotypical thinking about women or non-white people. But even in the social and human sciences, where the subject-matter at least allows room for them to operate, it is very doubtful that race, class, and gender interests are the root of all, or even most, error; and to imagine that the main sources of error in the natural sciences are racism, sexism, and bourgeois thinking is much, much farther-fetched.

This is not to deny that established scientists are sometimes slower than one might like to recognize, or quicker than one might like to dismiss, the worth of promising new ideas—"conservative" in a non-political sense; nor that scientists have sometimes accepted poorly warranted claims about women, or non-white people, because of prejudice. But to turn science into a tool for promoting the interests of women, or of non-white people, or of any class or group, is subject to precisely the same epistemological objections as Nazi and Soviet attempts to turn science into a political tool. And a community of advocates in which "people will have to negotiate with each other . . . about just whose perspective should prevail when 'facts' are being contested,"[39] would be a sorry substitute for a community of inquirers all of whom do their best to discover the truth of the questions that concern them, whether that truth advances their interests or not.

If science were to be transmuted into a tool for promoting a political agenda, or into negotiation among competing interest groups, it really *would* have come to an end. Aryan or proletarian or feminist "science" is not inquiry but advocacy, and "strong objectivity" is only politics in disguise. The demotion of science to politics really would constitute its annihilation; and so the perception, on the part of those scientists who take it seriously, that the New Cynicism is not merely mistaken, but dangerously so, is correct. For these mistaken ideas really could hinder, or even halt, the scientific enterprise.

THE THEME OF ANNIHILATION

Certain conditions required for scientific inquiry are, obviously enough, contingent, and might come to an end. The possibility cannot be ruled out that—whether through nuclear war or ecological disaster or some other catastrophe—the human race should die out, and the scientific enterprise with it.

The possibility cannot be ruled out, either, that the human race should survive a nuclear or ecological disaster but, bringing a new barbarism upon itself, be reduced to primitive conditions of life; nor that in such circumstances the scientific knowledge we now possess should be lost, or perhaps retained only as half-understood fragments. The same, interestingly enough, is true of art or literature: sufficiently catastrophic events could bring a new Dark Age in which the art or literature of the past is lost, or retained only as half-appreciated fragments. Thinking, however, of the gloriously vivid rock-painting of the Kalahari bushmen, and the subtlety of their music and dance, I suspect that art and music would soon revive even in such newly primitive conditions of life as I have imagined. A revival of science, however, would likely take much longer; for science seems to require more particular circumstances, a more specialized cultural niche, to arise and to flourish.

The possibility cannot be ruled out, either, that though the human race manages to avoid global catastrophe, the cultural conditions necessary for science to flourish might come to an end. Before modern science developed in Europe, there were many earlier anticipations which never quite reached critical mass. There must be some explanation of why modern science flourished when and where it did, and not at other times and places in societies no less rich in time and other resources—in the many non-scientific societies able to support a non-productive priestly caste, for example.[40]

The conditions that enabled the rise of modern science, and the conditions that enable it to continue, are doubtless different (think how unlike present-day North America, or Japan, or China, or India, or even modern Europe, seventeenth-century Europe was). Once well under way, it seems, science can flourish in quite varied circumstances. That was part of my argument that science is not inherently "Western"; though modern science arose in Europe, it now flourishes in many parts of the world, East as well as West. But it does not flourish in all circumstances whatsoever; and whatever exactly the conditions are for it to continue to flourish, they are not found always and everywhere.

In the light of Hitler's and Stalin's efforts to control and distort science, it is easily possible to imagine the rise and spread of totalitarian or theocratic regimes which, suppressing the freedom of thought and speech a flourishing science requires, are seriously inhospitable environments for the scientific enterprise. As Glashow reminds us, Arab astronomy, mathematics, and chemistry were once unrivalled; but with the rise, early in the second millennium, of *Taqlid*, the Islamic doctrine that there are no truths beyond those revealed in the Koran, scientists and scholars were banished. And so, rather than growing and flourishing, Arab science declined—leaving its footprints on our language, however, from "alkali" to "zenith." The ancient Chinese, Glashow continues, invented gunpowder and the compass, and were great navigators—until in the fifteenth century they decided that nothing beyond the Celestial Empire was worthy of discovery, and "burnt their great ships just before Columbus set forth with his tiny flotilla to discover a New World."[41]

It easily possible, also, to imagine a gradual disenchantment that makes our own society less and less willing to devote large resources to scientific research. ("I am not fond of predicting catastrophes," writes Sidney Smith; "but there are cracks in the universe.")[42] Among the factors that could contribute to this kind of disenchantment are concern about the ever-increasing cost of scientific research; unease about such issues as the use of embryonic or fetal tissue in medical research, or the possibility of cloning human beings; worries about the dangers of technology; a sense that science is growing too large and too powerful; resentment of scientists' perceived "elitism"; failures in science education; a rising interest in the mysterious, the miraculous, the occult; a growing fear that comfortable and comforting religious ideas are under threat; and—here is that shadowy but real connection between the themes of demotion and annihilation once again—the increasing popularity of the idea that the epistemic pretensions of the sciences are hollow.

Earlier, trying to understand why many people feel fear and resentment towards, as well as admiration for, science, I observed that after all, "it, and we, are only human." And thus far, in thinking about the conditions for science to continue, I haven't considered the possibility of non-human science. Though it is presumably a lot less likely that there is intelligent life elsewhere in the universe than it is that there is life elsewhere in the universe, it is nevertheless conceivable that there should be. Science as we know it bears the stamp of our human cognitive capacities and limitations; but it is conceivable that other creatures, with different cognitive capacities and limitations, might engage in something recognizably similar in its aims and its achievements to our, human science. If so, it is possible to imagine inessential human elements falling away as, through the history of human science, inessential cultural elements have done; even to imagine science surviving the extinction of human beings.

In acknowledging the possibility of non-human science, haven't I undermined what I said earlier about the value of science? It's a fair question, and it calls for an amendment; but I don't think the possibility of non-human science changes the spirit of what I said before. At its best, I said, science represents the most remarkable amplification and refinement of a distinctively human talent, the capacity to inquire. Perhaps, given the continuity of our capacity for inquiry with the curiosity of some animals, "characteristically human" would have been a wiser choice of words than "distinctively." Here on earth, anyway, there is no non-human science. But if other creatures elsewhere in the universe were to produce a science which is a refinement of their capacity to inquire as our science is of ours, it would be valuable, as human science is valuable, because it refines and amplifies this talent. And if there were non-human creatures capable of science, this would be, to my mind, at least as wonderful as there being other creatures capable of music or dance, or extra-terrestrial analogues of these human activities.

But I dare not stumble much further down this dark and very speculative path; and as I turn to questions about the limits of science—questions of culmination, diminishing returns, and completion—I shall confine myself to our, human, science.

THE LIMITS OF SCIENCE

Obviously enough, there are limits to science. There are plenty of questions—historical, legal, literary, logical, political, philosophical, commercial, culinary, etc., etc.—that simply fall outside the scope of the sciences (as that not-quite-translatable German word, *Wissenschaft*,[43] is broader than "science"). At any time, furthermore, there are many, many questions within the scope of the sciences which they aren't, at that time, able to answer, and many, many more questions which they aren't, at that time, even able to ask. It is possible, moreover, that there should be legitimate scientific questions which are beyond human cognitive capacities to answer, no matter how long science continues. For we humans have limited intellectual and other resources: limited intellectual integrity, respect for evidence; limited imaginative powers; limited capacity to reason; and limited sensory reach. (Were there extra-terrestrial scientists, they would be limited too, though no doubt in somewhat different ways—omniscient creatures wouldn't *need* science.)

However, as I have stressed all along, one of the most remarkable things about the natural sciences has been the ways they have found to overcome human limitations: instruments of every kind, extending innate human powers of observation; the calculus, statistics, computers, extending innate human powers of reasoning; metaphors, analogies, linguistic innovations, extending innate human powers of imagination. But in the nature of the case it is nearly impossible to predict what limitations might prove insuperable, and what limitations scientists will eventually find their way over, or around. Human computational power, for example, is certainly limited; but it is virtually impossible for us to predict how much further we might extend it artificially than we already have, as it would have been virtually impossible for a medieval mathematician struggling with long division in roman numerals even to conceive of the power of arabic numerals, let alone of the calculus or the computer.

Those who succeed in surpassing what previously seemed to be insuperable intellectual barriers often feel an extraordinary exhilaration. After discovering his third law on May 15, 1618, Kepler wrote: "What 25 years ago I had a vague premonition of; what I determined 16 years ago as the whole aim of research; what I devoted the best part of my life to, . . . *that* I have at last brought into the light . . . [I] yield freely to the sacred frenzy. . . . The die is cast, I am writing the book. . . . It can wait a century for a reader, as God himself has waited six thousand years for a witness."[44] But before they made their extraordinary breakthroughs, the great heroes of science often despaired of figuring out what seemed to be hopelessly intractable questions. In the early 1920s, Werner Heisenberg later wrote, quantum physics was in such difficulties that "we reached a state of despair"—but a state of despair out of which came a change of mind, a new approach, and a new resolution.[45]

Of course, it is not only our cognitive resources, but also our economic resources, that are limited. A reader of *The Double Helix* today will probably be struck by how modest the resources were that Watson and Crick's work on the structure of DNA required: a grant of a few thousand pounds to keep Watson in Cambridge, some bother about getting more templates made for the models, and so forth. By now, however, cutting-edge scientific work seems almost invariably to be astoundingly expensive. Probably there is at least a grain of truth in the thought that the apparatus (and mind-set) of grants-and-research-projects, the huge investments by governments and large industrial concerns, have themselves contributed to a sense that only really expensive work is worthwhile. Probably there is also at least a grain of truth in the thought that regulation of scientific research has imposed considerable bureaucratic costs. Nonetheless, there is also an important element of truth in the idea that, for reasons inherent in the scientific enterprise itself, we must expect the cost of new discoveries to rise as science advances.

By now, it seems, most of the important scientific discoveries that can be

made with tiny resources—with a candle and a piece of string, as I put it earlier—have been made; by now, furthermore, in physics especially, fundamental
research requires observations of the smaller and smaller and the faster and
faster, so that the equipment needed is likely to be more and more costly (only
"likely," because predictions of the cost of as-yet only imagined scientific equipment can prove quite mistaken). This is not to say that no important new discoveries will be made without vast sums of money—interesting work is still, sometimes, done on a shoe-string[46]—much less to say that vast sums of money will
guarantee important new discoveries; it is only to say that, on average, important
scientific discoveries are likely to be more expensive than those already in place.

Since the 1930s, when the first cyclotron was built in Berkeley, discoveries
about the structure of atoms and their nuclei have relied on using particles as pro-
jectiles to disintegrate some of the atomic nuclei with which they collided, by
means of machines designed to give particles very large amounts of energy. Particle accelerators have become more and more powerful, allowing the head-on
collision of two streams of particles; the Superconducting Super Collider—fifty-
four miles around, accelerating protons to almost the speed of light, and allowing
two beams of 20-trillion volt protons to collide head-on—would have been
capable of giving a pair of colliding protons a total energy of about 40 trillion
volts, and, it was hoped, would make it possible to determine how the symmetry
that relates the weak and electromagnetic interactions is broken. The cost of the
SSC was originally (1987) estimated at between $4 billion and $6 billion, then
(1992) at $8 billion, and finally, at the point at which Congress cancelled the
project (1995), at $11 billion.[47] Reflecting on the fate of the SSC, Weinberg
regretfully observes that the "tacit bargain" between science and society is unrav-
elling; society, he fears, is no longer willing to support pure science, or to
acknowledge that in the long run it is likely to produce the greatest benefits.[48]
Stent's prediction that society's willingness to support scientific work might
decline, or even be exhausted, was not entirely groundless.

What of his prediction of diminishing returns, in the sense of returns of diminishing scientific importance? After a large breakthrough in any area, there may be
a period of consolidation and application, or a cascade of new breakthroughs—as
with a crossword puzzle, where after filling a long, much-intersected entry we may
next complete lots of small, now-easy entries, or find other long, much-intersected
entries suddenly feasible. And by any standards, in the thirty-odd years since Stent
predicted that there would soon be nothing left for biologists to do but boring
"stamp-collecting" work, there has been a cascade of new discoveries in genetics
and molecular biology; and a whole vast range of new questions—some barely
conceivable in 1969—remains to be answered. Perhaps the student revolts of the

late 1960s made too strong an impression on him; in any case, Stent's book puts me in mind of Popper's metaphor of the scientist as climbing a mountain shrouded in fog, thinking he has reached the peak only because he can't see how far there is still to go.

By now, anyway, we know that there are higher peaks which in 1969 could barely be glimpsed. By 1998 Maddox could write that "[i]n recent times only the hectic period in which the foundations of quantum mechanics were laid between 1925 and 1930 compares in elan with present research in molecular biology." From the identification of reverse transcriptase in 1968, through the identification of cyclins in the late 1980s, to the identification of the molecules involved in the regulation of cell division, where, throughout the 1990s, a significant result was published at least once a week, the "naming of parts" has proceeded apace. And yet, Maddox suggests, in some ways "the application of molecular biology to the understanding of how the cell functions is still at the rudimentary stage of nineteenth century chemistry." Much remains unknown about DNA processing, RNA editing, protein assembly, the communication between a cell and the outside world, differentiation of cells; and despite the pioneering efforts of J. A. Winfree and, decades later, Albert Goldbeter, modelling of cellular processes is still rudimentary.[49]

"By the 1970s, after two decades of accumulating knowledge of the properties of DNA," Maddox continues, "it seemed only a matter of a short time before there would be an understanding of how the genome of a species orchestrates the fine details of inheritance. But that was not to be." Among the surprises he mentions are that there is horizontal transfer of genetic information between species, e.g., possibly from viruses to mammals; that there is inheritance via mitochondria (which is maternal inheritance, since mitochondria are found in ova but not in sperm); that the arrangements of individual genes themselves consist of pieces used in the eventual translation (exons) separated by stretches of DNA apparently irrelevant to the protein molecules eventually made from them (introns); that 95 percent or so of DNA seems to be functionally insignificant.[50] Nor is a solution to the puzzle of the origin of life yet in sight.

In other areas of science too we find a huge range of unanswered questions. The accretion of matter, the expanding universe, a consistent account of quantum particles—astonishing as all this is, it is nonetheless, Maddox writes, "an interim achievement only." There remain difficulties about the rate of expansion, obscurities in the idea of an inflationary universe, worries about why there is less matter than current cosmological theories indicate there should be,[51] about how to reconcile Einstein's theory of gravitation with quantum mechanics, and so on.[52] It is in fundamental physics, moreover, that the cost of basic research seems likeliest to hinder the work; to answer questions about the unification of strong, electroweak, and gravitational interactions would require energy concentrated on a single proton or electron about a hundred trillion times higher than the Superconducting Super Collider would have made available.[53]

But none of this yet touches the most abstract questions about the limits of science: Are there ultimate physical laws to be captured in the "final theory" of which Weinberg dreams? Is science really an "endless frontier," or only a vast but finite territory not yet exhaustively explored?

As I argued earlier, despite some striking similarities and parallels, science is in important respects quite different from art or literature. One such difference is now newly apropos: The idea cannot be dismissed out of hand that (if we don't run out of money or blow ourselves up first) the scientific enterprise might in principle reach culmination or even completion. Where art and literature are concerned, however, the idea of a final theory is as out of place as the idea of theory is;[54] and it is at least arguable that artists' and writers' explorations of ways and means of expressing emotions, conveying truths, and limning possibilities are inexhaustible in a way that scientists' investigations of things and events are not.

On the question of culmination, of ultimate laws of nature, Weinberg is up against Popper's insistence that there can be no final explanations. Since it is the aim of science to explain, Popper writes, "it will also be its aim to explain what so far has been accepted as an *explicans*; for example, a law of nature." Then comes the key claim for present purposes: that there can be no ultimate explanations neither capable of further explanation, nor in need of it; rather, "every explanation may be further explained, by a theory or conjecture of a higher degree of universality. There can be no explanation which is not in need of a further explanation."[55]

Whether or not "Always ask 'why?'" is a reasonable procedural maxim, the present concern is metaphysical. Popper's repudiation of the possibility of ultimate laws seems to be based primarily on the fact that, historically, this idea has been associated with the doctrine of essentialism, via the thesis that the ultimate laws of physics can be derived from the essence of matter (whether that essence was taken to be extension, as in Descartes, or inertia and the power to attract other matter, as in Roger Cotes).[56] But what is centrally at issue is whether there could be laws about the ultimate constituents of everything, laws of maximal generality not themselves susceptible of further explanation; whether a positive answer would or would not constitute "essentialism" is secondary.

Weinberg argues that the progress of physics thus far indicates that there *are* ultimate laws, that the arrows of explanation eventually converge on the ultimate constituents of everything. Granted, we don't yet know what these ultimate laws are; but to conclude from this that there are none "would be like a nineteenth-century explorer arguing that, because all previous arctic explorations . . . had always found that however far north they penetrated there was still more sea and ice left unexplored to the north, either there was no North Pole or in any case no one would ever reach it." Weinberg's main positive reason for believing that there are ultimate laws is, simply, that "it is very difficult to conceive of a regression of more and more fundamental theories becoming steadily more and more unified, without the arrows of explanation having to converge somewhere."[57]

As Popper says, given any law offered as an explicans, one may always ask "why?"; but it doesn't follow that all such questions are legitimate scientific questions—perhaps they are legitimate but metaphysical questions, or perhaps they are not legitimate questions at all. If so, it is possible that there should be ultimate laws. And if there are, the answer to the question, "and why are the constituents of everything thus and so?" may be: "they just are." Granted, this is not very satisfying intellectually; but it would be no more satisfying, surely, if there are no ultimate laws. Either way, our explanations will always appeal to something itself unexplained.

But there is no sharp line between cosmological and metaphysical questions; nor are there very clear criteria for identifying and individuating questions. Thinking of how cosmologists first transmuted, and then at least partially answered, the metaphysicians' question, "Why is there something rather than nothing?" when they developed their account of the accretion of matter, we might speculate that, if physicists were to find the ultimate laws, the demand for an explanation of those laws might eventually be transmuted into a kindred, yet different and more answerable, question. Alluding to Robert Nozick's suggestion that we look for something more satisfying than mere brute ultimate laws, Weinberg argues that it might be that though the final theory is not logically necessary, there is no way to modify it even slightly without arriving at logical absurdities; as, he continues, physicists have found it impossible to change the rules of quantum mechanics even slightly without such logical disasters as probabilities that come out as negative numbers.[58]

No one doubts, Weinberg continues, that in principle all the properties of DNA could be explained by solving the equations of quantum mechanics for electrons and nuclei of a few common elements, the properties of which are in turn explained by the standard model. However, that there are living things on earth which use DNA to pass on genetic information "depends on certain historical accidents: there is a planet like the earth; life and genetics somehow got started; and a long time has been available for evolution to do its work." The same goes for other sciences, such as geology and astronomy. Similarly, it is possible to calculate the nature of the matter formed in the first few minutes of the universe—about three-quarters hydrogen and one-quarter helium, with only a trace of other, mainly very light, elements: these were the raw materials out of which heavier elements were later formed in stars. But this whole scenario "depends on a historical assumption—that there was a more-or-less homogeneous big bang, with about ten billion photons for every quark."[59]

Acknowledging that what we now think of as initial conditions might come to be seen as part of the laws of nature, Weinberg wonders whether eventually the line between universal laws and historical contingencies might blur. What we now think of simply as initial conditions certainly might later prove to be susceptible of explanation; but, given that there are singularities, i.e., contingencies holding now but not then, here but not there, it seems that as a matter of logic the distinction between singular conditions and universal laws cannot evanesce altogether. If this is right, those ultimate laws would not be sufficient for the deduction of all natural phenomena without some ultimate singularity or singularities.

Beyond this, it gets less and less clear just what the question of in-principle completability amounts to. Presumably it would always be possible to calculate the value of this or that constant to yet more decimal places—but then we really would be in the realm of diminishing returns. More interestingly (even though, if there are ultimate laws and singularities, "why?" is not always an answerable question), perhaps there would nevertheless always be new questions—less fundamental than the questions to which the ultimate laws are answers, but not necessarily trivial questions—radiating outwards from those ultimate laws. I'm not sure. But as I think of what would be involved in deriving from ultimate laws and singularities, say, that at such-and-such a time this variety of beetle would evolve in the Amazon, which so-and-so many millennia later would become extinct, I am inclined to agree with Einstein and Weinberg that it is extraordinarily hard to imagine scientists' eventually completing every last answer to every legitimate scientific question.

The social sciences, of course, insofar as they concern themselves with local and contingent social roles, rules, and institutions, have a much more markedly historical aspect than the natural sciences. And that the historical contingencies of human societies might eventually be derivable from completely universal laws of nature seems, to put it mildly, much farther-fetched than the idea that cosmological events such as the big bang might be. Even if there are laws governing the universal aspects of human nature variously expressed in this society or that, and even if such laws were known, there would remain not only a vast array of details to be discovered, not only the ramifications of self-fulfilling (or self-undermining) predictions, but also the permanently open possibility of new manifestations of those laws in new social arrangements. So, though thus far the social sciences seem to lag far behind the natural sciences, the scope for future spurts and breakthroughs seems enormous, and the future prospects limited only by the possibility of the extinction of human societies.

AND, FINALLY, IN CONCLUSION

While the theme of demotion led us through a thicket of confusions, and the theme of the annihilation of science only to the conclusion that such an eventuality can by no means be ruled out a priori, the themes of culmination and completion drew us to extraordinarily challenging questions about the future of the scientific enterprise. It is possible that there are ultimate laws, and possible, if there are, that scientists will discover them. It is possible, furthermore, that what we now conceive as ineliminably historical elements in cosmology and biology should turn out to be derivable from completely universal laws of nature, though not that all historical singularities are eliminable.

We can't predict what currently accepted scientific ideas will turn out to need modification, minor or major; or what new discoveries will bring forth a cascade of new questions; or whether or which new questions will bring us up against the limitations of our human capacity to sense and reason, and to devise ways to overcome those natural limitations; or how far and how fast the cost of making new discoveries will escalate, or whether or when society's willingness to find the means to pursue them might end. But we can say that Spengler was quite wrong, that there is no danger that science will shortly "fall upon its own keen sword"; that Glashow and Maddox are almost certainly right, that even in those areas where culmination and perhaps completion may be in principle possible, gargantuan tasks remain; and that Yogi Berra was absolutely right—it ain't over till it's over.

NOTES

1. Spengler, *The Decline of the West*, 1:424.
2. Glashow, "The Death of Science!?" p. 23.
3. Maddox, *What Remains to Be Discovered*, p. 378.
4. Holton, "How to Think about the End of Science."
5. As Holton acknowledges in a longer, later version of his paper, "The Controversy over the End of Science."
6. Du Bois-Reymond's earlier lecture appears in English as "The Limits of Our Knowledge of Nature." Haeckel's book appears in English as *The Riddle of the Universe— at the Close of the Nineteenth Century*. My source is Rescher, *Scientific Progress*, pp. 20ff.

7. Rescher, *Scientific Progress*, p. 22.

8. Weinberg, *Dreams of a Final Theory*, pp. 13–14.

9. Quoted in *Physics Today* 21 (1968):56. My source is Rescher, *Scientific Progress*, p. 23.

10. Weinberg, *Dreams of a Final Theory*, pp. 14–15.

11. Holton, "How to Think about the End of Science," p. 65.

12. Spengler, *The Decline of the West*, 1:378, 380.

13. Ibid., 1:417 ("pure analysis," "the faith-forms of springtime"), 419 ("ruthlessly cynical"), 420 ("gently sloping route of decline").

14. Ibid., vol. 1, table III.

15. Einstein, "Principles of Research," pp. 225–26.

16. Stent, *The Coming of the Golden Age*, p. 94.

17. Elvee, "After Twenty-Five Years: The End of Science!" pp. x–xi.

18. Ibid., p. xi.

19. Harding, "Why Physics Is a Bad Model for Physics," p. 1.

20. Ibid., p. 16; as "authority" for this claim she refers to Smith, *The Everyday World as Problematic: A Feminist Sociology* and Hartsock, "The Feminist Standpoint: Developing the Ground for a Specifically Feminist Historical Materialism."

21. Harding, "Why Physics Is a Bad Model for Physics," pp. 18–19.

22. Glashow, "The Death of Science!?" pp. 24–25, 30.

23. Horgan is alluding to Bloom, *The Anxiety of Influence*.

24. Michelson, *Light Waves and Their Uses*, p. 163.

25. Weinberg, *Dreams of a Final Theory*, pp. 6, 17.

26. Einstein, "Principles of Research," p. 226.

27. Weinberg, *Dreams of a Final Theory*, p. 18.

28. See chapter 11, pp. 313 ff.

29. Holton, "How to Overcome the Limits of Science."

30. Ibid., p. 6 of the English typescript.

31. Glashow, "The Death of Science!?" p. 31.

32. Ibid.

33. Chapter 6, pp. 169 ff.

34. Peirce, *Collected Papers*, 1:45.

35. Ibid., 8:143.

36. The quotation from Hitler is reported by Hermann Rauschning, president of the Danzig Senate; my source is Holton, *Einstein, History, and Other Passions*, p. 31, and "Postmodernisms and the 'End of Science'." The quotation from Lenard is from his *Deutsche Physik*, reproduced in Clark, *Einstein: The Life and Times*, pp. 525–26; my source is Gross and Levitt, *Higher Superstition*, p. 129.

37. Harding, "Why Physics Is a Bad Model for Physics," p. 4. As "authority" for this, she refers to Fox Keller, "The Force of the Pacemaker Concept in Theories of Aggregation in Slime Mold," and "Cognitive Repression in Contemporary Physics"; Forman, "Weimar Culture, Causality, and Quantum Theory"; and Hintikka and Hintikka, "How Can Language be Sexist?"

38. Harding, "Why Physics Is a Bad Model for Physics," p. 8.

39. Ibid., p. 18.

40. I owe this thought to Meera Nanda.

41. Glashow, "The Death of Science!?" p. 26.

42. Gross, *The Oxford Book of Aphorisms*, p. 8 (no original source is given).

43. A rough approximation might be "rigorous, organized, systematic inquiry."

44. My source is Holton, "The Scientific Method Is Doing Your Damnedest, No Holds Barred" (the published, abridged, English version of "How to Overcome the Limits of Science"), p. 92.

45. My source for the Heisenberg quotation is Holton, "How to Overcome the Limits of Science," p. 6 of the typescript English version.

46. See Larson, "Science and Shoestring Technology," for a discussion of the difficulties faced by scientists without large resources, and a description of the research on phytoplankton and other limnological features at Crater Lake, Oregon, that he conducted using minimal equipment; and Yaukey, "The Sky's the Limit for Backyard Scientists" on data collected by amateur astronomers using a forty-year-old telescope and a homemade digital camera.

47. Maddox, *What Remains to Be Discovered*, pp. 65–67; Weinberg, *Dreams of a Final Theory*, chapter 12. Lemonick, "Superconductivity Heats Up"; "The $2 Billion Hole."

48. Weinberg, *Dreams of a Final Theory*, afterword to the Vintage edition, pp. 277–82.

49. Maddox, *What Remains to be Discovered*, chapter 5; the quotations are from p. 165.

50. Ibid., chapter 6; the quotation is from p. 199.

51. But see Kirshner, *The Extravagant Universe* (or, briefly, Begley, "Scientists Go on Hunt for the 'Dark Matter' Filling In the Universe") on relevant developments since 1998.

52. Maddox, *What Remains to Be Discovered*, pp. 367–68.

53. Weinberg, *Dreams of a Final Theory*, p. 234.

54. Which is not to deny that there might be other senses in which the idea of culmination would be appropriate to art or literature.

55. Popper, "The Aim of Science," pp. 194–95.

56. My source is Popper, "The Aim of Science," p. 194, where Cotes is described as a follower of Newton's.

57. Weinberg, *Dreams of a Final Theory*, pp. 231, 232.

58. Ibid., pp. 236 ff.; Nozick, *Philosophical Explanations*, chapter 2.

59. Weinberg, *Dreams of a Final Theory*, pp. 32–39; the quotations are from pp. 33 and 34.

BIBLIOGRAPHY

Note: Where there are references to more than one article in the same collection or anthology, the details of the collection or anthology are given separately, under the name(s) of the editor(s). For books and articles published more than once, where full publication details are given out of chronological order, the first source for which full details are given is the version cited.

Abrahamsen, David. *The Psychology of Crime*. New York: Columbia University Press, 1960.

Achinstein, Peter, and Stephen F. Barker, eds. *The Legacy of Logical Positivism: Studies in the Philosophy of Science*. Baltimore: Johns Hopkins Press, 1969.

Adams, Richard. *Watership Down*. 1972. Harmondsworth, Middlesex, U.K.: Penguin Books, Puffin Books, 1973.

"AIDS and the African." *The Boston Globe*. *See* Shillinger; Haygood.

Anderson, Martin. *Impostors in the Temple: A Blueprint for Improving Higher Education in America*. New York: Simon and Schuster, 1992. Reprint, Stanford, Calif.: Hoover Institution Press, 1996.

Andreski, Stanislav. *Social Sciences as Sorcery*. New York: St. Martin's Press, 1972.

Angell, Marcia. *Science on Trial: The Clash of Medical Evidence and the Law in the Breast Implant Case*. New York: W. W. Norton, 1996.

———. "Is Academic Medicine for Sale?" *New England Journal of Medicine* 342 (May 2000):1516–18.

Aronowitz, Stanley. *Science as Power: Discourse and Ideology in Modern Society*. Minneapolis: University of Minnesota Press, 1988.

Ashmore, Malcolm. *The Reflexive Thesis: Wrighting Sociology of Scientific Knowledge.* Chicago: University of Chicago Press, 1989.

Avery, Oswald T., Colin M. MacCleod, and Maclyn McCarty. "Studies of the Chemical Nature of the Substance Inducing Transformation in Pneumococcal Types." *Journal of Experimental Medicine* 79 (1944):137–58. Reprinted in *Conceptual Foundations of Genetics: Selected Readings,* ed. Harry O. Corwin and John B. Jenkins (Boston: Houghton Mifflin, 1976), 13–27.

Ayer, Alfred Jules, ed. *Logical Positivism.* Glencoe, Ill.: Free Press, 1959.

Babbage, Charles. *Reflections on the Decline of Science in England and Some of Its Causes.* London: B. Fellowes, 1830.

———. *Science and Reform: Selected Works.* Edited by Anthony Hyman. New York: Cambridge University Press, 1989.

Bacon, Francis. *The New Organon.* 1620. Edited by Fulton H. Anderson. New York: Liberal Arts Press, 1960.

Bailar, John C., III. "The Powerful Placebo and the Wizard of Oz." *New England Journal of Medicine* 344, no. 21 (May 2001):1630–32.

Bandow, D. "Keeping Junk Science out of the Courtroom." *Wall Street Journal,* 26 July 1999, A23.

Barash, David P. *Sociobiology and Behavior.* Amsterdam: Elsevier, 1977.

Barnes, Barry. *Scientific Knowledge and Sociological Theory.* London: Routledge and Kegan Paul, 1974.

———. "Natural Rationality: A Neglected Concept in the Social Sciences." *Philosophy of the Social Sciences* 6 (1976):115–26.

———. *Interests and the Growth of Knowledge.* London: Routledge and Kegan Paul, 1977.

———. "On the 'Hows' and 'Whys' of Social Change." *Social Studies of Science* 11 (1981):481–98.

Barnes, Barry, David Bloor, and John Henry. *Scientific Knowledge: A Sociological Analysis.* Chicago: University of Chicago Press, 1996.

Barnes, Barry, and Donald Mackenzie. "On the Role of Interests in Scientific Change." In Wallis, ed., *On the Margins of Science,* 49–66.

———. "Scientific Judgement: The Biometry-Mendelism Controversy." In Barnes and Shapin, eds., *Natural Order,* 191–210.

Barnes, Barry, and David Edge, eds. *Science in Context: Readings in the Sociology of Science.* Cambridge: MIT Press, 1982.

Barnes, Barry, and Steven Shapin, eds. *Natural Order: Historical Studies of Scientific Culture.* London: Sage, 1979.

Bartholet, Jeffrey. "The Plague Years." *Newsweek,* 17 January 2000, 32–37.

Barzun, Jacques. *Science: The Glorious Entertainment.* New York: Harper and Row, 1964.

Bauer, Henry H. *Scientific Literacy and the Myth of Scientific Method.* Urbana: University of Illinois Press, 1992.

Bazerman, Charles. *Shaping Written Knowledge: The Genre and Activity of the Experimental Article in Science.* Madison: University of Wisconsin Press, 1988.

Beattie, James. *An Essay on the Nature and Immutability of Truth.* 1776. New York: Garland, 1971.

Beecher, H. K. "The Powerful Placebo." *Journal of the American Medical Association* 159, no. 17 (1955):1602–1606.

Begley, Sharon. "The Search For Life." *Newsweek*, 6 December 1999, 54–61.

———. "The Ancient Mariners." *Newsweek*, 3 April 2000, 48–54.

———. "Culture Club." *Newsweek*, 26 March 2001, 48–50.

———. "The New Old Man." *Newsweek*, 2 April 2001, 46–47.

———. "Bickering over Old Bones." *Newsweek*, 23 July 2001, 40.

———. "In the Placebo Debate, New Support for Role of the Brain in Healing." *Wall Street Journal*, 10 May 2002, B1.

———. "Science Breaks Down When Cheaters Think They Won't Be Caught." *Wall Street Journal*, 27 September 2002, B1.

———. "Scientists Go on Hunt for the 'Dark Energy' Filling In the Universe." *Wall Street Journal*, 18 October 2002, B1.

Begley, Sharon, and Adam Rogers. "War of the Worlds." *Newsweek*, 10 February 1997, 56–58.

Begley, Sharon, with Anne Underwood. "Religion and the Brain." *Newsweek*, 7 May 2001, 48–57.

Behe, Michael J. *Darwin's Black Box: The Biochemical Challenge to Evolution.* New York: Free Press, 1996.

Belenky, Mary Field, et al. *Women's Ways of Knowing: The Development of Self, Voice, and Mind.* New York: Basic Books, 1986; tenth anniversary edition, 1996.

Bellow, Saul. *Ravelstein.* New York: Viking, 2000.

Benveniste, Jacques, et al. "Human Basophil Degranulation Triggered by Very Dilute Antiserum against IgE." *Nature* 330 (June 1988):816–18.

Benveniste, Jacques. "Reply [to Maddox et al.]." *Nature* 334 (July 1988):291.

Bergmann, Gustav. *Philosophy of Science.* Madison: University of Wisconsin Press, 1957.

Bernstein, David E. "Junk Science in the United States and the Commonwealth." *Yale Journal of International Law* 21 (1996):123–82.

Beyerchen, Alan. *Scientists under Hitler: Politics and the Physics Community in the Third Reich.* New Haven, Conn.: Yale University Press, 1977.

———. "What We Now Know about Nazism and Science." *Social Research* 59, no. 3 (fall 1992):615–41.

Bird, Caroline. *The Case Against College.* New York: David McKay Co., 1973. Reprint, New York: Bantam Books, 1975.

———. *Enterprising Women.* New York: W. W. Norton and Mentor Books, 1976.

Bird, Graham, ed. *Selected Writings of William James.* Everyman series. London: Dent; Rutland, Vt.: Charles E. Tuttle, 1995.

Black, Bert, Francisco J. Ayala, and Carol Saffran-Brinks. "Science and the Law in the Wake of *Daubert*: A New Search for Scientific Knowledge." *Texas Law Review* 72 (1994):715–802.

Bloom, Harold. *The Anxiety of Influence: A Theory of Poetry.* New York: Oxford University Press, 1973.

Bloor, David. *Knowledge And Social Imagery*. London: Routledge and Kegan Paul, 1976. 2d ed., Chicago: University of Chicago Press, 1992.

———. "The Strengths of the Strong Programme." *Philosophy of the Social Sciences* 11 (1981):199–213.

———. "Sociology of Knowledge." In *Companion to Epistemology*, edited by Jonathan Dancy and Ernest Sosa, 483–87. Oxford: Blackwell, 1992.

Bodenheimer, T. "Uneasy Alliance—Clinical Investigators and the Pharmaceutical Industry." *New England Journal of Medicine* 342 (May 2000):1539–44.

Bondi, Hermann. *The Universe at Large*. Garden City, N.Y.: Anchor Books, 1960.

Bonnischen, Robson, and Alan L. Schneider. "Battle of the Bones." *The Sciences*, 1 July 2000, 40–46.

Bounds, Wendy. "One Family's Search for a Faulty Gene." *Wall Street Journal*, 15 August 1996, sec. B.

Boyd, Richard. "Metaphor and Theory-Change; What Is 'Metaphor' a Metaphor For?" In *Metaphor and Thought*, edited by Andrew Ortony, 481–533. 2d ed. Cambridge: Cambridge University Press, 1993. First edition published 1979.

———. "Constructivism, Realism, and Philosophical Method." In *Inference, Explanation, and Other Frustrations: Essays in the Philosophy of Science*, edited by John Earman, 131–98. Berkeley and Los Angeles: University of California Press, 1992.

Bradbury, Malcolm. *Stepping Westward*. 1965. London and New York: Penguin Books, in association with Secker and Warburg, 1968.

Braithwaite, Richard. *Scientific Explanation: A Study of the Function of Theory, Probability and Law in Science*. Cambridge: Cambridge University Press, 1953.

Brannigan, Martha, and Ann Carrns. "Surprise Failure Dashes Hopes for HIV Product." *Wall Street Journal*, 16 June 2000, B1 and B6.

Bridgman, Percy W. "The Struggle for Intellectual Integrity." In Bridgman, *Reflections of a Physicist*, 361–79. Originally published in *Harper's Magazine*, December 1933.

———. "The Prospect for Intelligence." In Bridgman, *Reflections of a Physicist*, 526–52. Originally published in *Yale Review*, 1945.

———. "New Vistas for Intelligence." In Bridgman, *Reflections of a Physicist*, 553–68. Originally published in *Physical Science and Human Values*. Princeton, N.J.: Princeton University Press, 1947.

———. "On 'Scientific Method'." In Bridgman, *Reflections of a Physicist*, 81–83. Originally published in *The Teaching Scientist*, December 1949.

———. *Reflections of a Physicist*. 1950. 2d ed., New York: Philosophical Library, 1955.

Broad, William. "Evidence Builds That Mars Lander Is Source of Mystery Signal." *New York Times*, 1 February 2000, A17.

Broad, William, and Nicholas Wade. *Betrayers of the Truth: Fraud and Deceit in the Halls of Science*. New York: Simon and Schuster, Touchstone, 1982.

Brodbeck, May, ed. *Readings in the Philosophy of the Social Sciences*. New York: Macmillan, 1968.

Browne, E. Janet. *Voyaging*. Vol. 1 of *Charles Darwin*. New York: Knopf, 1995. Reprint, Princeton, N.J.: Princeton University Press, 1996.

Bruner, J. S., and Leo Postman. "On the Perception of Incongruity: A Paradigm." *Journal of Personality* 18 (1949):206–23.

Buckley, Michael J. "Religion and Science: Paul Davies and John Paul II." *Theological Studies* 51 (1990):310–24.

Burdick, Howard. "On Davidson and Interpretation." *Synthese* 80 (1989):321–45.

Burton, Thomas M. "In Trials, Potion of Herbs Slows Prostate Cancer." *Wall Street Journal*, 17 May 2000, B1.

———. "Unfavorable Drug Study Sparks Battle over Publication of Results," *Wall Street Journal*, 1 November 2000, B1 and B4.

Burton, Thomas M., and Gardiner Harris. "Note of Caution: Study Raises Specter of Cardiovascular Risk for Hot Arthritis Pills." *Wall Street Journal*, 22 August 2001, sec. A.

Butler, Samuel. *The Way of All Flesh*. 1903. Modern Library. New York: Knopf, 1998.

Byrd, R. C. "Positive Therapeutic Effects of Intercessory Prayer in a Coronary Care Unit." *Southern Medical Journal* 81 (1988):826–29.

Calvino, Italo. *Invisible Cities*. Translated from the Italian by William Weaver. New York: Harcourt Brace Jovanovich, 1974.

Carnap, Rudolf. *The Logical Structure of the World [and] PseudoProblems in Philosophy*. 1928. Translated by Rolf George. London: Routledge and Kegan Paul, 1967.

———. "The Old and the New Logic." Translated by Isaac Levi. In Ayer, ed., *Logical Positivism*, 133–46. Originally published as "Die alte und die neue Logik," *Erkenntnis* 1 (1930–31).

———. "Testability and Meaning." *Philosophy of Science* 3 (1936):419–71; 4 (1937):1–40.

———. "The Two Concepts of Probability." In Feigl and Sellars, eds., *Readings in Philosophical Analysis*, 330–48. Originally printed in *Philosophy and Phenomenological Research* 5 (1945):513–32.

———. *Logical Foundations of Probability*. 1950. 2d ed., Chicago: University of Chicago Press, 1962.

Carnes, Tony. "Houses of Worship: The Pentecostal City." *Wall Street Journal*, 13 April 2001, W17.

Cartwright, Nancy. *How the Laws of Physics Lie*. Oxford: Clarendon Press; New York: Oxford University Press, 1983.

Chalmers, Alan. *What is This Thing Called Science? An Assessment of the Nature and Status of Science and Its Methods*. Atlantic Highlands, N.J.: Humanities Press, 1976; St. Lucia, Australia: University of Queensland Press, 1977; and Milton Keynes, U.K.: Open University Press, 1978. 2d ed., St. Lucia, Australia: University of Queensland Press; Milton Keynes, U.K.: Open University Press; Indianapolis: Hackett, 1982. 3d ed., St. Lucia, Australia; University of Queensland Press, 1998; Indianapolis: Hackett, 1999.

———. Science and Its Fabrication. Minneapolis: University of Minnesota Press, 1990.

Chambers, Robert. *Vestiges of the Natural History of Creation*. 1844. 10th ed., London: J. Churchill, 1853. Reprint of the 1844 edition, New York: Humanities Press, 1969.

Chargaff, Erwin. "Chemical Specificity of Nucleic Acids and Mechanism of Their Enzymatic Degradation." *Experientia* 6 (1950):201–40.

Chatters, James C. *Ancient Encounters: Kennewick Man and the First Americans*. New York: Simon and Schuster, 2001.

Cheseboro, Kenneth. "Galileo's Retort: Peter Huber's Junk Scholarship." *American University Law Review* 42 (1993):1637–1726.

Chisholm, Roderick M. "Firth and the Ethics of Belief." *Philosophy and Phenomenological Research* 50, no. 1 (1991):119–28.

Churchland, Paul M. "The Ontological Status of Observables." In Churchland, *A Neurocomputational Perspective*, 139–51. Originally published in *Pacific Philosophical Quarterly* 63, no. 3 (1982):67–89.

———. "Folk Psychology and the Explanation of Human Behavior." In Churchland, *A Neurocomputational Perspective*, 111–27. Originally published in *Proceedings of the Aristotelian Society* suppl. 62 (1988):209–22. Also reprinted in *Philosophy of Mind and Action Theory*, ed. James E. Tomberlin, Philosophical Perspectives 3 (Atasadero, Calif.: Ridgeview, 1989), and in *The Future of Folk Psychology: Intentionality and Cognitive Science*, ed. John D. Greenwood (New York: Cambridge University Press, 1991), pp. 51–69.

———. "On the Nature of Theories." In Churchland, *A Neurocomputational Perspective*, 153–96. Originally published in *Scientific Theories*, ed. C. Wade Savage, Minnesota Studies in the Philosophy of Science 11 (Minneapolis: University of Minnesota Press, 1989), pp. 59–101.

———. *A Neurocomputational Perspective: The Nature of Mind and the Structure of Science*. Cambridge: MIT Press, Bradford Books, 1989.

Clark, Ronald W. *Einstein: The Life and Times*. New York: World Publishing, 1971.

Clifford, William Kingdon. "The Ethics of Belief." 1877. In *The Ethics of Belief and Other Essays*, edited by Leslie Stephen and Sir Frederick Pollock, 70–96. London: Watts and Co., 1947.

Cohen, Laurie P. "Reasonable Doubt." *Wall Street Journal*, 12 July 2000, sec. A.

Cohen, L. Jonathan. *The Probable and the Provable*. Oxford: Clarendon Press, 1977.

———. "Inductive Logic 1945–1977." In *Modern Logic—a Survey: Historical, Philosophical, and Mathematical Aspects of Modern Logic and Its Applications*, edited by Evandro Agazzi, 353–75. Dordrecht, The Netherlands: Reidel, 1981.

Collin, Finn. *Theory and Understanding: A Critique of Interpretive Social Science*. Oxford: Blackwell, 1985.

Collins, Harry M. "Stages in the Empirical Programme of Relativism." *Social Studies of Science* 11 (1981):3–10 (introduction to Collins, ed., *Knowledge and Controversy*).

———. "What is TRASP? The Radical Programme as a Methodological Imperative." *Philosophy of the Social Sciences* 11 (1981):215–24.

———. "Special Relativism: The Natural Attitude." *Social Studies of Science* 12 (1982):139–43.

———. "An Empirical Relativist Programme in the Sociology of Scientific Knowledge." In Knorr-Cetina and Mulkay, eds., *Science Observed*, 85–113.

———. *Changing Order: Replication and Induction in Scientific Practice*. London: Sage, 1985.

———, ed. *Knowledge and Controversy: Studies of Modern Natural Science*. Special issue of *Social Studies of Science* 11, no. 1 (1981).

Collins, Harry M., and Trevor J. Pinch. "The Construction of the Paranormal: Nothing Unscientific Is Happening." In Wallis, ed., *On the Margins of Science*, 237–70.

———. *Frames of Meaning: The Social Construction of Extraordinary Science*. London: Routledge and Kegan Paul, 1982.

———. *The Golem: What Everyone Should Know about Science*. Cambridge: Cambridge University Press, 1993.

Collins, Harry M., and Steven Yearley. "Epistemological Chicken." In Pickering, ed., *Science as Practice and Culture*, 301–26.

———. "Journey into Space." In Pickering, ed., *Science as Practice and as Culture*, 369–89.

Conant, James B. *Science and Common Sense*. New Haven, Conn.: Yale University Press, 1951.

Cooper, William [H. S. Hoff]. *The Struggles of Albert Woods*. 1952. Harmondsworth, Middlesex, U.K.: Penguin Books, 1966.

Counts, George S., and Nucia Lodge. *The Country of the Blind: The Soviet System of Mind Control*. Boston: Houghton Mifflin, 1949.

Cowley, Geoffrey. "Fighting the Disease: What Can Be Done." *Newsweek*, 17 January 2000, 38.

———. "Cannibals to Cows: The Path of a Deadly Disease." *Newsweek*, 12 March 2001, 53–61.

Cowley, Geoffrey, and Anne Underwood. "What's 'Alternative'?" *Newsweek*, 23 November 1998, 68.

Cowley, Geoffrey, with Anne Underwood. "Alzheimer's: Unlocking the Mystery." *Newsweek*, 31 January 2000, 46–51.

Craig, Elizabeth. "Molecules, Chaperone." In *Molecular Biology and Biotechnology: A Comprehensive Desk Reference*, edited by Robert A. Meyers, 162–65. New York: VCH, 1995.

Crick, Francis. "The Genetic Code." In *The Molecular Basis of Life: An Introduction to Molecular Biology (Readings from Scientific American)*, 198–206 and 217–23. With introductions by Robert H. Haynes and Philip C. Hanawalt. San Francisco: W. H. Freeman, 1968. Originally published in *Scientific American* (October 1962):66–74.

———. *What Mad Pursuit: A Personal View of Scientific Discovery*. New York: Basic Books, 1988.

Crossen, Cynthia. "The Treatment: A Medical Researcher Pays for Challenging Drug-Industry Funding." *Wall Street Journal*, 3 January 2001, A1 and A6.

Cunningham, Stephen. "This Story Has Legs: Centipede Beats Odds, Baffles Researchers." *Wall Street Journal*, 4 November 1999, B1.

Daly, Martin, and Margo Wilson. *Sex, Evolution, and Behavior: Adaptation for Reproduction*. North Scituate, Mass.: Duxbury Press, 1978. Reprint, Boston: Willard Grant Press, 1983.

Darwin, Charles. *The Origin of Species by Means of Natural Selection: Or, The Preservation of Favoured Races in the Struggle for Life*. 1859. Edited by Burrow, J. W., Harmondsworth, Middlesex, U.K., and Baltimore: Penguin Books, Pelican Classics,

1968. The excerpt from the conclusion of the second edition quoted in the text is from Gardner, ed., *Great Essays in Science*, 3–13.

———. *Autobiography and Letters*. Edited by Francis Darwin. New York: D. Appleton and Company, 1893. Reprint, New York: Dover, 1952.

———. *On Evolution: The Development of the Theory of Natural Selection*. Edited by Thomas F. Glick and David Kohn. Indianapolis: Hackett, 1996.

Davies, Paul C. W. *The Mind of God: The Scientific Basis for a Rational World*. New York: Simon and Schuster, Touchstone, 1992.

Dawkins, Richard. *The Selfish Gene*. Oxford: Oxford University Press, 1976, reprint 1989.

———. *The Blind Watchmaker*. New York: W. W. Norton, 1986.

———. "When Religion Steps on Science's Turf." *Free Inquiry*, 18, no. 2 (spring 1998):18–19.

———. *Unweaving the Rainbow: Science, Delusion, and the Appetite for Wonder*. Boston and New York: Houghton Mifflin, 1998.

Dean, J. "Controversy over Classification: A Case Study from the History of Botany." In Barnes and Shapin, eds., *Natural Order*, 211–30.

Dean, William D. Foreword to Elvee, *The Death of Science?*

Dennett, Daniel C. "Faith in the Truth." In *The Values of Science*, edited by Wes Williams, 95–109. Boulder, Colo.: Westview Press, 1999. Paper originally delivered at the Amnesty Lectures, Oxford, 1997. Reprinted in *Free Inquiry* 21, no. 1 (winter 2000–01):40–42.

"Defining Disciplines: Anthropology Becomes Two Departments." *Humanities and Sciences Quarterly* (Stanford University), summer 1998.

"A Devastated Continent." Newsweek, 1 January 2000, 41.

Dewey, John. *Logic, The Theory of Inquiry*. New York: Henry Holt, 1938.

Dickson, David. *The New Politics of Science*. New York: Pantheon Books, 1984.

Diderot, Denis. *Additions aux pensées philosophiques*. In *Rameau's Nephew and Other Works*. Translated from the French by Jacques Barzun and Ralph H. Bowen. Garden City, N.Y.: Doubleday, 1956. Originally published c. 1762.

"Did the Mars Lander Crash in a Grand Canyon?" *Newsweek*, 17 January 2000, 57.

Dose, Klaus. "The Origin of Life: More Questions Than Answers." *Interdisciplinary Science Reviews* 13 (1988):348–56.

Du Bois-Reymond, Emil. "The Limits of Our Natural Knowledge." *Popular Scientific Monthly* 5 (1874):17–32. Originally published as *Über die Grenzen des Naturerkennens: Die sieben Welträtsel—zwei Vorträge* (includes lectures of 1872 and 1880).

Duhem, Pierre. *The Aim and Structure of Physical Theory*. Translated by Philipp P. Weiner. Princeton, N.J.: Princeton University Press, 1954. Originally published as *La théorie physique* (1914).

Dummett, M. A. E. "Ought Research to Be Unrestricted?" In *Science and Ethics*, edited by Rudolf Haller. *Grazer Philosophische Studien* 12–13 (1981):281–98.

Durkheim, Emile. *The Rules of Sociological Method*. 1895. 8th ed., translated by Sarah A. Solovay and John H. Mueller. Edited by George E. Catlin. Chicago: University of Chicago Press, 1938. Reprint, Glencoe, Ill.: Free Press; London: Collier-Macmillan, 1964.

Dyson, Freeman. *Imagined Worlds*. Cambridge: Harvard University Press, 1997.

Earman, John. "The SAP Also Rises: A Critical Examination of the Anthropic Principle." *American Philosophical Quarterly* 24, no. 4 (1987):307–17.

Easterbrook, Gregg. "The New Fundamentalism." *Wall Street Journal*, 8 August 2000, A22.

Eccles, John, and Karl R. Popper. See Popper, Karl R., and John Eccles.

Eddington, Sir Arthur. *New Pathways in Science*. Cambridge: Cambridge University Press, 1935. Reprint, Ann Arbor: University of Michigan Press, 1959.

Einstein, Albert. "Principles of Research." In *Ideas and Opinions*, 224–27. Paper originally delivered as lecture in celebration of Max Planck's sixtieth birthday in 1918. First published in *Mein Weltbild* (Amsterdam: Querido Verlag, 1934).

———. "The Religious Spirit of Science." In *Ideas and Opinions*, 40. Originally published in *Mein Weltbild* (Amsterdam: Querido Verlag, 1934).

———. "Physics and Reality." In *Ideas and Opinions*, 290–323. Originally published in *The Journal of the Franklin Institute* 221, no. 3 (1936).

———. "Science and Religion." In *Ideas and Opinions*, 41–52. Originally in part from an address at Princeton Theological Seminary, 1939, and in part from a symposium on science, philosophy, and religion, 1950.

———. *Ideas and Opinions of Albert Einstein*. Translated from the German by Sonja Bargmann. New York: Crown Publishers, 1954.

Eliot, George. *Adam Bede*. 1859. With a foreword by F. R. Leavis, New York: Signet Classics, 1961.

———. *Romola*. 1863. Edited by Andrew Sanders, Harmondsworth, Middlesex, U.K.: Penguin Books, 1980.

———. *Daniel Deronda*. 1876. Edited by Barbara Hardy, Harmondsworth, Middlesex, U.K.: Penguin Books, 1967.

Ellison, Sarah. "Ads for a Holocaust Exhibit in London Cause a Stir." *Wall Street Journal*, 2 November 2000, sec. B.

Elvee, Richard Q. "After Twenty-Five Years: The End of Science!" Introduction to Elvee, ed., *The End of Science?*

———, ed. *The End of Science? Attack and Defense*. Lanham, Md.: University Press of America, 1992.

Faigman, David. "Annual Report on Science and the Law." National Institute of Justice, National Conference on Science and the Law. 2001.

Feigl, Herbert. "The Origin and Spirit of Logical Positivism." In Achinstein and Barker, eds., *The Legacy of Logical Positivism*, 3–24.

Feigl, Herbert, and Wilfrid Sellars, eds. *Readings in Philosophical Analysis*. New York: Appleton-Century Crofts; London: G. A. Noble, 1949.

Festinger, Leon. *A Theory of Cognitive Dissonance*. Evanston, Ill.: Row and Peterson, 1957.

Festinger, Leon, Henry W. Riecken, and Stanley Schachter. *When Prophecy Fails: A Social and Psychological Study of a Modern Group That Predicted the Destruction of the World*. Minneapolis: University of Minnesota Press, 1956. Reprint, New York: Harper and Row, 1964.

Feyerabend, Paul K. *Against Method: Outlines of an Anarchistic Theory of Knowledge*.

London: New Left Press; Atlantic Highlands, N.J.: Humanities Press, 1975. Reprint, London: Verso, 1978.

———. *Farewell to Reason*. London: Verso, 1987.

———. *Killing Time: The Autobiography of Paul Feyerabend*. Chicago: University of Chicago Press, 1995.

Fine, Arthur. "The Natural Ontological Attitude." In *Scientific Realism*, edited by Jarrett Leplin, 83–107. Berkeley and Los Angeles: University of California Press, 1984.

Fischman, Joshua. "New Clues Surface about the Making of the Mind." *Science* 262 (1993):1517.

Fish, Stanley. "Professor Sokal's Bad Joke." In *Lingua Franca*, editors of, eds., *The Sokal Hoax*, 81–84. Originally published in the *New York Times*, 22 May 1996, A23.

Forman, Paul. "Weimar Culture, Causality, and Quantum Theory, 1918–1927: Adaptation by German Physicists and Mathematicians to a Hostile Intellectual Environment." *Historical Studies in the Physical Sciences* 3 (1971):1–115.

Foucault, Michel. *Power/Knowledge: Selected Interviews and Other Writings, 1972–1977*. Edited by Colin Gordon et al. New York: Pantheon Books, 1980.

Fox, John. "The Ethnomethodology of Science." In Nola, ed., *Realism and Relativism in Science*, 59–80.

Fox Keller, Evelyn. "The Force of the Pacemaker Concept in Theories of Aggregation in Cellular Slime Mold." In Fox Keller, *Reflections on Gender and Science*, 150–57.

———. "Cognitive Repression in Contemporary Physics." In Fox Keller, *Reflections on Gender and Science*, 139–49.

———. *Reflections on Gender and Science*. New Haven, Conn.: Yale University Press, 1985.

Frankel, Mark S. "The Role of Science in Making Good Decisions." Testimony before the U.S. House Committee on Science, 10 June 1998. American Association for the Advancement of Science [online]. www.aaas.org/spp/sfrl/projects/testim/mftest.htm.

Frege, Gottlob. "On Sense and Reference." 1892. Translated by Max Black, in *Translations from the Philosophical Writings of Gottlob Frege*, edited by Peter Geach and Max Black, 56–78. Oxford: Basil Blackwell, 1966.

Frey-Wyssling, Albert Friedrich. "Frühgeschichte und Ergebnisse der submikroskopischen Morphologie." *Mikroskopie* 19 (1964):2–12.

Friedman, Michael. "Explanation and Scientific Understanding." *Journal of Philosophy* 71 (1974):5–19.

Fuchs, Victor. "What Every Philosopher Should Know about Health Economics." Proceedings of the American Philosophical Society 140, no. 2 (1996):186–95.

Fuller, Steve. *Philosophy, Rhetoric and the End of Knowledge*. Madison: University of Wisconsin Press, 1993.

Futuyma, Douglas J. *Science on Trial: The Case for Evolution*. New York: Pantheon Books, 1983.

Galison, Peter. *How Experiments End*. Chicago: University of Chicago Press, 1987.

Gardner, Martin. *Science: Good, Bad, and Bogus*. Amherst, N.Y.: Prometheus Books, 1981.

———. "The Religious Views of Stephen Gould and Charles Darwin." *Skeptical Inquirer* 23, no. 4 (July–August 1999):8–13.

———, ed. *Great Essays in Science*. New York: Pocket Books, 1957.

Garfinkel, Harold. *Studies in Ethnomethodology*. Englewood Cliffs, N.J.: Prentice-Hall, 1967.

Garfinkel, Harold, Michael Lynch, and Eric Livingston. "The Work of a Discovering Science Construed with Materials from the Optically Discovered Pulsar." *Philosophy of the Social Sciences* 11 (1981):131–58.

Gergen, Kenneth. "Feminist Critique of Science and the Challenge of Social Epistemology." In Gergen, ed., *Feminist Thought and the Structure of Knowledge*, 27–48.

Gergen, Mary M., ed. *Feminist Thought and the Structure of Knowledge*. New York: New York University Press, 1988.

Giannelli, Paul, "The Admissibility of Scientific Evidence: *Frye* v. *United States*, a Half-Century Later," *Columbia Law Review* 80 (1980):1197–1250.

Giddens, Anthony. "Nine Theses on the Future of Sociology." In Giddens, *Social Theory and Modern Sociology*. Cambridge, Mass.: Polity Press, 1987.

Giere, Ronald N. "The Feminism Question in the Philosophy of Science." In Nelson and Nelson, eds., *Feminism, Science, and the Philosophy of Science*, 3–15.

———. *Science Without Laws*. Chicago: University of Chicago Press, 1999.

Gilbert, G. Nigel, and Michael J. Mulkay. *Opening Pandora's Box: A Sociological Analysis of Scientists' Discourse*. Cambridge: Cambridge University Press, 1984.

Gilbert, William. *On the Loadstone and Magnetic Bodies and on the Great Magnet the Earth*. 1600. In Hutchins, ed., *Gilbert, Galileo, Harvey*, 1–121.

Gillespie, Charles Coulston. *The Edge of Objectivity: An Essay in the History of Scientific Ideas*, Princeton, N.J.: Princeton University Press, 1960.

Gillie, Oliver. "Crucial Data Was Faked by Eminent Psychologist." *Sunday Times* (London), 24 October 1976.

Glashow, Sheldon. "The Death of Science!?" In Elvee, ed., *The End of Science?* 23–32.

Glick, Thomas F., and David Kohn, eds. *See* Darwin, Charles.

Glymour, Clark. *Theory and Evidence*. Princeton, N.J.: Princeton University Press, 1980.

Goodman, Nelson. "The New Riddle of Induction." In Goodman, *Fact, Fiction and Forecast*, 59–83. 2d ed. Indianapolis: Bobbs-Merrill, 1965. First edition published 1954 (London: University of London, Athlone Press).

———. *Ways of Worldmaking*. Hassocks, U.K.: Harvester, 1978.

Goodstein, David. "Scientific Fraud." *American Scholar* 60 (fall 1991):505–15. Reprinted in *Engineering and Science* 54, no. 2 (winter 1991):10–19.

———. "Conduct and Misconduct in Science." In Gross, Levitt, and Lewis, eds., *The Flight from Science and Reason*, 31–38.

Goldberg, Cary. "Judges' Unanimous Verdict on DNA Lessons: Wow!" *New York Times*, 24 April 1999, A10.

Goldman, Alvin I. *Epistemology and Cognition*. Cambridge: Harvard University Press, 1986.

Gosse, Philip Henry. *Omphalos: An Attempt to Untie the Geological Knot*. London: Van Voorst, 1857. Reprint, Woodbridge, Conn.: Ox Bow Press, 1998.

Gottesman, Michael. "From *Barefoot* to *Daubert* to *Joiner*: Triple Play or Double Error?" *Arizona Law Review* 40 (1998):753–80.

Gould, Stephen Jay. "Morton's Ranking of Races by Cranial Capacity." *Science* 200 (1978):503–509.

———. *The Panda's Thumb: More Reflections on Natural History*. New York: W. W. Norton, 1980.

———. *The Mismeasure of Man*. New York: W. W. Norton, 1981.

———. *Rocks of Ages: Science and Religion in the Fullness of Life*. New York: Ballantine, 1999.

Grafton, Sue. *O is for Outlaw*. New York: Ballantine, 2000.

Graham, Michael H. *Federal Rules of Evidence in a Nutshell*. St. Paul, Minn.: West Publishing, 1987.

Green, Michael D. "Expert Witnesses and Sufficiency of Evidence in Toxic Substance Litigation: The Legacy of Agent Orange and Bendectin Litigation." *Northwestern Law Review* 86 (1992):643–99.

Greeno, James. "Explanation and Information." In *Statistical Explanation and Statistical Relevance*, edited by Wesley Salmon, 89–104. Pittsburgh: University of Pittsburgh Press, 1971.

Gross, Alan G. *The Rhetoric of Science*. Rev. ed. Cambridge: Harvard University Press, 1996. First edition published 1990.

———. "Learned Ignorance." *2B: A Journal of Ideas* 13 (1998):94–96.

Gross, John, ed. *The Oxford Book of Aphorisms*. New York: Oxford University Press, 1983.

Gross, Paul R. *Politicizing Science Education*. Washington, D.C.: Thomas B. Fordham Foundation, 2000.

Gross, Paul R., and Norman Levitt. *Higher Superstition: The Academic Left and Its Quarrels with Science*, Baltimore: Johns Hopkins University Press, 1994.

Gross, Paul R., Norman Levitt, and Martin W. Lewis, eds. *The Flight from Science and Reason*. Annals of the New York Academy of Sciences 775, 1996. Baltimore: Johns Hopkins University Press, 1997.

Grove, J. W. *In Defence of Science: Science, Technology and Politics in Modern Society*. Toronto: University of Toronto Press, 1989.

Grünbaum, Adolf. "A New Critique of Theological Interpretations of Physical Cosmology." *British Journal for the Philosophy of Science* 51 (2000):1–43.

Guterl, Fred, with Mary Carmichael. "Water, Water Everywhere." *Newsweek*, 10 June 2002, 41.

Guth, Alan H. *The Inflationary Universe: The Quest for a New Theory of Cosmic Origins*. Reading, Mass.: Addison Wesley; London: Cape, 1997.

Haack, Susan. *Deviant Logic*. Cambridge: Cambridge University Press, 1974. 2d, exp. ed. published as *Deviant Logic, Fuzzy Logic: Beyond the Formalism* (Chicago: University of Chicago Press, 1996).

———. *Philosophy of Logics*. Cambridge: Cambridge University Press, 1978.

———. "Epistemology *with* a Knowing Subject." *Review of Metaphysics* 33, no. 2, issue 130 (December 1979):309–35.

————. "'Realism'." *Synthese* 73 (1987):275–99.

————. *Evidence and Inquiry: Towards Reconstruction in Epistemology.* Oxford: Blackwell, 1993.

————. "Knowledge and Propaganda: Reflections of an Old Feminist." *Partisan Review* 60, no. 4 (fall 1993):556–63. Reprinted in Haack, *Manifesto of a Passionate Moderate*, 123–36.

————. "Dry Truth and Real Knowledge: Epistemologies of Metaphor and Metaphors of Epistemology." In *Aspects of Metaphor*, edited by Jaakko Hintikka, 1–22. Dordrecht, The Netherlands: Kluwer, 1994. Reprinted in Haack, *Manifesto of a Passionate Moderate*, 69–89.

————. "Puzzling Out Science," *Academic Questions* 8, no. 2 (spring 1995):20–31. Reprinted in Haack, *Manifesto of a Passionate Moderate*, 90–103.

————. "Preposterism and Its Consequences." In *Social Philosophy and Policy* 13, no. 2 (1996). Reprinted in *Scientific Innovation, Philosophy, and Public Policy*, edited by Ellen Frankel Paul et al. (Cambridge: Cambridge University Press, 1996), 196–315; and in Haack, *Manifesto of a Passionate Moderate*, 188–208.

————. "Reflections on Relativism: From Momentous Tautology to Seductive Contradiction." In *Metaphysics*, edited by James E. Tomberlin. Philosophical Perspectives 10. Oxford: Blackwell, 1996, and *Noûs*, Supplement, 1996, 297–315. Reprinted in Haack, *Manifesto of a Passionate Moderate*, 149–166.

————. "As for That Phrase 'Studying in a Literary Spirit' . . . ." *Proceedings and Addresses of the American Philosophical Association* 70, no. 2 (1996):57–75. Reprinted in Haack, *Manifesto of a Passionate Moderate*, 48–68.

————. "Science as Social? — Yes and No." In Nelson and Nelson, eds., *Feminism, Science, and Philosophy of Science*, 79–93. Reprinted in Haack, *Manifesto of a Passionate Moderate*, 104–22.

————. "'The Ethics of Belief' Reconsidered." In *The Philosophy of Roderick M. Chisholm*, edited by Lewis Hahn, 130–44. Library of Living Philosophers. La Salle, Ill.: Open Court, 1997. Reprinted in *Knowledge, Truth and Duty*, ed. Matthias Steup (New York: Oxford University Press, 2001), pp. 21–33.

————. "Confessions of an Old-Fashioned Prig." In Haack, *Manifesto of a Passionate Moderate*, 7–30.

————. *Manifesto of a Passionate Moderate: Unfashionable Essays.* Chicago: University of Chicago Press, 1998.

————. "Staying for an Answer." *Times Literary Supplement* (London), July 1999, 12–14.

————. "'La teoría de la coherencia de la verdad y el conocimiento' de Davidson." In *Ensayos sobre Davidson*, edited by Carlos Caorsi, 165–87. Montevideo, Uruguay: Press of Universidad de la Republica, 1999.

————. "After My Own Heart: Dorothy Sayers's Feminism." *The New Criterion* 19, no. 9 (May 2001):10–14.

————. "Viejo y Nuevo Pragmatismo." *Diánoia* 46, no. 47 (2001):21–59. English translation in *Pragmatism and American Philosophy*, edited by Paulo Ghiraldeli, Jr. and John Shook (forthcoming).

———. "Worthwhile Lives." *Free Inquiry* 22, no. 1 (winter 2001–2002):50–51.

———. "Realisms and Their Rivals: Recovering Our Innocence." *Facta Philosophica* 4 (2002):67–88.

———. "Trial and Error: The Supreme Court's Philosophy of Science." Paper presented at the Coronado Conference on Scientific Evidence and Public Policy. 2003. Abridged version, entitled "Disentangling *Daubert*," forthcoming in *APA* [American Philosophical Association] *Newsletter on Philosophy and Law* (fall 2003).

Hacking, Ian. *Representing and Intervening: Introductory Topics in the Philosophy of Natural Science.* New York: Cambridge University Press, 1983.

———. "The Participant Irrealist at Large in the Laboratory." *British Journal for the Philosophy of Science* 39 (1988):277–94.

———. *The Social Construction of What?* Cambridge: Harvard University Press, 1999.

Haeckel, Ernst. *The Riddle of the Universe—at the Close of the Nineteenth Century.* Translated by Joseph McCabe. New York and London: Harper, 1901. Originally published as *Die Welträtsel* (Bonn, 1889).

Halloran, S. Michael. "Towards A Rhetoric of Scientific Revolution." In *Proceedings of the 31st Conference on College Composition and Communication*, 229–36. 1980. An extended version of the paper was published as "The Birth of Molecular Biology: An Essay in the Rhetorical Criticism of Scientific Discourse," *Rhetoric Review* 3, no. 1 (September 1984):70–83.

Hamilton, W. D. "The Genetical Evolution of Social Behaviour." Parts 1 and 2. *Journal of Theoretical Biology* 7 (1964):1–16, 17–52.

Hammersley, Martyn. *The Politics of Social Research.* London: Sage, 1995.

Hand, Learned. "Historical and Practical Considerations Regarding Expert Testimony." *Harvard Law Review* 15 (1901):40–58.

Hansen, Mark. "The Great Detective." *ABAJ* [American Bar Association Journal] 87 (April 2001):36–44, 77.

Hanson, Norwood Russell. *Patterns of Discovery: An Inquiry into the Conceptual Foundations of Science.* Cambridge: Cambridge University Press, 1958.

Hanson, Sandra L. *Lost Talent: Women in the Sciences.* Philadelphia: Temple University Press, 1996.

Harding, Sandra. *The Science Question in Feminism.* Ithaca, N.Y.: Cornell University Press, 1986.

———. *Whose Science? Whose Knowledge? Thinking From Women's Lives.* Ithaca, N.Y.: Cornell University Press, 1991.

———. "Why Physics Is a Bad Model for Physics." In Elvee, ed., *The End of Science?* 1–22.

Harding, Sandra, and Merrill Hintikka, eds. *Discovering Reality: Feminist Perspectives on Epistemology, Metaphysics, Methodology, and Philosophy of Science.* Dordrecht, The Netherlands: Kluwer, 1983.

Hare, Richard M. *The Language of Morals.* Oxford: Oxford University Press, 1952.

Harries, Karsten. *The Broken Frame.* Washington, D.C.: Catholic University of America Press, 1989.

Harris, Henry. "Rationality in Science." In Heath, *Scientific Explanation*, 36–52.

Hartsock, Nancy. "The Feminist Standpoint: Developing the Ground for a Specifically Feminist Historical Materialism." In Harding and Hintikka, eds., *Discovering Reality*, 283–310.

Harvey, William. "Motion of the Heart." In Hutchins, ed., *Gilbert, Galileo, Harvey*, 265–304.

Harwood, J. "Professional Factors." Part 1 of "The Race-Intelligence Controversy: A Sociological Approach." *Social Studies of Science* 6 (1976):369–94.

Hawking, Stephen. *A Brief History of Time*. New York: Bantam Books, 1988. Updated, tenth anniversary edition, 1998.

Hayden, Thomas. "A Message, but Still No Answer." *Newsweek*, 6 December 1999, 60.

Haygood, Will, and *Globe* staff. "AIDS and the African." *The Boston Globe*, 11 October 1999, A1 and A25, and 13 October 1999, A1, city edition.

Hayles, N. Katherine. *Chaos Bound: Orderly Disorder in Contemporary Literature and Science*. Ithaca, N.Y.: Cornell University Press, 1990.

Hearnshaw, Leslie Spencer. *Cyril Burt, Psychologist*. Ithaca, N.Y.: Cornell University Press, 1979.

Heath, A. F., ed. *Scientific Explanation*. Oxford: Clarendon Press, 1981.

Heilbroner, Robert L. "Economics by the Book." *The Nation*, 20 October 1997, 16–19.

Hempel, Carl G. "A Purely Syntactical Definition of Confirmation." *Journal of Symbolic Logic* 8 (1943):122–143.

———. "Studies in the Logic of Confirmation." In Hempel, *Aspects of Scientific Explanation*, 3–47. Originally published in two parts in *Mind* 54 (1945):1–26, 97–121.

———. "Postscript (1964) on Confirmation." In Hempel, *Aspects of Scientific Explanation*, 47–51.

———. *Aspects of Scientific Explanation*. New York: Free Press; London: Collier-Macmillan, 1965.

———. *Philosophy of Natural Science*. Englewood Cliffs, N.J.: Prentice-Hall, 1966.

———. "The Irrelevance of the Concept of Truth for the Critical Appraisal of Scientific Theories." 1990. In *Selected Philosophical Essays [by] Carl G. Hempel*, edited by Richard Jeffrey, 75–84. Cambridge: Cambridge University Press, 2000.

Hempel, Carl G., and Paul Oppenheim. "A Definition of 'Degree of Confirmation'." *Philosophy of Science* 12 (1945):98–115.

———. "Studies in the Logic of Explanation." In Hempel, *Aspects of Scientific Explanation*, 245–90. Originally published in *Philosophy of Science* 15 (1948):135–75.

Henderson, Willie, Tony Dudley-Evans, and Roger Backhouse, eds. *Economics and Language*. London: Routledge, 1993.

Hensley, Scott. "Alzheimer's Cause May Be Metals Buildup," *Wall Street Journal*, 21 June 2001, B4.

Heritage, John. *Garfinkel and Ethnomethodology*. Cambridge: Polity Press; Oxford: Blackwell, 1984.

Herschel, John Frederick William. *A Preliminary Discourse on the Study of Natural Philosophy*. London: Longman, 1830–31.

Hershey, A. D., and Martha Chase. "Independent Functions of Viral Protein and Nucleic Acid in Growth of Bacteriophage." *Journal of General Physiology* 36 (1952):39–56.

Hesse, Mary. *Models and Analogies in Science*. London: Sheed and Ward, 1963. Reprint, Notre Dame, Ind.: University of Notre Dame Press, 1966.

――――. "Positivism and the Logic of Scientific Theories." In Achinstein and Barker, eds., *The Legacy of Logical Positivism*, 85–114.

――――. *The Structure of Scientific Inference*. Berkeley and Los Angeles: University of California Press, 1974.

――――. "How to Be Postmodern without Being a Feminist." *The Monist* 77, no. 4 (1994):445–61.

Himmelfarb, Gertrude. *Darwin and the Darwinian Revolution*. New York: W. W. Norton, 1968. Originally published 1959 (Garden City, N.Y.: Doubleday, reprint 1962).

Hintikka, J. J. K. "Towards a Theory of Inductive Generalisation." In *Proceedings of the 1964 International Congress for Logic, Methodology and Philosophy of Science*, edited by Yehoshua Bar-Hillel, 274–88. Amsterdam: North-Holland, 1964.

Hintikka, J. J. K., and Merrill Hintikka. "How Can Language Be Sexist?" In Harding and Hintikka, eds., *Discovering Reality*, 139–48.

Hixson, Joseph. *The Patchwork Mouse*. Garden City, N.Y.: Anchor Press, 1976.

Holbach, Baron Paul-Henri-Dietrich d'. *The System of Nature: Or, Laws of the Moral and Physical World*. 1770. Translated from the French by H. D. Robinson, New York: B. Franklin Press, 1970.

Holton, Gerald. "Subelectrons, Presuppositions, and the Millikan-Ehrenhoft Dispute." *Historical Studies in the Physical Sciences* 9 (1978):161–224.

――――. "How to Think About the End of Science." In Elvee, ed., *The End of Science?* 63–74.

――――. "The Controversy over the End of Science." In Holton, *Science and Anti-Science*, 126–44.

――――. *Science and Anti-Science*. Cambridge: Harvard University Press, 1993.

――――. *Einstein, History, and Other Passions: The Rebellion Against Science at the End of the Twentieth Century*. Woodbury, N.Y.: American Institute of Physics Press, 1995. Reprint, New York: Addison-Wesley, 1996.

――――. "The Scientific Method Is Doing Your Damnedest, No Holds Barred." *The Reading Room* 2 (2000):91–98. Abridged version of a longer paper originally published in Italian in *Limiti e frontiere della scienza*, ed. Pino Donghi (Laterza, Italy: Editori Gius, 1999), pp. 3–19.

――――. "The Rise of Postmodernisms and the 'End of Science'." *Journal of the History of Ideas* 61, no. 2 (April 2000):327–41.

Holton, Gerald, and Gerhard Sonnert. *See* Sonnert, Gerhard, and Gerald Holton.

Hooper, Laural L., Joe S. Cecil, and Thomas A. Willging. "Assessing Causation in Breast Implant Litigation: The Role of Scientific Panels." *Law and Contemporary Problems* 64, no. 4 (fall 2001):139–89.

Horgan, John. *The End of Science: Facing the Limits of Knowledge in the Twilight of the Scientific Age*. New York: Addison-Wesley, Helix Books, 1996.

Horwich, Paul, ed. *World Changes: Thomas Kuhn and the Nature of Science*. Cambridge: MIT Press, 1993.

Hoyningen-Huene, Paul. "Context of Discovery and Context of Justification." *Studies in the History and Philosophy of Science* 18, no. 4 (1987):501–15.

Hrobjartsson, Asbjorn, and Peter Gotzsche. "Is the Placebo Powerless? An Analysis of Clinical Trials Comparing Placebo with No Treatment." *New England Journal of Medicine* 344, no. 21 (May 2001):1594–1602.

Hubbard, Ruth. "Some Thoughts about the Masculinity of the Natural Sciences." In Gergen, ed., *Feminist Thought and the Structure of Knowledge*, 1–15.

Huber, Peter W. *Galileo's Revenge: Junk Science in the Courtroom*. New York: Basic Books, 1991.

———. "Junk Science in the Courtroom." *Valparaiso University Law Review* 26 (1992):723–55.

Huber, Peter W., and Kenneth R. Foster, eds. *Judging Science: Scientific Knowledge and the Federal Courts*. Cambridge: MIT Press, 1997.

Hull, David. *Science as Process*. Chicago: University of Chicago Press, 1988.

Hunt, Morton. *The New Know-Nothings: The Political Foes of the Scientific Study of Human Nature*. New Brunswick, N.J.: Transaction Publishers, 1999.

Hutchins, Robert Maynard, ed. *Gilbert, Galileo, Harvey*. Great Books of the Western World 28. Chicago: William Benton, for Encyclopaedia Britannica, 1952.

Huxley, Thomas H. Letter to F. Dysert, 30 January 1859.

———. "On the Advisableness of Improving Natural Knowledge." 1866. In Huxley, *Collected Essays*, 1:18–41.

———. *On a Piece of Chalk*. 1868. Edited by Loren Eiseley, New York: Charles Scribner's Sons, 1967.

———. "Possibilities and Impossibilities." 1891. In Huxley, *Collected Essays*, 5:192–208.

———. *Collected Essays*. 9 vols. New York: Appleton, 1894. Reprint, Bristol, U.K.: Thoemmes, 2001. Originally published 1893–94 (London: Macmillan).

———. "Science and Culture." In Gardner, ed., *Great Essays in Science*, 128–44. Originally delivered as address at the opening of Mason College (now the University of Birmingham), U.K., in 1880, and published in Huxley, *"Science and Culture" and Other Essays* (New York: Appleton, 1884), pp. 7–30.

Hyman, Anthony, ed. *See* Babbage, Charles.

Irigaray, Luce. "Sujet de la science, sujet sexué?" In *Sens et place de connaissance dans la société*, 95–121. Paris: Centre National de Recherche Scientifique, 1987. Cited in Sokal and Bricmont, *Fashionable Nonsense*, 109, 288.

James, William. "The Methods and Snares of Psychology." In Bird, ed., *Selected Writings*, 232–48. Originally published in *Principles of Psychology* (New York: Henry Holt, 1890).

———. "The Will to Believe." In *The Will to Believe and Other Essays in Popular Philosophy*, 1–31. New York: Dover, 1956, 1–31. Collection originally published in 1897; essay in 1896.

———. "On a Certain Blindness in Human Beings." In Bird, ed., *Selected Writings*, 320–40. Originally published in *Talks to Teachers on Psychology and to Students on Some of Life's Ideals* (1899).

———. *A Pluralistic Universe*. In *"Radical Empiricism" and "A Pluralistic Universe,"* edited by Ralph Barton Perry, with an introduction by Richard J. Bernstein. New York: Dutton, 1971. Reprint, Cambridge: Harvard University Press, 1977. Originally delivered as the Hibbert Lectures at Manchester College, Oxford; originally published 1909 (New York: Longman's, Green).

Jensen, Arthur R. "Kinship Correlations Reported by Sir Cyril Burt." *Behavioral Genetics* 4, no. 1 (1974):1–28.

John Paul II, Pope. "Message on Evolution to the Pontifical Academy of Sciences." *Acta Apostolicae Sedia* 88 (1996).

Johnson, George. "In Silica Fertilization: All Science Is Computer Science." *New York Times*, 25 March 2001, sec. 4.

Johnson, Phillip E. *Darwin on Trial*, Washington, D.C.: Regnery Gateway, 1991. 2d ed., Downers Grove, Ill.: InterVarsity Press, 1993.

Jonakait, Randolph N. "Forensic Science: The Need for Regulation." *Harvard Journal of Law and Technology* 4 (1991):109–91.

Jones, Malcolm. "Odd Outing." *Newsweek*, 7 February 2000, 69.

Joubert, Elsa. *Poppie*. London: Coronet Books and Hodder and Stoughton, 1981. Originally published in Afrikaans in 1979.

Joynson, Robert B. *The Burt Affair*. London: Routledge, 1989.

Judson, Horace Freeland. *The Eighth Day of Creation: Makers of the Revolution in Biology*. New York: Simon and Schuster; Harmondsworth, Middlesex, U.K.: Penguin Books, 1979.

Kalb, Claudia. "We Have to Save Our People." *Newsweek*, 24 July 2000, 29.

Kamin, Leon J. *The Science and Politics of I.Q.* Potomac, Md.: Lawrence Erlbaum; Harmondsworth, Middlesex, U.K.: Penguin Books, 1974.

Kaufmann, Felix. "The Logical Rules of Scientific Procedure." *Philosophy and Phenomenological Research* 2 (June 1942):457–71.

Kendrew, John. *The Thread of Life: An Introduction to Molecular Biology*. Cambridge: Harvard University Press; London: Bell, 1966.

Kesan, Jay P. "An Autopsy of Scientific Evidence in a Post-*Daubert* World." *Georgetown Law Journal* 84 (1996):1985–2041.

Kevles, Bettyann Holtzmann. *Naked to the Bone: Medical Imaging in the Twentieth Century*. New Brunswick, N.J.: Rutgers University Press, 1997. Reprint, Reading, Mass.: Addison-Wesley, 1998.

Kirshner, Robert. *The Extravagant Universe: Exploding Stars, Dark Energy, and the Accelerating Cosmos*. Princeton, N.J.: Princeton University Press, 2002.

Kitcher, Philip. "Explanatory Unification and the Causal Structure of the World." In Kitcher and Salmon, eds., *Scientific Explanation*, 410–505. Reprinted in abridged form as "The Unification Model of Scientific Explanation," in Klee, ed., *Scientific Inquiry*, 181–89.

———. *The Advancement of Science: Science without Legend, Objectivity without Illusions*. New York: Oxford University Press, 1993.

Kitcher, Philip, and Wesley Salmon, eds. *Scientific Explanation*. Minnesota Studies in the Philosophy of Science 13. Minneapolis: University of Minnesota Press, 1989.

Kitto, H. D. F. *The Greeks*. Harmondsworth, Middlesex, U.K.: Penguin Books, 1951.

Klee, Robert, ed. *Scientific Inquiry: Readings in the Philosophy of Science*. New York: Oxford University Press, 1999.

Knorr-Cetina, Karin D. *The Manufacture of Knowledge: An Essay on the Constructivist and Contextual Nature of Science*. Oxford: Pergamon Press, 1981.

Knorr-Cetina, Karin D., Roger G. Krohn, and Richard D. Whitley, eds. *The Social Process of Scientific Investigation*, Dordrecht, The Netherlands: Reidel, 1981.

Knorr-Cetina, Karin D., and Michael Mulkay, eds. *Science Observed: Perspectives on the Social Study of Science*. London: Sage, 1983.

Koehler, Jonathan J. "DNA Matches and Statistics: Important Questions, Surprising Answers." *Judicature* 76 (1993):222–29.

Koehler, Jonathan J., Audrey Chia, and Samuel Lindsey. "The Random Match Probability in DNA Evidence: Irrelevant and Prejudicial?" *Jurimetrics Journal* 35 (1995):201–219.

Koertge, Noretta. "Wrestling with the Social Constructor." In Gross, Levitt, and Lewis, eds., *The Flight from Science and Reason*, 266–73.

———, ed. *A House Built on Sand: Exposing Postmodernist Myths about Science*. New York: Oxford University Press, 1998.

Kolata, Gina. "Placebo Effect Is More Myth Than Science, Study Says." *New York Times*, 24 May 2001, A20.

———. "Putting Your Faith in Science?" *New York Times*, 27 May 2001, sec. 4, p. 2.

Kuhn, T. S. *The Structure of Scientific Revolutions*. Chicago: University of Chicago Press, 1962. 2d, enl. ed., 1970; 3d ed., with index, 1996.

———. "Logic of Discovery or Psychology of Research?" In Lakatos and Musgrave, eds., *Criticism and the Growth of Knowledge*, 1–23.

———. "Reflections on Receiving the John Desmond Bernal Award." *4S [Society for the Social Study of Science] Review* 1, no. 4 (1983):26–30.

———. "Afterwords." In Horwich, ed., *World Changes*, 311–42.

Kurtz, Paul, ed. *Moral Problems in Contemporary Society: Essays in Humanistic Ethics*. Englewood Cliffs, N.J.: Prentice-Hall, 1969. Reprint, Amherst, N.Y.: Prometheus Books, 1973.

Kyburg, Henry. "Reply to Wesley C. Salmon's 'The Status of Prior Probabilities in Statistical Explanation'." *Philosophy of Science* 32 (1965):147–51.

Kyburg, Henry, and Choh Man Teng. *Uncertain Inference*. Cambridge: Cambridge University Press, 2001.

Lacan, Jacques. "Of Structure as an Inmixing of an Otherness Prerequisite to Any Subject Whatever." In *The Languages of Criticism and the Sciences of Man*, edited by Richard Macksey and Eugenio Donato, 186–95. Baltimore: Johns Hopkins University Press, 1970.

Lakatos, Imre. "Falsification and the Methodology of Scientific Research Programmes." In Lakatos, *The Methodology of Scientific Research Programmes*, 8–93. Originally published in Lakatos and Musgrave, eds., *Criticism and the Growth of Knowledge*, 91–195.

————. *The Methodology of Scientific Research Programmes.* Vol. 1 of *Philosophical Papers,* edited by J. Worrall and G. Currie. Cambridge: Cambridge University Press, 1978.

Lakatos, Imre, and Alan Musgrave, eds. *Criticism and the Growth of Knowledge.* Cambridge: Cambridge University Press, 1970.

Lamarck, Jean-Baptiste. *Philosophie Zoologique.* Paris: Duminil-Leseur, 1809. Translated as *Zoological Philosophy.* Chicago: University of Chicago Press, 1984.

Lamb, Trevor, and Janine Bourriau, eds. *Universe: Art and Science.* New York: Cambridge University Press, 1995.

Langreth, Robert, and Chris Adams. "NIH Parley to Scrutinize Gene-Therapy Death." *Wall Street Journal,* 26 November 1999, B6.

Landsman, Stephan. "Of Witches, Madmen, and Product Liability: An Historical Survey of the Use of Expert Testimony." *Behavioral Sciences and the Law* 13 (1995):131–57.

Larson, Douglas W. "Science and Shoestring Technology." *National Forum* 80, no. 4 (fall 2000):6–7.

Latour, Bruno. *Science in Action: How to Follow Scientists and Engineers Through Society.* Cambridge: Harvard University Press; Milton Keynes, U.K.: Open University Press, 1987.

Latour, Bruno, and Steve Woolgar. *Laboratory Life: The Social Construction of Scientific Facts.* Beverly Hills, Calif., and London: Sage Library of Scientific Research, 1979. Reprint, Princeton, N.J.: Princeton University Press, 1986.

Laudan, Larry. "The Pseudo-Science of Science?" In Laudan, *Beyond Positivism and Relativism,* 183–209. Originally published in *Philosophy of the Social Sciences* 11 (1981):173–98.

————. "The Demise of the Demarcation Problem." In Ruse, ed., *But Is It Science?* 337–50. Originally published in *Physics, Philosophy, and Psychoanalysis,* edited by R. S. Cohen and Larry Laudan (Dordrecht, The Netherlands: Reidel, 1983), 111–28.

————. "Progress or Rationality? The Prospects for Normative Naturalism." In Laudan, *Beyond Positivism and Relativism,* 125–41. Originally published in *American Philosophical Quarterly* 24 (1987):19–33.

————. "Demystifying Underdetermination." In Laudan, *Beyond Positivism and Relativism,* 29–54. Originally published in *Scientific Theories,* ed. C. Wade Savage, Minnesota Studies in the Philosophy of Science 14 (Minneapolis: University of Minnesota Press, 1990), pp. 267–97. Reprinted, abridged, under the title "A Critique of Underdetermination," in Klee, ed., *Scientific Inquiry,* 83–99.

————. "'The Sins of the Fathers . . .'; Positivist Origins of Postpositivist Relativisms." In Laudan, *Beyond Positivism and Relativism,* 3–25.

————. *Beyond Positivism and Relativism: Theory, Method and Evidence.* Boulder, Colo.: Westview Press, 1996.

Laudan, Larry, and Jarrett Leplin. "Empirical Equivalence and Underdetermination." In Laudan, *Beyond Positivism and Relativism,* 55–73. Originally published in *Journal of Philosophy* 88 (1991):449–72.

Lavoisier, A. *Oeuvres*. Paris: Impresse Imperiale, 1862.

Lawson, Hilary, and Lisa Appignanesi, eds. *Dismantling Truth: Reality in the Post-Modern World*. New York: St. Martin's Press, 1989.

Lemonick, Michael D. "Superconductivity Heats Up." *Time*, 2 March 1987, 62.

———. "The $2 Billion Hole." *Time*, 1 November 1993, 69.

Levchine [Levshin], A. *Description des hordes et des steppes des Kirghiz-Kazaks ou Kirghiz-kaissaks*. Translated into French from the Russian by Ernest Charriere. Paris: Imprimerie Royale, 1840.

Levitt, Norman. *Prometheus Bedeviled: Science and the Contradictions of Contemporary Culture*. New Brunswick, N.J.: Rutgers University Press, 1999.

Lewontin, Richard. "Honest Jim's Big Thriller About DNA." Review of *The Double Helix*, by James Watson. *Chicago Sun-Times*, 25 February 1968, 1–2. Reprinted in the Stent edition of Watson, *The Double Helix*, 185–87.

Limon, John. "The Double Helix as Literature." *Raritan* 5, no. 3 (1986):26–47.

Lingua Franca, editors of, eds. *The Sokal Hoax: The Sham that Shook the Academy*. Lincoln: University of Nebraska Press, 2000.

Llewellyn, Karl. *Bramble Bush: Our Law and Its Study*. 2d ed. New York: Oceana, 1951. First edition published 1930.

Locke, John. *An Essay Concerning Human Understanding*. 1690.

Lodge, David. *Changing Places: A Tale of Two Campuses*. 1975. Harmondsworth, Middlesex, U.K.: Penguin Books, 1978.

Lowe, Adolph. "Comment" [on Hans Jonas]. In Natanson, ed., *Philosophy of the Social Sciences*, 152–57.

Lucretius [Titus Lucretius Carus]. *De rerum natura*. Translated by Ronald Melville as *On the Nature of the Universe*. Oxford: Oxford University Press, 1997. Also translated by Alban Dewes Winspar (New York: Harbor Press, 1956).

Lumsden, C. J., and E. O. Wilson. *Promethean Fire*. Cambridge: Harvard University Press, 1983.

Lurie, Alison. *Imaginary Friends*. 1967. New York: Owl Books, 1998.

Machlup, Fritz. "Are the Social Sciences Really Inferior?" In Natanson, ed., *Philosophy of the Social Sciences*, 158–80. Originally published in *The Southern Economic Journal* 27, no. 3 (1961):163–84.

MacKenzie, Donald. "Statistical Theory and Social Interests: A Case Study." *Social Studies of Science* 8, no. 1 (1978):35–83.

———. *Statistics in Britain 1865–1930*. Edinburgh, Scotland: Edinburgh University Press, 1981.

Maddox, John. *What Remains to Be Discovered: Mapping the Secrets of the Universe, the Origins of Life, and the Future of the Human Race*. New York: Simon and Schuster, 1998, and London: Macmillan, 1998.

Maddox, John, et al. "'High-Dilution' Experiments an Illusion." *Nature* 334 (July 1988):287–90.

Maki, Uskali. "Two Philosophies of the Rhetoric of Economics." In Henderson, Dudley-Evans, and Backhouse, eds., *Economics and Language*, 23–50.

————. "Diagnosing McCloskey." *Journal of Economic Literature* 33, no. 3 (1995):1300–18.

Manier, Edward. "Levels of Reflexivity: Unnoted Differences within the 'Strong Programme' in the Sociology of Knowledge." *PSA* [Proceedings of the Philosophy of Science Association] 1 (1980):197–207.

Marglin, Frédérique Apffel. "Smallpox in Two Systems of Knowledge." In *Dominating Knowledge: Development, Culture and Resistance*, edited by Frédérique Apffel Marglin and Stephen A. Marglin, 102–44. Oxford: Clarendon Press, 1990.

Maxwell, Grover. "The Ontological Status of Theoretical Entities." In *Scientific Explanation, Space, and Time*, edited by Herbert Feigl and Grover Maxwell, 3–27. Minnesota Studies in the Philosophy of Science 3. Minneapolis: University of Minnesota Press, 1962. Reprinted, abridged, in Klee, ed., *Scientific Inquiry*, 30–44.

Mayo, Deborah G. *Error and the Growth of Experimental Knowledge*, Chicago: University of Chicago Press, 1996.

McCloskey, Donald N. [now Deidre]. *The Rhetoric of Economics*. Madison: University of Wisconsin Press, 1985, reprint 1988.

————. *Knowledge and Persuasion in Economics*. Cambridge: Cambridge University Press, 1994.

McCormick, Charles. *McCormick on Evidence*. Edited by John W. Strong et al. St. Paul, Minn.: West Publishing, 1992, reprint 1999. Originally published as *Handbook of the Law of Evidence*, 1954.

McGinn, Colin. "Philosophical Materialism." *Synthese* 44 (1980):173–206.

McGinnis, John. "The Politics of Cancer Research." *Wall Street Journal*, 28 February 1997, A14.

McIver, Tom. *Anti-Evolution: A Reader's Guide to Writings before and after Darwin.* 1988. Baltimore: Johns Hopkins University Press, 1992.

McMullin, Ernan, ed. *The Social Dimensions of Science*. Notre Dame, Ind.: University of Notre Dame Press, 1992.

Medawar, Peter B. "Is the Scientific Paper a Fraud?" In Medawar, *The Threat and the Glory*, 228–33. Originally published 1963.

————. *Induction and Intuition in Scientific Thought*. Memoirs of the American Philosophical Society 75. London: Methuen, 1969. Based on the Jayne lectures of 1968.

————. "The Strange Case of the Spotted Mice." In Medawar, *The Threat and the Glory*, 71–82. Reprinted in *The Strange Case of the Spotted Mice and Other Classic Essays on Science*. New York: Oxford University Press, 1996. Originally published 1976.

————. "Scientific Fraud." In Medawar, *The Threat and the Glory*, 64–70. Originally published 1983.

————. *The Threat and the Glory: Reflections on Science and Scientists*. Oxford: Oxford University Press, 1990. Reprint, New York: Harper Collins, 1998.

Meehl, Paul E. *Clinical versus Statistical Prediction: A Theoretical Analysis and a Review of the Evidence*. Minneapolis: University of Minnesota Press, 1954. Reprinted, with new preface, Northvale, N.J.: Jason Aronson, 1996.

————. "Corroboration and Verisimilitude: Against Lakatos's 'Sheer Leap of Faith'." *Minnesota Center for Philosophy of Science* 90, no. 1 (1990):1–49.

Melville, Hermann. *Moby-Dick: Or, the Whale*. 1851. New York: Signet, 1998.

Mencken, H. L. *Treatise on the Gods*. New York: Knopf, 1930.

Merton, Robert K. "Science and the Social Order." In Merton, *Social Theory and Social Structure*, 591–603. Originally published 1937.

————. "The Normative Structure of Science." In *The Sociology of Science: Theoretical and Empirical Investigation*, edited by Norman W. Storer, 267–78. Chicago: University of Chicago Press, 1973. Originally published 1942.

————. *Social Theory and Social Structure*. Enlarged ed. New York: Free Press, 1968. Originally published 1957 (Glencoe, Ill.: Free Press).

"Meteorite—or Wrong." *Newsweek*, 26 January 1998, 8.

Michelson, Albert. *Light Waves and Their Uses*. Chicago: University of Chicago Press, 1903. Lectures delivered at the Lowell Institute in 1899.

Michener, James. *The Covenant*. 1980. London: Transworld Publishers, Corgi Books, 1982.

Millikan, Robert Andrews. "Alleged Sins of Science." *Scribner's Magazine* 87, no. 2 (1930):119–30.

Milton, Katherine. "Civilization and Its Discontents: Amazonian Indians." *Natural History* 101, no. 3 (March 1992):36–42.

Monod, Jacques. *Chance and Necessity: An Essay on the Natural Philosophy of Modern Biology*. Translated from the French by Austryn Wainhouse. New York: Knopf, 1971.

Montefiore, Hugh. *The Probability of God*. London: SCM [Student Christian Movement] Press, 1985.

Mulkay, Michael J. *Science and the Sociology of Knowledge*. London: George Allen and Unwin, 1979.

————. "Noblesse Oblige." In Mulkay, *Sociology of Science*, 161–82.

————. *Sociology of Science: A Sociological Pilgrimage*. Bloomington: Indiana University Press, 1991.

Murr, Andrew. "Final Answer: It Crashed." *Newsweek*, 10 April 2000, 46.

Musgrave, Alan. "Idealism and Antirealism." In Klee, ed., *Scientific Inquiry*, 344–52.

Musgrave, Alan, and Imre Lakatos, eds. See Lakatos, Imre, and Alan Musgrave, eds.

Nagel, Ernest. *The Structure of Science*. New York: Harcourt, Brace, and World, 1961.

————. "The Value-Oriented Bias of Social Inquiry" (excerpt from *The Structure of Science*). In Brodbeck, ed., *Readings in the Philosophy of the Social Sciences*, 98–113.

Nance, Dale. "The Best Evidence Principle." *Iowa Law Review* 73, no. 2 (January 1988):227–97.

Nanda, Meera. "The Epistemic Charity of Social Constructivist Critics of Science and Why the Third World Should Refuse the Offer." In Koertge, ed., *A House Built on Sand*, 286–311.

"NASA Scientists Seem Close to Confirming the Likelihood . . . ." *Wall Street Journal*, 19 December 1999, A1.

Natanson, Maurice, ed. *Philosophy of the Social Sciences: A Reader*. New York: Random House, 1963.

National Academy of Sciences. *Responsible Science: Ensuring the Integrity of the Research Process*. Washington, D.C.: National Academy of Sciences Press, 1992.

Nelson, Jack, and Lynn Hankinson Nelson, eds. *Feminism, Science, and the Philosophy of Science*. Dordrecht, The Netherlands: Kluwer, 1996.

Nicod, Jean. *Foundations of Geometry and Induction*. Translated by P. P. Weiner. London: Trench, Trubner, and Co.; New York: Harcourt and Brace, 1930.

Nietzsche, Friedrich. *Daybreak: Thoughts on the Prejudices of Morality*. 1881. Translated by R. J. Hollingdale. Edited by Maudemarie Clark and Brian Leiter. Cambridge: Cambridge University Press, 1997.

Nola, Robert, ed. *Relativism and Realism in Science*. Dordrecht, The Netherlands: Kluwer, 1988.

Notturno, Mark. "Popper, *Daubert*, and Kuhn." Washington, D.C.: Interactivity Foundation, 2000.

Nozick, Robert. *Philosophical Explanations*. Cambridge: Harvard University Press, 1981.

Nylenna, Magne, et al. "Handling of Scientific Dishonesty in the Nordic Countries." *The Lancet* 354, no. 9172 (July 1999):57–61.

Office of Inspector General of the National Science Foundation. Semiannual report to Congress, no. 9, 1993 (on fraud in science).

Olby, Robert C. *The Path to the Double Helix*. Seattle: University of Washington Press, 1974.

Olding, Alan. *Modern Biology and Natural Theology*. London: Routledge, 1991.

———. "Popper for Afters." *Quadrant* 43, no. 2 (December 1999):19–22.

———. "Maker of Heaven and Microbiology." *Quadrant* 44, nos. 1–2 (January–February 2000):62–68.

Oppenheim, Paul, and Carl G. Hempel. *See* Hempel, Carl G., and Paul Oppenheim.

Oppenheimer, J. Robert. "Physics in the Modern World." In Gardner, ed., *Great Essays in Science*, 186–204. Originally published 1947.

Packard, Vance. *The Status-Seekers: An Exploration of Class Behavior in America and the Hidden Barriers that Affect You, Your Community, Your Future*. Harmondsworth, Middlesex, U.K.: Penguin Books; New York: D. McKay, 1959.

Paley, William. *Natural Theology: Or, Evidence of the Existence and Attributes of the Deity Collected from the Appearances of Nature*. London: R. Faulder, 1802. Reprint, abridged, Indianapolis: Bobbs-Merrill, 1963.

Parker-Pope, Tara. "New Tests Go beyond Cholesterol to Find Heart-Disease Risks." *Wall Street Journal*, 22 June 2001, B1.

Peirce, Charles Sanders. *Collected Papers*. 8 vols. Edited by C. Hartshorne, P. Weiss, and A. Burks. Cambridge: Harvard University Press, 1931–58. References are by volume and paragraph number.

Pennock, Robert T. *Tower of Babel: The Evidence against the New Creationism*. Cambridge: MIT Press, Bradford Books, 1999.

Percy, Walker. "The Fateful Rift: The San Andreas Fault in the Modern Mind." In Percy, *Signposts in a Strange Land*, 271–91. New York: Farrar, Straus, and Giroux, 1991. Originally published 1990.

Perutz, Max. *Is Science Necessary? Essays on Science and Scientists*. New York: E. P. Dutton, 1989.

———. "The Pioneer Defended." Review of Gerald Geison, *The Private Science of Louis Pasteur*, by Gerald Geison. *The New York Review of Books*, 21 December 1995, 54–58.

Peters, Eric. "Benchmark Victory for Sound Science." *Washington Times*, 30 April 1999, A19.

Pickering, Andrew. "The Role of Interests in High-Energy Physics: The Choice between Charm and Color." In Knorr-Cetina, Krohn, and Whitley, eds., *The Social Process of Scientific Investigation*, 107–38.

———, ed. *Science as Practice and Culture*. Chicago: University of Chicago Press, 1992.

Pigliucci, Massimo. "Where Do We Come From? A Humbling Look at the Biology of Life's Origin." *Skeptical Inquirer* 23, no. 5 (September–October 1999):21–27.

Pinch, Trevor J., and Trevor J. Pinch. "Reservations about Reflexivity and New Literary Forms or Why Let the Devil Have All the Good Tunes?" In Woolgar, ed., *Knowledge and Reflexivity*, 178–97.

Pius XII, Pope. "Humani Generis." *Acta Apostolicae Sedis* 42 (1950).

Planck, Max. *Scientific Autobiography and Other Papers*. Translated by F. Gaynor. New York: Philosophical Library, 1949. Reprint, New York: Greenwood Press, 1968.

Poincaré, Henri. *Electricité et optique*. 1890–91.

"Pointer Rules Fed'l Science Panel Report Not Tainted by Payments to Panelist." *Medical Legal Aspects of Breast Implants* 7, no. 5 (April 1999):1, 4, 5.

Polanyi, Michael. *Personal Knowledge: Towards a Post-Critical Philosophy*. Chicago: University of Chicago Press, 1958.

———. *The Republic of Science: Its Political and Economic Theory*. Chicago: Roosevelt University, 1962. Reprinted in *Knowing and Being*, edited by Marjorie Grene, 49–62. Chicago: University of Chicago Press, 1969.

Popper, Karl R. *The Logic of Scientific Discovery*. English translation with the assistance of Dr. Julius Freed and Lou Freed. London: Hutchinson; New York: Basic Books, 1959. Reprint, London: Routledge, 1997. Originally published in German (Vienna: Springer Verlag, 1934).

———. "The Aim of Science." In Popper, *Objective Knowledge*, 191–205. Originally published in *Ratio* 1, no. 1 (December 1957): 24–35.

———. "Science: Conjectures and Refutations." In *Conjectures and Refutations: The Growth of Scientific Knowledge*, by Karl R. Popper, 33–65. London: Routledge and Kegan Paul, 1962. Originally published as "Philosophy of Science: A Personal Report," in *British Philosophy in Mid-Century*, edited by C. A. Mace (1957). Reprinted, abridged, under the title "Falsificationism," in Klee, ed., *Scientific Inquiry*, 65–71.

———. *The Poverty of Historicism*. London: Routledge and Kegan Paul; Boston: Beacon Press, 1957. Reprint, London and New York: Routledge, 1997. Originally published in three parts in *Economica* 11, nos. 42, 43 (1944), and 46 (1945).

———. "Normal Science and Its Dangers." In Lakatos and Musgrave, eds., *Criticism and the Growth of Knowledge*, 51–58.

———. "Epistemology without a Knowing Subject." In Popper, *Objective Knowledge*,

106–52. Originally published in *Proceedings of the Third International Congress for Logic, Methodology and Philosophy of Science*, edited by B. van Rooteslaar and J. F. Staal (Amsterdam: North-Holland, 1968), 333–73.

———. *Objective Knowledge: An Evolutionary Approach.* Oxford: Clarendon Press, 1972. 9th ed. Oxford: Oxford University Press, 1995.

———. "The Problem of Demarcation." In *A Pocket Popper*, edited by David Miller, 118–30. Oxford: Fontana, 1983. Originally published 1974.

———. "Subjective Experience and Linguistic Formulation." In Schilpp, ed., *The Philosophy of Karl Popper*, 1111–14.

———. "The Verification of Basic Statements." In Schilpp, ed., *The Philosophy of Karl Popper*, 1110–11.

Popper, Karl R., and John Eccles. *The Self and Its Brain.* Berlin, Heidelberg, London, and New York: Springer International, 1977. Paperback edition, London: Routledge, 1993.

Portugal, Franklin H., and Jack S. Cohen. *A Century of DNA: A History of the Discovery of the Structure and Function of the Genetic Substance.* Cambridge: MIT Press, 1977.

Press, Frank, ed. *Science and Creationism: A View from the National Academy of Sciences.* Washington, D.C.: National Academy of Sciences Press, 1984.

Price, H. H. *Belief.* London: Allen and Unwin; New York: Humanities Press, 1969.

Putnam, Hilary. "Is Logic Empirical?" In *Boston Studies in the Philosophy of Science*, vol. 1, edited by R. S. Cohen and M. R. Wartofsky, 216–41. New York: Humanities Press, 1969.

———. *Mathematics, Matter and Method.* 2 vols. Cambridge: Cambridge University Press, 1975.

Quine, W. V. "Two Dogmas of Empiricism." In Quine, *From a Logical Point of View*, 20–46. Originally published in *Philosophical Review* 60 (1951):20–43.

———. *From a Logical Point of View.* New York and Evanston, Ill.: Harper and Row, Harper Torchbooks, 1963. Originally published 1953 (Cambridge: Harvard University Press).

———. *Word and Object.* Cambridge: MIT Press, 1960.

———. "Epistemology Naturalized." In *Ontological Relativity*, 69–90.

———. "Natural Kinds." In *Ontological Relativity*, 114–38.

———. *Ontological Relativity and Other Essays.* New York: Columbia University Press, 1969.

———. "Grades of Theoreticity." In *Experience and Theory*, edited by Lawrence Foster and J. W. Swanson, 1–17. Amherst: University of Massachusetts Press, 1970.

———. "On Empirically Equivalent Systems of the World." *Erkenntnis* 9 (1975):313–28.

———. *From Stimulus to Science.* Cambridge: Harvard University Press, 1995, 1998.

Quine, W. V., and Joseph Ullian. *The Web of Belief.* New York: Random House, 1970. 2d ed., 1978.

Ramon y Cajal, Santiago. *Advice for a Young Investigator.* 1923. Translated by Neely Swanson and Larry W. Swanson. Cambridge: MIT Press, Bradford Books, 1999.

Ramsey, Frank P. *On Truth: Original Manuscript Materials (1927–29).* Edited by N. Rescher and U. Majer. Dordrecht, The Netherlands: Kluwer, 1990.

Rauch, Jonathan. *Kindly Inquisitors: The New Attacks on Free Thought*. Chicago: University of Chicago Press, 1993.

Read, Eileen White. "For Parched Lawns, a Patch of Blue." *Wall Street Journal*, 18 August 2002, W10.

Regalado, Antonio. "Bell Labs Scientists to Withdraw Papers Based on Fraudulent Data." *Wall Street Journal*, 1 November 2002, B7.

Reichenbach, Hans. *Theory of Probability*. Translated by Ernest H. Hutton and Maria Reichenbach. Berkeley and Los Angeles: University of California Press, 1949. Originally published as *Wahrscheinlichkeitslehre* (Leiden: A. W. Sijthoff's Uiteversmaatschappi, 1935).

———. *Experience and Prediction: An Analysis of the Foundations and the Structure of Knowledge*. Chicago: University of Chicago Press, 1938.

———. "On the Justification of Induction." *Journal of Philosophy* 37 (1940):97–103.

Reid, Thomas. *Essays on the Intellectual Powers*. 1785. In *The Works of Thomas Reid*, edited by G. N. Wright. Vol. 2. London: Thomas Tegg, 1843. Reprint, Charlottesville, Va.: Lincoln-Rembrandt, 1986.

Rescher, Nicholas. *Scientific Progress: A Philosophical Essay on the Economics of Research in Natural Science*. Oxford: Blackwell, 1978.

Reuben, David. "Realism in the Social Sciences." In Lawson and Appignanesi, eds., *Dismantling Truth*, 58–78.

Richards, I. A. Science and Poetry. London: Kegan Paul, Trench, Trubner and Co., 1926. Reprint, New York: W. W. Norton, 1970.

Ridley, Mark. *Evolution*. Oxford and Boston: Blackwell Scientific, 1993.

Robinson, Richard. *An Atheist's Values*. Oxford: Clarendon Press, 1964. Reprint, Oxford: Blackwell, 1975.

Rogers, Adam. "Come In, Mars." *Newsweek*, 19 August 1996, 56–57.

Roll-Hansen, Nils. "Studying Natural Science without Nature? Reflections on the Realism of So-Called Laboratory Studies." *Studies in the History and Philosophy of Biology and Biomedical Science* 29, no. 1 (1998):165–87.

Rollinson, Neil. "Quantum." *London Review of Books*, 25 November 1999, 28.

Root-Bernstein, Robert. "Darwin's Rib: Religious Beliefs of Students Encountered While Teaching Introductory Evolution." *Discover*, September 1995, 38–41.

Rorty, Richard. *Consequences of Pragmatism*. Hassocks, U.K.: Harvester Press; Minneapolis: University of Minnesota Press, 1982.

———. "Science as Solidarity." In *The Rhetoric of the Human Sciences*, edited by John S. Nelson, Allan Megill, and Deidre [formerly Donald] M. McCloskey, 38–52. Madison: University of Wisconsin Press, 1987.

Roth, Paul. "What Does the Sociology of Science Explain? Or, When Epistemological Chickens Come Home to Roost," *History of the Human Sciences* 7, no. 1 (1994):95–108.

Routley, Richard, et al. *Relevant Logics and Their Rivals*. Atascadero, Calif.: Ridgeview, 1982.

Ruder, Debra Bradley. "Opening the Drawer One More Time." *Newsweek*, 10 July 2000, 12.

Ruse, Michael. *From Monad to Man: The Concept of Progress in Evolutionary Biology.* Cambridge: Harvard University Press, 1996.

———, ed. *But Is It Science? The Philosophical Question in the Creation/Evolution Controversy.* Amherst, N.Y.: Prometheus Books, 1988, 1996.

Russell, Bertrand. *Human Knowledge, Its Scope and Limits.* New York: Simon and Schuster, 1948. Reprint, London: Allen and Unwin, Routledge, 1992.

Rutherfurd, Edward. *London: The Novel.* 1997. London: Ballantine and Arrow; New York: Fawcett Crest, 1998.

Sager, Ryan H. "A Profession in Crisis, a World Still in Misery." Reviews of *From Pathology to Politics*, by James T. Bennett and Thomas J. DiLorenzo, and *Betrayal of Trust*, by Laurie Garrett. *Wall Street Journal*, 29 August 2000, A24 and A25.

Salmon, Wesley. "Statistical Explanation." In *The Nature and Function of Scientific Theories*, edited by R. G. Colodny, 173–231. Pittsburgh: University of Pittsburgh Press, 1970. Reprinted in *Statistical Explanation and Statistical Relevance*, edited by Wesley Salmon, Richard Jeffrey, and James G. Greeno, (Princeton, N.J.: Princeton University Press, 1971), 29–87.

———. *Scientific Explanation and the Causal Structure of the World.* Princeton, N.J.: Princeton University Press, 1984.

———. "Four Decades of Scientific Explanation," in Kitcher and Salmon, eds., *Scientific Explanation*, 3–195.

Salzberg, Stephen A., and Kenneth R. Redden. *Federal Rules of Evidence Manual: A Complete Guide to the Federal Rules of Evidence.* Charlottesville, Va.: Michie, 1982. 8th ed., with co-editors Michael M. Martin and Daniel J. Capra, 2002.

Sapp, Jan. *Where the Truth Lies.* Cambridge and New York: Cambridge University Press, 1990.

Savage-Rumbaugh, E. Sue, and Roger Lewin. *Kanzi: The Ape at the Brink of the Human Mind.* New York: Wiley, 1994.

Saxe, John Godfrey, "The Blind Men and the Elephant." In *The Oxford Treasury of Children's Poems*, edited by Michael Harrison and Christopher Stuart-Clark, 116–17. Oxford: Oxford University Press, 1988.

Sayers, Dorothy L. *Gaudy Night.* 1936. Seven Oaks, U.K.: Coronet Books, Hodder and Stoughton, 1990.

———. "The Human-Not-Quite-Human." In Sayers, *Unpopular Opinions: Twenty-One Essays*, 142–49. London: V. Gollancz, 1946. Reprint, New York: Harcourt and Brace, 1947.

Sayre, Anne. *Rosalind Franklin and DNA.* New York: W. W. Norton, 1975.

Schilpp, Paul Arthur, ed. *The Philosophy of Karl Popper.* La Salle, Ill.: Open Court, 1974.

Schmitt, R. B. "Witness Stand: Who Is an Expert? In Some Courtrooms, the Answer Is 'Nobody'." *Wall Street Journal*, 17 June 1997, A1 and A8.

Schoeck, Helmut. *Envy: A Theory of Social Behaviour.* Translated by Michael Glenny and Betty Ross. Indianapolis: Liberty Fund, 1987. Originally published as *Der Neid* (Freiburg: Alber, 1966).

Schoofs, Mark. "Undermined: African Gold Giant Finds History Impedes a Fight Against AIDS." *Wall Street Journal*, 26 June 2001, A1 and A10.

Schulman, Kevin, et al. "A National Survey of Provisions in Clinical-Trial Agreements between Medical Schools and Industry Sponsors." *New England Journal of Medicine* 347, no. 17 (October 2002):1335–41.

Schutz, Alfred. "Concept and Theory Formation in the Social Sciences." In Natanson, ed., *Philosophy of the Social Sciences*, 210–30. Originally published in *Journal of Philosophy* 51 (1954):257–72.

Schwartz, Adina. "A 'Dogma of Empiricism' Revisited: *Daubert* v. *Merrell Dow Pharmaceuticals, Inc.*, and the Need to Resurrect the Philosophical Insight of *Frye* v. *United States.*" *Harvard Journal of Law and Technology* 10, no. 2 (1997):149–237.

Scriven, Michael. "Definitions, Explanations and Theories." In *Concepts, Theories, and the Mind-Body Problem*, edited by Herbert Feigl, Michael Scriven, and Grover Maxwell, 99–105. Minnesota Studies in the Philosophy of Science 2. Minneapolis: University of Minnesota Press, 1958.

———. "Explanations, Predictions, and Laws." In *Scientific Explanation, Space and Time*, edited by Herbert Feigl and Grover Maxwell, 170–230. Minnesota Studies in the Philosophy of Science 3. Minneapolis: University of Minnesota Press; Oxford: Oxford University Press, 1962.

Searle, John R. *The Construction of Social Reality*. New York: Free Press, 1995.

Sellars, Wilfred. "Scientific Realism or Irenic Instrumentalism?" In *In Honor of Philipp Frank*, edited by Robert Cohen and Marx Wartofsky, 171–204. Boston Studies for the Philosophy of Science 2. New York: Humanities Press, 1965.

Shapin, Steven. "Phrenological Knowledge and the Social Structure of Early Nineteenth-Century Edinburgh." *Annals of Science* 32 (1975):219–33.

———. "The Politics of Observation: Cerebral Anatomy and Social Interests in the Edinburgh Phrenology Debate." In Wallis, ed., *On the Margins of Science*, 139–78.

———. "Homo Phrenologicus: Anthropological Perspectives on an Historical Problem." In Barnes and Shapin, eds., *Natural Order*, 41–71.

———. *A Social History of Truth: Civility and Science in Seventeenth-Century England*. Chicago: University of Chicago Press, 1994.

———. *The Scientific Revolution*. Chicago: University of Chicago Press, 1996.

Shapin, Steven, and Simon Schaffer. *Leviathan and the Air-Pump: Hobbes, Boyle and the Experimental Life*. Princeton, N.J.: Princeton University Press, 1985.

Shattuck, Roger. *Forbidden Knowledge: From Prometheus to Pornography*. New York: St. Martin's Press and Harcourt and Brace, Harvest Books, 1996.

Shermer, Michael. *How We Believe: The Search for God in an Age of Science*. New York: W. H. Freeman, 2000.

Shillinger, Kurt. "AIDS and the African." *The Boston Globe*, 10 October 1999, A1; 12 October 1999, A1 and A20–21; 13 October 1999, A17, city edition.

Sismondi, Sergio. "Some Social Constructions." *Social Studies of Science* 23 (1993):515–53.

Skrabanek, Petr, and James McCormick. *Follies and Fallacies in Medicine*. 1989. Amherst, N.Y.: Prometheus Books, 1990, 1997.

Slezak, Peter. "Bloor's Bluff: Behaviourism and the Strong Programme." *International Studies in the Philosophy of Science* 5, no. 3 (1991):241–56.

―――. "The Social Construction of Social Constructivism." *Inquiry* 37, no. 2 (1994):139–57.

Smith, Dorothy. *The Everyday World as Problematic: A Feminist Sociology.* Boston: Northeastern University Press; Toronto: Toronto University Press; Milton Keynes, U.K.: Open University Press, 1987.

Smith, H. Irvin. "Leviticus Contains Many Proscriptions." *Wall Street Journal*, 20 June 2000, A27.

Snow, C. P. *The Search.* 1934. New York: Charles Scribner's Sons, 1958.

Snyder, Laurence H. *The Principles of Heredity.* Boston: D. C. Heath, 1935. 4th ed., 1951.

Soifer [Soyfer], Valerii, *The Tragedy of Soviet Science.* Translated from the Russian by Leo Gruliow and Rebecca Gruliow. New Brunswick, N.J.: Rutgers University Press, 1994.

Sokal, Alan. "Transgressing the Boundaries: Toward a Transformative Hermeneutics of Quantum Gravity." *Social Text* 46–47 (spring-summer 1996):217–52. Reprinted in *Lingua Franca*, editors of, eds., *The Sokal Hoax*, 11–48.

Sokal, Alan, and Jean Bricmont. *Fashionable Nonsense: Postmodern Intellectuals' Abuse of Science.* London: Profile Books; New York: Picador, 1998. Originally published as *Impostures intellectuelles* (Paris: Editions Odile Jacob, 1997).

Sonnert, Gerhard, with Gerald Holton. *Gender Differences in Science Careers: The Project Access Study.* New Brunswick, N.J.: Rutgers University Press, 1995.

Spengler, Oswald. *The Decline of the West.* Translated by Charles Francis Atkinson. New York: Knopf, 1926–28. Reprint, London: Allen and Unwin, 1932. Originally published as *Der Untergang des Abendlandes*, 2 vols. (Vienna: Braumüller, 1918–22; Munich: Back, 1920–22).

Starrs, James E. "A Still-Life Water-Color: *Frye* v. *United States*." *Journal of Forensic Science* 27, no. 3 (1982):684–94.

Stendhal, Marie Henri Beyle. *The Red and the Black.* Translated by Catherine Slater and Roger Pearson. Oxford and New York: Oxford University Press, 1991.

Stent, Gunther S. *The Coming of the Golden Age: A View of the End of Progress.* Garden City, N.Y.: Natural History Press, 1969.

Stich, Stephen P. *From Folk Psychology to Cognitive Science: The Case against Belief.* Cambridge: MIT Press, Bradford Books, 1983.

Stove, David C. *Popper and After: Four Modern Irrationalists.* Oxford: Pergamon Press, 1982. Republished under the title *Anything Goes: Origins of the Cult of Scientific Irrationalism* (Paddington, Australia: Macleay Press, 1998).

―――. "D'Holbach's Dream." In Stove, *Against the Idols of the Age*, 81–91. Originally published in *Quadrant* (December 1989):28–31.

―――. "Cole Porter and Karl Popper: Philosophy of Science in the Jazz Age." In Stove, *The Plato Cult and Other Philosophical Follies*, 1–26. Oxford: Blackwell, 1991.

―――. *Against the Idols of the Age.* Edited by Roger Kimball. New Brunswick, N.J.: Transaction Publishers, 1999.

Stowe, Harriet Beecher. *Uncle Tom's Cabin.* 1852. New York and London: Bantam Books, 1981.

Sturrock, John. "Le pauvre Sokal." Review of Sokal and Bricmont, *Fashionable Nonsense. London Review of Books,* 16 July 1998, 8–9.

Suppe, Frederick. *The Semantic Conception of Scientific Theories and Scientific Realism.* Urbana: University of Illinois Press, 1989.

———, ed. *The Structure of Scientific Theories.* Urbana: University of Illinois Press, 1974. 2d ed., 1977.

Swinburne, Richard. *The Existence of God.* New York: Oxford University Press, 1979. Rev. ed., 1991.

———. *Is There a God?* New York: Oxford University Press, 1996.

"Symposium on Science and Rules of Evidence." In *Federal Rules Decisions,* vol. 99, 188–234. St. Paul, Minn.: West Publishing, 1984.

Szasz, Thomas. *The Second Sin.* Garden City, N.Y.: Anchor Books; London: Routledge and Kegan Paul, 1974.

Tanenhaus, Sam. "Bellow, Bloom and Betrayal." *Wall Street Journal,* 2 February 2000, A26.

Tarski, Alfred. "The Semantic Conception of Truth and the Foundations of Semantics." In Feigl and Sellars, eds., *Readings in Philosophical Analysis,* 52–84. Originally published in *Philosophy and Phenomenological Research* 4 (1944):341–76.

Tessman, Irwin, and Jack Tessman. "Efficacy of Prayer: A Critical Examination of Claims." *Skeptical Inquirer* 24, no. 2 (March–April 2000):31–33.

Thagard, Paul. "The Best Explanation: Criteria for Theory Choice." *Journal of Philosophy* 75 (1978):76–92.

———. "Explanatory Coherence." *Behavioral and the Brain Sciences* 12, no. 3 (1989):435–502.

———. "The Dinosaur Debate: Explanatory Coherence and the Problem of Competing Hypotheses." In *Philosophy and AI: Essays at the Interface,* edited by J. Pollock and R. Cummins, 279–300. Cambridge: MIT Press, Bradford Books, 1991.

———. *Conceptual Revolutions.* Princeton, N.J.: Princeton University Press, 1992.

———. *How Scientists Explain Disease.* Princeton, N.J.: Princeton University Press, 1999.

———. *Coherence in Thought and Action.* Cambridge: MIT Press, Bradford Books, 2000.

Theocharis, Theo, and M. Psimopoulos. "Where Science Has Gone Wrong." *Nature* 329 (October 1987):595–98.

Thomas, Evan, Mark Hosenball, and Michael Isikoff. "The JFK-Marilyn Hoax." *Newsweek,* 6 June 1997, 36–37.

Thomson, Judith. "Grue." *Journal of Philosophy* 63 (1966):289–309.

Tomlinson, Hugh. "After Truth: Post-Modernism and the Rhetoric of Science." In Lawson and Appignanesi, eds., *Dismantling Truth,* 43–57.

Tooze, John, and James D. Watson. *See* Watson, James D., and John Tooze.

Tressell, Robert. *The Ragged-Trousered Philanthropists.* 1914. St. Albans, Herts, U.K.: Granada Publishing, Panther Books, 1965.

Trivers, Robert. *Social Evolution.* Menlo Park, Calif.: Benjamin/Cummings Publishing, 1985.

Twain, Mark. *Life on the Mississippi*. 1883. New York: Oxford University Press, 1996. Originally serialized in the *Atlantic Monthly* in 1875.

Tzara, Tristan. "The Dada Manifesto." *Seven Dada Manifestos and Lampisteries*. Translated by Barbara Wright. London: Calder.

"An Unnatural Disaster." Editorial, *Wall Street Journal*, 11 December 1998, A22.

Van Fraassen, Bas C. *The Scientific Image*. Oxford: Clarendon Press, 1980.

Velikovsky, Immanuel. *Worlds in Collision*. New York: Macmillan; London: Gollancz, 1950. Reprint, New York: Quality Paperback Book Club, 1997.

———. *Ages in Chaos*. Garden City, N.Y.: Doubleday, 1952. Reprint, London: Abacus, 1973.

Vickers, Brian. *In Defense of Rhetoric*. Oxford: Clarendon Press, 1988, 1997.

Vladislav, Jan, ed. *Vaclav Havel, or Living in Truth*. London: Faber and Faber, 1987.

Vonnegut, Kurt, Jr. *Wampeters, Foma & Granfaloons*. New York: Delecourt Press, 1974.

Waal, Franz B. M. de. *The Ape and the Sushi Master: Cultural Reflections by a Primatologist*. New York: Basic Books, 2001.

Wade, Nicholas. "Microbiology: Hazardous Profession Faces New Uncertainties," *Science*, n. s., 182 (November 1973):566–67.

———. *The Ultimate Experiment: Man-Made Evolution*. New York: Walker, 1977.

Waismann, Friedrich. "Logische Analyse des Wahrscheinlichkeitsbegriffs." *Erkenntnis* 1 (1930–31):228–48.

Walker, Laurens, and John Monahan. "Scientific Authority: The Breast Implant Litigation and Beyond." *Virginia Law Review* 86, no. 4 (May 2000):801–33.

Wallace, Alfred Russel. *Man's Place in the Universe: A Study of the Results of Scientific Research in Relation to the Unity or Plurality of Worlds*. New York: McClure and Phillips, 1903. Reprint, London: Chapman and Hall, 1904.

Wallace, Irving. *The Prize*. 1962. London: New English Library, 1967.

———. *The Man*. 1964. London: Cassell, 1965, 1966.

Wallis, Roy, ed. *On the Margins of Science: The Social Construction of Rejected Knowledge*. Sociological Review Monograph. Keele, Staffordshire, U.K.: University of Keele, 1979.

Wang, Hao. *Beyond Analytic Philosophy: Doing Justice to What We Know*. Cambridge: MIT Press, Bradford Books, 1986.

Watkins, J. W. N. *Science and Scepticism*. London: Hutchinson; Princeton, N.J.: Princeton University Press, 1984.

Watson, James D. *The Double Helix: A Personal Account of the Discovery of DNA*. Edited by Gunther Stent, with commentary, reviews, and Watson and Crick's original scientific papers. New York: W. W. Norton, 1980. Originally published 1967 (New York: Atheneum).

———. *Molecular Biology of the Gene*. New York: W. A. Benjamin, 1965. 3d ed., Menlo Park, Calif.: W. A. Benjamin, 1976.

Watson, James D., and Francis Crick. "Molecular Structure of Nucleic Acids: A Structure for Deoxyribonucleic Acid." *Nature* 171 (25 April 1953):737–38.

———. "Genetic Implications of the Structure of Deoxyribose Nucleic Acid," *Nature* 171 (30 May 1953):964–67.

————. "The Structure of DNA." In *Viruses. Cold Spring Harbor Symposia on Quantitative Biology* 18 (1953):123–131.

Watson, James D., and John Tooze, eds. *The DNA Story: A Documentary History of Gene Cloning*. San Francisco: W. H. Freeman, 1981.

Weber, Max. *The Theory of Social and Economic Organization*. Glencoe, Ill.: Free Press; New York: Oxford University Press; London: W. Hodge, 1947. Originally published as *Wirtschaft und Gesellschaft* (1922).

————. *The Methodology of the Social Sciences*. Edited by Edward Shils and Henry A. Finch. New York: Free Press, 1949. Includes two essays originally published in *Archiv für Sozialwissenschaft und Sozialpolitik*, 1904 and 1905, and a third originally published in *Logos*, 1917.

————. "The Interpretive Understanding of Social Action." In Brodbeck, ed., *Readings in the Philosophy of the Social Sciences*, 19–33. Excerpted from *The Theory of Social and Economic Organization*.

————. "'Objectivity' in Social Science." In Brodbeck, *Readings in the Philosophy of the Social Sciences*, 85–97. Excerpted from *The Methodology of the Social Sciences*.

Weinberg, Steven. *Dreams of a Final Theory*. New York: Pantheon Books, 1992. Reprint, New York: Vintage, 1993.

————. "Sokal's Hoax." *New York Review of Books*, 8 August 1996, 11–12. Reprinted in *Lingua Franca*, editors of, eds., The *Sokal Hoax*, 148–59, and in Weinberg, *Facing Up*, 138–54.

————. "A Designer Universe?" *New York Review of Books*, 21 October 1999, 46–48. Reprinted in Weinberg, *Facing Up*, 230–42.

————. *Facing Up: Science and Its Cultural Adversaries*. Cambridge: Harvard University Press, 2001.

Whelan, Richard. *The Book of Rainbows: Art, Literature, Science, and Mythology*. Cobb, Calif.: First Glance Books, 1997.

Whewell, William. *Philosophy of the Inductive Sciences*. 1840. In *Selected Writings on the History of Science*, edited by Yehuda Elkana, 121–254. Chicago: University of Chicago Press, 1984.

White, Andrew Dickson. *A History of the Warfare of Science with Theology in Christendom*. 2 vols. 1896. New York: Dover, 1960. Also reprinted 1997 (Bristol, U.K.: Thoemmes).

White, Gregory L. "GM Takes Advice from Disease Sleuths to Debug Cars." *Wall Street Journal*, 8 April 1999, B1 and B4.

Wicker, W. "The Polygraphic Truth Test and the Law of Evidence." *Tennessee Law Review* 22, no. 6 (1953):711–42.

Wilford, John Noble. "2 New Chemical Studies Find Meteorite Samples Show No Traces of Past Life on Mars." *New York Times*, 16 January 1998, A22.

Wilson, Catherine. "Instruments and Ideologies: The Social Construction of Knowledge and Its Critics." *American Philosophical Quarterly* 33, no. 2 (1996):167–81.

Wilson, E. O. *Consilience: The Unity of Knowledge*. New York: Knopf, 1998. Reprint, New York: Vintage, 1998.

Windschuttle, Keith. *The Killing of History: How Literary Critics and Social Theorists are Murdering Our Past.* Rev. ed. New York: Free Press, 1997. Originally published 1994 (Sydney: Macleay Press).

"Wish You Were Here." *Oxford Today* 10, no. 3 (1998):40.

Woodward, Kenneth. "The Changing Face of the Church." *Newsweek,* 16 April 2001, 46–52.

Woolgar, Steve. "Interests and Explanation in the Social Study of Science," *Social Studies of Science* 11 (1981):365–94.

———. "Critique and Criticism: Two Readings of Ethnomethodology," *Social Studies of Science* 11, no. 3 (1981):504–14.

———. "The Ideology of Representation and the Role of the Agent." in Lawson and Appignanesi, eds., *Dismantling Truth,* 131–44.

———. *Science: The Very Idea.* London and New York: Ellis Horwood/Tavistock Publications, 1988. Reprint, London: Routledge, 1993.

———, ed. *Knowledge and Reflexivity: New Frontiers in the Sociology of Knowledge.* London: Sage, 1988.

Woolf, Patricia, K. *Project on Scientific Fraud and Misconduct: Report on Workshop Number One.* Washington, D.C.: American Association for the Advancement of Science, 1988.

Wrangham, Richard W., et al., eds. *Chimpanzee Culture.* Cambridge: Harvard University Press, 1994.

Wright, Charles Alan, and Kenneth W. Graham, Jr. *Federal Practice and Procedure (Evidence) Sections 5161–5270.* Vol. 22. St. Paul, Minn.: West Publishing, 1978.

Wynne, Brian. "C. G. Barkla and the J Phenomena: A Case Study in the Treatment of Deviance in Physics." *Social Studies of Science* 6, nos. 3–4 (1976):307–47.

Yaukey, John. "The Sky's the Limit for Backyard Scientists." *USA Today,* 2 May 2000, 9D.

Zimmerman, Rachel. "Wary of HIV, Some Vendors Stop Selling N–9 Spermicide." *Wall Street Journal,* 25 September 2002, sec. B.

———. "Medical Schools' Research Pacts Are Criticized." *Wall Street Journal,* 24 October 2002, D8.

———. "AIDS's Spread Inflames Other Crises." *Wall Street Journal,* 27 November 2002, D3.

INDEX